MICRO-DOPPLER CHARACTERISTICS OF RADAR TARGETS

ELSEVIER

science & technology books

Companion Web Site:

http://booksite.elsevier.com/9780128098615/

Micro-Doppler Characteristics of Radar Targets

张群(Qun Zhang) 罗迎(Ying Luo) 陈永安(Yong-an Chen) 著

Available Resource:

- MATLAB codes discussed in chapters 2, 3 and 4

ELSEVIER

B H

MICRO-DOPPLER CHARACTERISTICS OF RADAR TARGETS

QUN ZHANG
YING LUO
YONG-AN CHEN

国防工业出版社
National Defense Industry Press

ELSEVIER

Amsterdam • Boston • Heidelberg • London
New York • Oxford • Paris • San Diego
San Francisco • Singapore • Sydney • Tokyo
Butterworth-Heinemann is an imprint of Elsevier

Butterworth-Heinemann is an imprint of Elsevier
The Boulevard, Langford Lane, Kidlington, Oxford OX5 1GB, United Kingdom
50 Hampshire Street, 5th Floor, Cambridge, MA 02139, United States

Library of Congress Cataloging-in-Publication Data
A catalog record for this book is available from the Library of Congress

British Library Cataloguing-in-Publication Data
A catalogue record for this book is available from the British Library

ISBN: 978-0-12-809861-5

For information on all Butterworth-Heinemann publications visit our website at https://www.elsevier.com/

Publisher: Jonathan Simpson
Acquisition Editor: Simon Tian
Editorial Project Manager: Vivi Li
Production Project Manager: Debbie Clark
Designer: Greg Harris

Typeset by TNQ Books and Journals

CONTENTS

PREFACE

Radar target recognition technology has been significant in radar systems. Meanwhile, it is one of the most active areas in the development of radar technologies. With the rapid development of wideband/ultra-wideband signal processing technology, semiconductor technology, and computer technology, radar that can only position and orbit measure has developed into radar with multifeatures measuring. The development of radar has changed its connotation profoundly. The traditional observing and tracking radar used for detection, ranging, and angle measurement has developed into feature-measuring radar, which can be used for fine structures and movement features extraction of the target. In general, the information of the radar target features is implied in radar echoes, such as radar cross-section (RCS) and its fluctuation statistical model, the Doppler spectrum, high-resolution imaging, and the polarization scattering matrices of the target. These features are obtained by the specific design of the wave and the processing, analysis, and transformation of echo's magnitude and phase.

However, with the rapid development of target feature control technology, the false target and the decoy can imitate the features, such as the RCS, the track, geometric structures, and surface material of the real target accurately. Thus, the accurate imitation makes the radar target recognition based on traditional features more difficult and even invalid.

Micromotion is regarded as the unique movement status and the fine feature of radar targets. It is induced by the specific forces acting on the specific structures of the target. Micromotion is of small amplitude, which leads to the low controllability and great difficulty to be imitated. Thus, the target recognition technology based on microfeatures is regarded as one of the most promising technologies in the field of radar target recognition. It has drawn wide attention of researchers both domestically and abroad in recent years.

With the support of the National Natural Science Foundation of China "Study on micro-Doppler feature extraction and imaging techniques of moving targets in MIMO radar" (Grant No. 60971100), "Distinguishing and micro-motion feature extraction of space group targets based on image sparse decomposition theory" (Grant No. 61471386), "Research on interferometric three-dimensional imaging theory and feature extraction techniques for micromotional space targets" (Grant No. 61571457), and "Cognitive extraction methods of radar target's micro-Doppler features based on resource-optimal scheduling" (Grant No. 61201369), the authors have made a comprehensive study on the radar target micro-Doppler effect theories and relative technologies in recent years. Based on these recent study results, the authors illustrate the principle of the radar target micro-Doppler effect from two aspects, that is, micro-Doppler effect analysis and micro-Doppler feature extraction. The analysis of the micro-Doppler effect of different targets in different radar systems is analyzed by the author. Meanwhile, multiple different micro-Doppler feature extraction methods and the three-dimensional micro-Doppler feature reconstruction methods are discussed. This book provides a reference to scholars interested in radar target micro-Doppler effect.

This book consists of seven chapters. Chapter 1 is an introduction to the concepts, basic principles, application fields, and the research history and present situation of micro-Doppler effect in radar. In Chapters 2–4, the micro-Doppler effect in different radar systems is analyzed. The micro-Doppler effect in narrowband radar and wideband radar is discussed in Chapters 2 and 3, respectively. The micro-Doppler effect in bistatic radar is discussed in Chapter 4. In Chapters 5 and 6, the micro-Doppler feature extraction methods are introduced. The multiple micro-Doppler feature analysis and extraction methods, including time-frequency analysis, image processing, orthogonal matching pursuit decomposition, empirical-model decomposition, and high-order moment function analysis are discussed in detail in Chapter 5. In Chapter 6, three-dimensional feature reconstruction methods are introduced. Chapter 7 summarizes the book and lists some perspectives in micro-Doppler theory research.

We would like to express sincere thanks to graduate student Dong-hu Deng, for his work on the writing of Chapter 5.5, and graduate students Jian Hu, Yi-jun Chen, Jia-cheng Ni, Yu-xue Sun, Qi-fang He, Tao-yong Li, Chen Yang, Di Meng, Yi-shuai Gong, who participated in the programming, figure making, and proofreading of this book. In the MATLAB files of this book, time-frequency analysis toolbox is utilized. Due to intellectual property considerations, the codes of this toolbox cannot be listed in the book, however, they can be accessed on the Website: http://tftb.nongnu.org/. We would like to show special thanks for the authors of time-frequency analysis toolbox for their programming work.

The research scope for many aspects of micro-Doppler effect is limited. However the study of radar target micro-Doppler effect is a rapidly developing field, with new theories and more research on engineering technologies and practices. We would welcome readers' input, should they notice any unintended errors in this book.

Authors

December, 2015

ABBREVIATIONS

BRF	Burst repetition frequency
DFT	Discrete Fourier transform
EMD	Empirical-mode decomposition
FT	Fourier transform
HOMF	High-order moments function
HT	Hough transform
IFT	Inverse Fourier transform
ISAR	Inverse synthetic aperture radar
JEM	Jet engine modulation
LFM	Linear frequency modulation
LFMCW	LFM continuous wave
LOS	Line of sight
m-D effect	Micro-Doppler effect
MF	Modulus function
MIMO	Multi-input multi-output
MP	Matching pursuit
OFDM-LFM	Orthogonal frequency division multiplexing-LFM
OMP	Orthogonal matching pursuit
PD radar	Pulse Doppler radar
PMF	Product of modulus function
PRF	Pulse repetition frequency
PRI	Pulse repetition interval
PWVD	Pseudo Wigner–Ville distribution
RCS	Radar cross-section
RSPWVD	Reassigned SPWVD
RVP	Residual video phase
SAR	Synthetic aperture radar
SFCS	Stepped-frequency chirp signal
SFM	Sinusoidal frequency modulation
SNR	Signal-to-noise ratio
SPWVD	Smoothed pseudo Wigner–Ville distribution
STFT	Short-time Fourier transform
SWVD	Smoothed Wigner–Ville distribution
WT	Wavelet transform
WVD	Wigner–Ville distribution

CHAPTER ONE

Introduction

Radar has been proven to be a powerful detection and recognition tool in modern high-tech warfare. With the development of radar technology, radar is not just for target detection and positioning; the automatic recognition for noncooperative target has been one of its important functions, where the selection of target features and how to extract them are the core problems to be solved. In recent years, radar target recognition technology has been greatly developed along with the advances in modern signal processing technology and wideband radar technology. However, the target antirecognition technology progressed at the same time. Active and passive interference technologies, which provide high fidelity to the real target, along with the application of illusory target digital synthesis technique, have brought complications to the detection and recognition of targets.

For the past few years, micro-Doppler effect of radar target has been a hotspot in the field of target feature extraction and identification. Target recognition technique based on micro-Doppler effect has been considered to have the most potential of further development in the area of radar target recognition. In the following subsections, the scattering center model is introduced first, followed by basic concepts of micro-Doppler effect. Both national and international current situations of radar target micro-Doppler effect research and application are then explained. Finally, the content organization of this book is given.

1.1 SCATTERING CENTER MODEL

The generation mechanism of backscattered electromagnetic signal from targets is quite complicated, as it is related to several factors like radar line-of-sight angle, polarization, wavelength, target structure and material, etc. Although the scattering mechanism of electromagnetic waves has been determined by Maxwell equation, it is still difficult to critically describe for a complex target in real application. In radar signal processing, especially in radar imaging, the scattering center model is usually adopted for target backscattering approximation. When target size is far larger than wavelength, the target can be considered as a sum of discrete scatterers; therefore the backscattered signal can be regarded as generated by a series of scatterers on the target. In this case, when the transmitted signal is $p(t)$, the received echo signal can be found as

$$s(t) = \sum_{n} \sigma_n p\left(t - \frac{2R_n}{c}\right) \tag{1.1}$$

Micro-Doppler Characteristics of Radar Targets
ISBN 978-0-12-809861-5, http://dx.doi.org/10.1016/B978-0-12-809861-5.00001-1

where σ_n is the backscattered amplitude of the n-th scattering center, and R_n is the radial distance of the n-th scattering center to the radar.

The scattering center model critically follows the first law of electromagnetic scattering theory, which is the approximation of Maxwell equation in high frequency, under the assumption that target size is far larger than wavelength and thus is scattering in the optical region. When the assumption fails, the scattering center model fails as well. For example, when target size is in the same scale as wavelength, the radar target is scattering in the resonant region, where the scattering magnitude varies with the change of wavelength, and the maximum scattering magnitude can be larger than the minimum value by more than 10 dB.

In modern radar, the assumption that target size is far larger than wavelength usually holds true. Take X band radar as an example—the wavelength is in centimeters, which is far smaller than most man-made objects. It is clear that if radar transmits pulse signals and there are several scatterers on the target in radial direction, the echo return will be composed of a series of pulses with the scattering center model applied, where different time instances that the pulses arrive at the receiver can tell the different positions of corresponding scatterers. The backscattered wave is actually the projection of target scattering centers on the radar line-of-sight direction, which is the theoretical basis of forming a one-dimensional range profile of a target. The wide applications of high-resolution one-dimensional range profile have proven the effectiveness of the scattering center model.

All the following discussions in this book have been conducted with the scattering center model applied. To explain it in more detail, the micro-Doppler effect is sensitive to the radar wavelength of operation, which means the micro-Doppler effect is more remarkable with smaller wavelength. In this way, the precondition of scattering center model can be satisfied in most cases in the analysis of micro-Doppler effect.

1.2 CONCEPT OF MICRO-DOPPLER EFFECT

1.2.1 Doppler Effect

Radar detection is mainly for targets in motion, such as satellites, ballistic missiles, debris in space, aerial planes, ships on the sea, and vehicles on the ground. The relative radial motion between signal source and receiver will induce a frequency shift on the received signal, and this physical phenomenon is called the Doppler effect. The Doppler effect was found by Christian Doppler in 1842, and has been put into application in electromagnetics since 1930. The first-generation Doppler radar was produced in 1950. Many important functions, like velocity measurement, moving target identification, and synthetic aperture radar imaging, are based on the Doppler effect. When a target moves toward radar with a certain radial velocity, the echo signal frequency will be shifted, and this shift in frequency is called the Doppler frequency shift, which is determined by the wavelength and target's radial velocity. Suppose the transmitted signal is

$$p(t) = u(t)\exp(\mathrm{j}\omega_0 t) \tag{1.2}$$

where $u(t)$ is the signal envelope, j is an imaginary number unit, ω_0 is the angular frequency of transmitted signal, and t represents time. Take the case of monostatic radar for analysis. The backscattered echo signal $s(t)$ can be expressed as

$$s(t) = \sigma p(t - t_r) = \sigma u(t - t_r)\exp(\mathrm{j}\omega_0(t - t_r)) \tag{1.3}$$

where t_r is the time delay of echo signal with respect to the transmitted signal and σ is the target reflection coefficient. When the target location does not vary with time with respect to the radar, there are fixed time delays in both the envelope and phase of echo signal. When the target moves with a certain radial velocity v_r with respect to radar, the echo signal received at t instance is actually transmitted at $t - t_r$. The time instant that the target is illuminated by signal can be calculated as $t' = t - t_r/2$, and at that moment the distance from radar to target is

$$R(t') = R_0 - v_r t' \tag{1.4}$$

where R_0 is the initial distance between radar and target when $t = 0$. The time consumed by the two-way propagation for a distance of $R(t')$ is actually the echo signal time delay t_r, that is

$$t_r = \frac{2R(t')}{c} \tag{1.5}$$

where c is the propagation speed of electromagnetic wave, which equals the velocity of light in free space. By combining Eqs. (1.4) and (1.5), it can be figured out that

$$t_r = \frac{2(R_0 - v_r t)}{c - v_r} \tag{1.6}$$

Substituting Eq. (1.6) into Eq. (1.3), the formula of the echo signal can be found as

$$s(t) = \sigma u\left(\frac{c + v_r}{c - v_r}t - \frac{2R_0}{c - v_r}\right)\exp\left[\mathrm{j}\omega_0\left(\frac{c + v_r}{c - v_r}t - \frac{2R_0}{c - v_r}\right)\right] \tag{1.7}$$

From expression Eq. (1.7), two properties of echo signal can be analyzed as follows:

1. There is a shift in the angular frequency of echo signal. From the phase term of expression Eq. (1.7), it is clear that the angular frequency is shifted from ω_0 to $\frac{c+v_r}{c-v_r}\omega_0$;
2. There is a scaling change for the echo signal envelope in terms of time. The signal is scaled by the factor $\frac{c+v_r}{c-v_r}$ in signal envelope term. When the target moves toward radar,

v_r is a positive value, which means the signal is narrowed in time and widened in frequency.

In most cases of radar applications, the envelope change of echo signal can be ignored. Even for the signal of large time-bandwidth product, the scaling change of echo signal envelope will not greatly influence the process of phase term. Since in radar imaging signal processing the key point actually lies in the process of phase term, the change of echo signal angular frequency should be mainly considered. Generally, the relative motion velocity v_r of radar and target is far smaller than electromagnetic wave propagation speed c; therefore the shift in angular frequency can be approximated by

$$\frac{c+v_r}{c-v_r}\omega_0-\omega_0=\frac{2v_r}{c-v_r}\omega_0\approx\frac{2v_r}{c}\omega_0=2\pi\frac{2v_r}{\lambda}, \tag{1.8}$$

where λ is the wavelength of transmitted signal. The corresponding frequency shift is

$$f_\mathrm{d}=\frac{2v_r}{\lambda}, \tag{1.9}$$

which is the so-called Doppler frequency. It is proportional to the target radial velocity v_r with respect to radar and inversely proportional to the wavelength λ. When the target moves toward radar, the Doppler frequency is a positive value, and the echo signal frequency is higher than the transmitted signal frequency. In contrast, when the target moves away from radar, the Doppler frequency is a negative value, and the echo signal frequency is lower than the transmitted signal frequency.

1.2.2 Micro-Doppler Effect

Doppler effect reveals the specificity of a target integral motion's modulation on radar signal. By extracting the Doppler frequency shift of the echo signal, the radial motion velocity of a target can be obtained. In fact, in real radar applications, targets with single-motion pattern hardly exist. For targets like satellites, airplanes, vehicles, and pedestrians, their motion dynamics are all in complex forms. Simple measurements of velocity and distance can no more satisfy the needs of practical applications than the characteristics of complex motion dynamics, which can provide critical information for delicate target recognition.

In 2000 Victor C. Chen at the US Navy Research Laboratory successfully introduced the concept of micromotion and micro-Doppler into radar research, which opened up the new research field of target micromotion characteristics extraction based on radar signal processing [1,2]. The tiny motions like vibration, rotation, and accelerated motion of target or part of a target besides the target translation as a whole, are called micromotions. As an objective feature of target motion, micromotion is common in nature, such as the precession of a ballistic missile in midcourse, the rotation of antennas, turbine blades, and helicopter rotor wings, the rotation of wheels in motor vehicles like tanks and

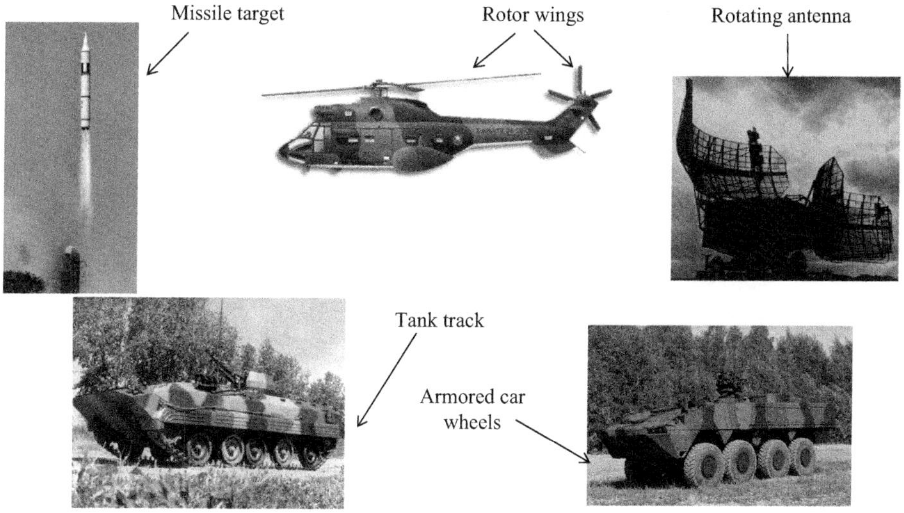

Figure 1.1 Typical targets with micromotion characteristics.

trucks, engine-induced vehicle vibrations, the pitch and yaw motions of ships, and the swing motions of pedestrian arms and legs, etc. Fig. 1.1 illustrates some typical military targets with micromotion characteristics. Micromotion characteristics are commonly accepted as unique motion features of radar targets. Using high-resolution radar and modern signal processing technique to extract these kinds of tiny motion features can provide new solutions for noncooperative radar target detection and recognition. For example, information about wheel rotation and engine vibration can be utilized for recognition of military vehicles or tanks on the ground; knowledge about helicopter rotor wings and jet aircraft engine blades' rotation can be adopted for a higher level of accuracy of recognition of aerial targets; the characteristics of self-rotation warheads' precession can be used to increase the effectiveness of warhead recognition. Micromotion characteristics' extraction opens up a new field of target feature analysis, providing features of favorable stability and good distinguishability for target identification.

The main theoretical basis and technological means of radar target micromotion characteristics extraction is the micro-Doppler effect of a radar target. According to the aforementioned Doppler effect theory, the micromotion of a target or part of a target will induce frequency modulation on a radar echo signal, which is in fact a Doppler sideband besides the main Doppler frequency shift induced by target major structure motion. This kind of frequency modulation induced by micromotion is called micro-Doppler effect. Micro-Doppler effect can be regarded as the result of interaction between target part and target major structure. It reflects the modulation of radar target micromotion on echo signal. Through analyzing the micromotion characteristics of the

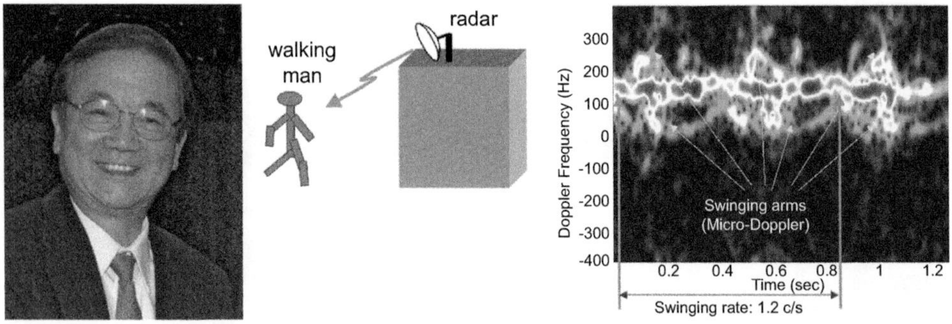

Figure 1.2 Victor C. Chen in the US Navy Research Laboratory and the experimental result of pedestrian target micro-Doppler effect analysis that he published.

echo signal and extracting the internal micromotion feature, we can determine target structure and motion dynamics, thus realizing the classification, recognition, and imaging of radar targets.

The concept of micro-Doppler was first put forward in a laser radar system. As is known, a distance change of half wavelength in radial direction will induce 2π shift in phase term. The wavelength of laser is very short: taking wavelength $\lambda = 2\ \mu m$ as an example, just 1 μm radial distance change can generate 2π shift in phase. Therefore the sensitivity of laser radar is very high, and thus even very tiny vibration of a target can be easily observed. To validate the micro-Doppler effect in microwave radar, Chen used X band radar to detect trigonometric scatterer target with vibration amplitude 1 mm and vibration frequency 10 Hz, and successfully extracted micro-Doppler frequency shift in echo signal through time-frequency analysis method. It clearly shows that although the operational wavelength of microwave radar is far larger than that of laser radar, the micro-Doppler effect induced by target micromotion can still be effectively observed with the help of modern time-frequency analysis technique. In addition, he provided the micro-Doppler signal analysis result (as shown in Fig. 1.2) of a pedestrian with X band radar, which further demonstrated the potential application prospect of micro-Doppler effect in radar target recognition [3–5]. After that, research institutions published many related papers, which further discussed the promising potential of micro-Doppler effect in radar target recognition. The research on radar-based micro-Doppler effect gradually attracted increasing attention and focus.

1.3 MICRO-DOPPLER EFFECT RESEARCH AND APPLICATIONS

Since micromotion characteristics are unique motion features of radar targets, using high-resolution radar and modern signal processing technique to extract these kinds of tiny motion features can provide new solutions for noncooperative radar target detection and recognition. In recent years, the research on radar target micro-Doppler effect has been furthered with many valuable findings.

For various kinds of target micromotion dynamics in nature, rotation and variation are the two most basic micromotion forms. Therefore current research on micro-Doppler effect are mainly focused on them. Besides, complex forms of target micro-motion, like micro-Doppler effect induced by precession and nutation of warhead, human body walking, etc., have been thoroughly investigated. For the development history of micro-Doppler effect, it can be divided into two stages: narrowband radar micro-Doppler effect analysis and corresponding feature extraction; and wideband radar micro-Doppler effect analysis and corresponding feature extraction.

Researchers first turned their attention to the micro-Doppler effect in narrowband radar. Time-frequency analysis is the most widely used technique in it. Scholars like Chen conducted systematic research on micro-Doppler frequency shift of target acceleration, rotation, vibration, and roll motion, and provided detailed discussion on the use of high-resolution time-frequency analysis method in micro-Doppler feature extraction. T. Thayaparan applied self-adaptive joint time-frequency analysis and wavelet transform theory in the detection and separation of narrowband radar micro-Doppler signal, and successfully extracted micro-Doppler signal from helicopter and human body echo signals [6–8]. As the result of time-frequency analysis is the distribution of micro-Doppler signal in time-frequency plane, the micro-Doppler time-frequency spectrum through time-frequency analysis is usually called "time-frequency image" [9]. In narrowband radar, the micro-Doppler effect of a target can be observed, but the accuracy of micro-Doppler parameter estimation is not satisfying due to the limits of time-frequency analysis resolution. Especially for multicomponent micro-Doppler signal, the parameter estimation is even more difficult.

With the increase of radar signal bandwidth, the extraction of the target tiny motion feature becomes possible. In wideband radar, thanks to the high resolution of wideband signal in range, the micro-Doppler effect within each range bin can be separately analyzed for micro-Doppler feature extraction of higher accuracy. Meanwhile, motion parameters of target micromotion in part can be effectively extracted through analyzing the change of one-dimensional range profiles. After conduct translation compensation on the echo signal, the echo signal envelope of target main scatterers is aligned in range direction if the range migration over range bins is ignored. Therefore each main scatterer corresponds to a horizontal line in range–slow time plane (suppose that the vertical axis represents range and the horizontal axis represents slow time). Meanwhile, the motion of micromotion scatterer with respect to the main scatterer cannot be compensated, where obvious range migration can be observed. Therefore each micromotion scatterer corresponds to a curve determined by its equation of motion in range–slow time plane. Different from the aforementioned time-frequency image, the micro-Doppler effect of this case can be described by range–slow time image.

As an effective representation of target tiny motion characteristics, micro-Doppler effect has promising application prospects in the target recognition field. According to the present situation, the main applications of micro-Doppler effect can be summarized as follows.

1.3.1 Space Target Recognition

Space targets generally refer to targets like satellites, ballistic missiles, space debris, etc. Currently, micro-Doppler theory has been most thoroughly investigated and applied in midcourse warhead target identification [10,11]. The flight course of ballistic missiles can be generally divided into boost phase, midcourse, and reentry phase. The midcourse flight of a ballistic missile can take over 70% of the total flight time, and possesses the greatest difficulty of recognition and tracking at the same time. Therefore the recognition of midcourse target has become a current research hotspot. Ballistic missiles usually experience separation of warhead and rocket, bait release, coasting flight, and reentry in the whole flight from launching to attacking. Since target and bait are traveling nearly the same speed in coasting flight, and share similar size and scattering properties, recognition based on structure feature can hardly work. After entering midcourse flight, a warhead usually adopts self-rotation to maintain a stable pose state. According to rigid body attitude dynamics, a warhead target can be easily influenced by disturbance to generate precession and nutation when in self-rotation. In contrast, bait or a fake warhead will be displayed in roll motion or yaw motion as they have no pose control system. Besides, the self-rotation frequency of a warhead is usually just several hertz due to its relatively large height, while that of relatively light bait is generally larger to keep its stability. As discussed previously, there are significant differences between the micromotion characteristics of baits, fake warheads, and true warheads, which can be utilized for separation and recognition of different targets.

As early as March 29 and October 20 in 1990, the United States carried out two experiments called "Firefly" to verify the feasibility of midcourse ballistic missile true and fake warhead target recognition based on micromotion characteristics. In the experiment, baits are stored in a storage tank and expand to cone balloons after launched out in self-rotation. The controlled cone balloons simulated several different kinds of precession, and were observed by "Filepond" laser radar. Finally, the motion of bait over 700 km away was successfully observed, and several target motion parameters were estimated. MIT Lincoln Laboratory summarized the development of monitoring radar and related techniques in ballistic missile defense, and pointed out that ballistic missile recognition technique has evolved from narrowband radar-based radar cross-section analysis to wideband radar-based joint feature recognition including recognition of target structure and size, ballistic coefficient, and micromotion characteristics [12,13].

The US Navy missile defense committee conducted verification on ballistic missile defense based on sea-based radar, and put forward that micromotion characteristics are the unique characteristics of high-threat targets for missile defense radar systems, which can help radar to distinguish warhead targets from bait and debris. According to the report, the ground-based X band radar can measure the micromotion characteristics of threatening targets with great precision with an operational distance of over 2000 km and range resolution of 15 cm. Therefore the micromotion characteristics extraction of ballistic missile targets can provide a promising technique for ballistic missile target recognition.

1.3.2 Aerial Target Recognition

Aerial target usually refers to man-made targets like helicopters, propeller aircraft, jet aircraft, and biological targets like birds. For the observation of targets like helicopters, propeller aircraft, and jet aircraft, there are usually periodic micro-Doppler frequency components in echo signals when the target is in a certain pose, as the rotation parts of these aircraft targets will induce modulation on echo signal. Take a helicopter target as example: its main wing and tail wing will generate micro-Doppler signal [14]. It provides significant feature information for recognition of helicopter targets. To prove the micro-Doppler effect in helicopter targets, the US Army Research Laboratory used 92 GHz millimeter—wave band continuous wave radar to measure the micro-Doppler signal of an Mi-24 "Hind" D-type helicopter, analyzed its micro-Doppler spectrum, including the spectrum of jet engine, main wing, and tail wing, and measured the micro-Doppler features under different angle views. The experiment results showed that the recognition of helicopter targets can be effectively achieved through extraction of micro-Doppler spectrum feature in echo signal [15].

In fact, the additional modulation of rotor wings and engine blade rotation on echo signal, that is, "jet engine modulation (JEM)" [16], had been found in inverse synthetic aperture radar (ISAR) experiments before the concept of micro-Doppler was introduced into the radar target recognition field. JEM phenomenon refers to the fact that with regular ISAR imaging methods, micro-Doppler spectrum will influence the focuses of micromotion scatterers' corresponding range bins in azimuth and lead to interference stripes in the final ISAR image. It can be seen that the tail of an airplane in the ISAR image is almost submerged by the interference stripes of micro-Doppler effect [17]. This means that regular ISAR imaging methods are not suitable or even not applicable at all for radar targets with strong micro-Doppler scatterers. For this type of target, on the one hand, the micromotion characteristics can be extracted for target recognition; on the other hand, the micro-Doppler signal in echo signal can be cleared to obtain final imaging results of better quality. With this thought in mind, research [17] put forward a micro-Doppler signal separation method based on chirplet decomposition, realized the separation of micro-Doppler signal and target main body echo signal, and obtained clear imaging results of the jet aircraft.

Bird targets are a significant kind of aerial targets. The acquisition and processing of avian situation information is an essential problem in avian ecology [18–20], which is of significant essence for ecological protection, environment security monitoring, and wild animal epidemic prevention [21,22]. Avian situation detection and monitoring is also the basis of bird strike avoidance and aerial security guarantee. Radar becomes a powerful tool for avian situation monitoring thanks to its all-weather, all-time long-distance detection ability. Bird targets will cause faint micro-Doppler effect due to their wing flap during flight, which provides a critical basis for distinguishing bird targets from man-made targets [23–25]. There are significant differences in the micro-Doppler characteristics of bird wing vibration and other biological or nonbiological targets. Meanwhile, the micro-Doppler effects of different bird targets represent themselves differently due to the differences in flight features. By extracting the micro-Doppler characteristics, particular information about bird targets can be obtained, like flap frequency, wingspan, tail length, specific flap features, the change patterns of flapping and gliding, etc., which are quite useful for bird identification. For example, in the flight of birds of relatively large size, there are glide and flap two patterns with low flap frequency, while for birds with relatively small size, they keep flapping to maintain stability, and the flap frequency is usually higher.

1.3.3 Ground Motion Target Recognition

Micro-Doppler effect is also of great importance in the detection and recognition of ground targets like cars, tanks, armored cars, pedestrians, etc. Motor vehicles like cars, tanks, and armored cars are the most commonly known ground motion targets, whose micromotion contains several micromotion dynamics such as the rotation of wheels, the rotation of track, the engine-induced car body vibration, etc. The validation experiments of car micro-Doppler effect conducted in some countries showed that the micro-Doppler effect induced by wheel rotation will lead to interference stripes in the final ISAR image [26]. It can provide valuable features for vehicle target recognition. The man-portable surveillance tracking radar based on micro-Doppler feature made by Thales company from Great Britain, focusing on the Doppler spectrum of echo signal, is able to recognize three main kinds of ground targets (wheeled vehicles, tracked vehicles, and pedestrians) by Fisher linear criteria classifier, whose comprehensive recognition rate is over 80% [27]. Besides, the engine of motor vehicles can induce the vibration of the whole body when in working. The US Georgia engineering institute carried out experiments on the vibration extraction of a trailer bus with the funding of the US Department of Transportation. X band radar was used in the experiment to measure the frequency of vehicle-bridge system coupling vibration caused by a large trailer bus in heavy load when passing through the bridge. The experiment reached the conclusion that the vibration frequency of the trailer bus is usually 1−5 Hz, which is specifically determined by the load weight and load stack types [28]. The research findings in this field are applied in bridge health monitoring and vibration control to extend the service life of bridges. According to statistics, there are over 40 bridge collapse events in the world due to bridge damage, overload, and extreme loads (not including earthquakes). Vehicle−bridge system coupling vibration caused by large trailer load vehicles when passing through the bridge is one of the key factors related to bridge service life. If the vibration information of vehicles can be obtained before they pass through the bridge for active control and tuning of the bridge vibration frequency, the damage of vehicle−bridge system coupling vibration on bridges can possibly be reduced to extend the service life of bridge. Therefore there are quite promising prospects of vehicle vibration feature extraction based on micro-Doppler effect in civil applications.

Another important application of micro-Doppler effect is human body gait feature recognition. Compared to optics and infrared devices, radar possesses the advantages of all-weather, all-time, and good penetrability. The human body gait recognition based on micro-Doppler signal is of great significance in territory border monitoring and street battles. The research of human body micromotion characteristics mainly focuses on two aspects: human body micro-Doppler signal modeling and human body micro-Doppler recognition based on radar observation. Currently, many research institutes have participated in this work and collected large amounts of measured data. The research findings have pointed out that the intelligence recognition of human body targets with different gaits is quite possible through micro-Doppler characteristics analysis. Compared to the motion of man-made targets, the motion representation of human body targets is quite complex. In order to model the micro-Doppler signal of human body targets, division of the human body into head, body, big-handed arm (left/right), forearm (left/right), thigh (left/right), shank (left/right), and foot (left/right), etc. is usually adopted. According to the motion state of different parts, their specific motion is modeled to analyze the overall micro-Doppler feature [45]. On the basis of detailed division of human body motion, the micro-Doppler features caused by motion of different parts can be analyzed to provide the basis for human body target intelligent recognition.

After relevant in-depth study of these years, the recognition technique based on micro-Doppler feature has been accepted as an effective way to target precise identification. Compared to recognition based on imaging results, the micro-Doppler feature can provide better robustness. For example, the micromotion frequency feature does not vary with the change of radar line-of-sight angle, while regular imaging results vary significantly with the change of azimuth angle. As micromotion can be commonly found in natural targets and man-made targets, micro-Doppler effect analysis and micromotion feature extraction are expected to have wide applications, and will certainly attract more researchers.

1.4 THE ORGANIZATION OF THIS BOOK

This book is devoted to the conceptual description of radar target micro-Doppler effect basic theories and the explanation of corresponding micro-Doppler feature extraction technique. The main content can be divided into two parts: the first part covers micro-Doppler effect analysis, including Chapters 2–4; the second part introduces several micro-Doppler feature extraction techniques, including Chapters 5 and 6. The specific content in each chapter is as follows.

In Chapter 2, micro-Doppler effect in narrowband radar is discussed. Focusing on common motion dynamics like rotation, vibration, and precession, it establishes a mathematical model of target micromotion and analyzes their micro-Doppler effect in narrowband radar. Rotation, vibration, and precession are typical micromotion dynamics extracted from nature. Analyzing their micro-Doppler effects can provide a basic foundation for further discussions on micro-Doppler effects of complicated motions.

In Chapter 3, micro-Doppler effect in wideband radar is explained. The parametric expressions of micro-Doppler effect in range–slow time image for several common wideband signals, ie, linear frequency modulation (LFM) signal, stepped-frequency chirp signal (SFCS), and LFM continuous wave (LFMCW), are induced. LFM pulse signal is the most common signal in radar systems, whose micro-Doppler effect is quite typical, and similar to that of some other wideband signals like phase-coding pulse signal and frequency-coding pulse signal, etc. The micro-Doppler effect of SFCS can represent the micro-Doppler effect of other similar stepped-frequency signals. The conclusion obtained can also apply for stepped-frequency pulse signal. LFM CW radar is usually deployed in millimeter-wave seeker of antimissile missiles, whose micro-Doppler effect is quite essential for true-fake warhead identification in air defense and antimissile operation.

In Chapter 4, micro-Doppler effect in bistatic radar is investigated. The aforementioned analyses of micro-Doppler effect are all based on monostatic radar. A bistatic (multistatic) radar system is an important trend of future radar technology development. Separation of transmitter and receiver enables bistatic (multistatic) radar system some unique advantages compared to monostatic radar, which can also be supported by the ever-increasing in-depth research and wide applications. Therefore the research of micro-Doppler effect in bistatic (multistatic) radar is of great value. Compared to the micro-Doppler effect in monostatic radar, that of bistatic (multistatic) radar is special to some extent, but their basic generation mechanisms of micro-Doppler effect are similar. Hence, the micro-Doppler effect in bistatic radar is discussed in a single chapter in this book. The discussion is conducted with rotation as an example, where the micro-Doppler effects of bistatic radar with narrowband and wideband signal are analyzed, respectively. Chapter 4 is the basis of discussion of micro-Doppler effect in multi-input multi-output (MIMO) radar and three-dimensional micromotion characteristics reconstruction in Chapter 6.

In Chapter 5, micro-Doppler feature analysis and extraction is discussed. To obtain the micro-Doppler feature of target to achieve target identification through analyzing and extracting micromotion characteristics is the fundamental objective of radar target micro-Doppler effect research. Currently, micro-Doppler feature analysis and extraction has been widely investigated, and there are many examples in the literature of micro-Doppler feature analysis and extraction method for different applications. Time-frequency analysis method, image processing method, orthogonal matching pursuit decomposition method, empirical-mode decomposition method, and high-order moment function analysis method are introduced in this chapter; they have different unique advantages in micro-Doppler feature analysis and extraction. Chapter 5 is also the basis of three-dimensional micromotion characteristics reconstruction discussed in Chapter 6.

In Chapter 6, three-dimensional micromotion characteristics reconstruction methods are specified. In monostatic radar, the micro-Doppler characteristic parameters are determined by the projection of target micromotion part motion vector on line-of-sight direction; therefore only the structure and motion feature of target micromotion part on radar line-of-sight direction can be extracted. Due to the complexity of motion target posture, the target micro-Doppler feature will differ from each other significantly under different radar angle view, which will influence the accuracy of target recognition. On the basis of Chapter 5, micro-Doppler effect is extended into a multistatic radar case in Chapter 6. Rotation target three-dimensional micromotion characteristics reconstruction methods in narrow band and wideband MIMO radar are introduced, respectively, and precession target three-dimensional micromotion characteristics reconstruction method in wideband radar is also discussed.

Chapter 7 summarizes the content of this book and presents the application and development prospects of micro-Doppler effect theory.

REFERENCES

[1] Chen VC. Analysis of radar micro-Doppler signature with time-frequency transform. In: Proceedings of statistical signal and array processing, Pocono Manor, PA, USA; 2000. p. 463–6.

[2] Chen VC. Time-frequency signatures of micro-Doppler phenomenon for feature extraction. In: Proceedings of SPIE in wavelet applications VII, vol. 4056. Orlando, FL, USA: SPIE Press; 2000. p. 220–6.

[3] Chen VC, Li F. Analysis of micro-Doppler signatures. IEE Proceedings - Radar, Sonar and Navigation 2003;150(4):271–6.

[4] Chen VC, Li FY, Ho SS, et al. Micro-Doppler effect in radar: phenomenon, model and simulation study. IEEE Transactions on Aerospace and Electronic Systems 2006;42(1):2–21.

[5] Chen VC. Micro-Doppler effect of micro-motion dynamics: a review. In: Proceedings of SPIE on independent component analyses, wavelets, and neural networks, vol. 5102. Orlando, USA: IEEE Press; 2003. p. 240–9.

[6] Thayaparan T, Abrol S, Riseborough E. Micro-Doppler feature extraction of experimental helicopter data using wavelet and time-frequency analysis. In: RADAR 2004, proceedings of the international conference on radar systems; 2004.
[7] Thayaparan T, Abrol S, Riseborough E, et al. Analysis of radar micro-Doppler signatures from experimental helicopter and human data. IET Radar, Sonar & Navigation 2007;1(4):289–99.
[8] Thayaparan T, Lampropoulos G, Wong SK, et al. Application of adaptive joint time-frequency algorithm for focusing distorted ISAR images from simulated and measured radar data. IEE Proceedings - Radar, Sonar and Navigation 2003;150(4):213–20.
[9] Marple SL. Sharpening and bandwidth extrapolation techniques for radar micro-Doppler feature extraction. In: Proceedings of the international conference on radar. Adelaide, Australia: IEEE Press; 2003. p. 166–70.
[10] Gao H, Xie L, Wen S, et al. Micro-Doppler signature extraction from ballistic target with micro-motions. IEEE Transactions on Aerospace and Electronic Systems 2010;46(4):1969–82.
[11] Liang M, Jin L, Tao W, et al. Micro-Doppler characteristics of sliding-type scattering center on rotationally symmetric target. Science China Information Sciences 2011;54(9):1957–67.
[12] Lemnios William Z, Grometstein Alan A. Overview of the Lincoln laboratory ballistic missile defense program. The Lincoln Laboratory Journal 2002;13(1):9–32.
[13] Camp WW, Mayhan JT, O'Donnell RM. Wideband radar for ballistic missile defense and range-Doppler imaging of satellites. The Lincoln Laboratory Journal 2000;12(2):267–80.
[14] Chen VC, M William J, Braham H. Micro-Doppler analysis in ISAR – review and perspectives. In: International radar conference-surveillance for a safer world, Bordeaux, France; October 2009. p. 1–6.
[15] Nalecz M, Andrianik RR, Wojtkiewicz A. Micro-Doppler analysis of signal received by FMCW radar. In: Proceedings of international radar symposium. Dresden, Germany: IEEE Press; 2003. p. 231–5.
[16] Bell MR, Grubbs RA. JEM modeling and measurement for radar target identification. IEEE Transactions on Aerospace and Electronic Systems 1993;29(1):73–87.
[17] Li J, Ling H. Application of adaptive chirplet representation for ISAR feature extraction from targets with rotating parts. IEE Proceedings - Radar, Sonar and Navigation 2003;150(4):284–91.
[18] Jones J. Habitat selection studies in avian ecology: a critical review. The Auk 2001;118(2):557–62.
[19] Bruderer B, Peter D, Steuri T. Behaviour of migrating birds exposed to X-band radar and a bright light beam. The Journal of Experimental Biology 1999;202:1015–22.
[20] Ruth JM, Barrow WC, Sojda RS, et al. Advancing migratory bird conservation and management by using radar: an interagency collaboration. 2005. p. 1173. Open-File Report.
[21] Furness RW, Greenwood JJD. Birds as monitors of environmental change. London: Chapman Press; 1993.
[22] Liu J, Xiao H, Lei F, et al. Highly pathogenic H5N1 influenza virus infection in migratory birds. Science 2005;309(5738):1206.
[23] Ozcan AH, Baykut S, Sahinkaya DSA, Yalcin IK. Micro-Doppler effect analysis of single bird and bird flock for linear FMCW radar. In: 2012 20th signal processing and communications applications conference. Mugla: IEEE Press; 18–20 April, 2012. p. 1–4.
[24] Zhu F, Luo Y, Zhang Q, Feng Y-Q, Bai Y-Q. ISAR imaging for avian species identification with frequency-stepped chirp signals. IEEE Geoscience and Remote Sensing Letters 2010;7(1):151–5.
[25] Zhang Q, Zeng Y-S, He Y-Q, Luo Y. Avian detection and identification with high-resolution radar. In: 2008 IEEE radar conference. Rome, Italy; 26–30 May, 2008. p. 2194–9.
[26] Ghaleb A, Vignaud L, Nicolas JM. Micro-Doppler analysis of wheels and pedestrians in ISAR imaging. IET Signal Processing 2008;2(3):301–11.
[27] Stove AG, Sykes SR. A Doppler-based automatic target classifier for a battlefield surveillance radar. In: Proceedings of IEEE international conference on radar, Edinburgh, UK; October 2002. p. 419–23.
[28] Greneker G, Geisheimer J, Asbell D. Extraction of micro-Doppler from vehicle targets at X-band frequencies. In: Proceedings of SPIE on radar sensor technique, vol. 4374. Orlando, USA: SPIE Press; 2001. p. 1–9.

CHAPTER TWO

Micro-Doppler Effect in Narrowband Radar

Micro-Doppler effects in narrowband radar are introduced in this chapter. We mainly analyze the micro-Doppler effects induced by rotations, vibrations, and precessions of targets. Rotations, vibrations, and precessions are typical micromotion dynamics abstracted from a target's motion in the real world. Typical rotations include rotations of helicopter rotors, rotations of mechanical scanning radar antennas on a ship, rotations of turbine blades, etc. Typical vibrations include engine-induced car surface vibrations, mechanical oscillations of a bridge, etc. Precession is a major movement form of spatial targets, for example, precessions usually go with a ballistic missile in its midcourse flight. Analyzing the micro-Doppler effects of rotations, vibrations, and precessions provides a basic foundation for further discussions on micro-Doppler effects of complicated motions.

2.1 MICRO-DOPPLER EFFECT OF TARGETS WITH ROTATION

For better representation of the relative three-dimensional (3D) rotations of a target to the radar, three coordinate systems should be established: radar coordinate system, target-local coordinate system, and reference coordinate system. Among them, the radar coordinate system's origin is radar and keeps stationary; however, the target changes its pose; the target-local coordinate system's origin is the target center and moves with the target; the reference coordinate system is adopted to provide a coordinate transform relation between radar coordinate system and target-local coordinate system, whose origin is also the target center, and it has parallel axes with those of radar coordinate system and the same translation with that of the target.

As shown in Fig. 2.1, the radar is stationary and located at the origin Q of the radar coordinate system (U,V,W). The origin O of the reference coordinate system (X,Y,Z), which is parallel to the radar coordinate system, is located at a distance $\boldsymbol{R}_0$ from the radar with azimuth angle α and elevation angle β in the radar coordinate system. Suppose that the target can be modeled as a rigid body with a translation velocity $\boldsymbol{v}$ with respect to the radar. The target-local coordinate system (x,y,z) shares the same origin O with the reference coordinate system. The target rotates about the x-axis, y-axis, and z-axis with the angular rotation velocity ω_x, ω_y, and ω_z, respectively, when undergoing a translation,

Micro-Doppler Characteristics of Radar Targets
ISBN 978-0-12-809861-5, http://dx.doi.org/10.1016/B978-0-12-809861-5.00002-3

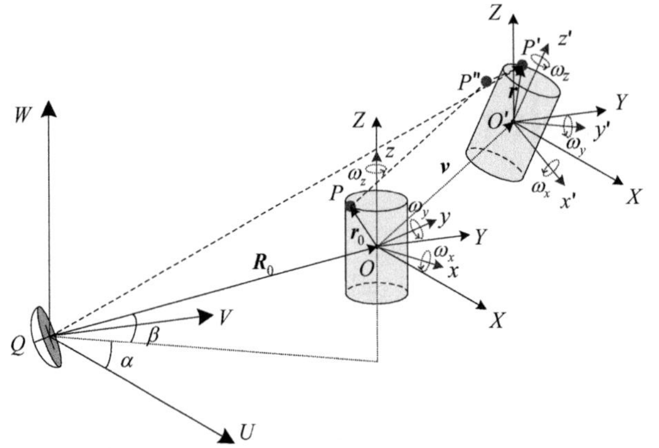

Figure 2.1 Geometry of the radar and a target with three-dimensional rotation.

which can be either represented in the target-local coordinate system as $\boldsymbol{\omega} = (\omega_x, \omega_y, \omega_z)^{\mathrm{T}}$ (the superscript "T" means transposition), or represented in the reference coordinate system as $\widehat{\omega} = (\omega_X, \omega_Y, \omega_Z)^{\mathrm{T}}$. A point scatterer P of the target is located at $\boldsymbol{r}_0 = (r_{x0}, r_{y0}, r_{z0})^{\mathrm{T}}$ in the target-local coordinate system and $\widehat{\boldsymbol{r}}_0 = (r_{X0}, r_{Y0}, r_{Z0})^{\mathrm{T}}$ in the reference coordinate system at the initial time. A single vector will be represented by different mathematical formulas in target-local coordinate system and reference coordinate system, whose transform relation is decided by Euler rotation matrix.

Now we introduce the concept of Euler rotation matrix. Euler rotation matrix is a transformation between the target-local coordinate system and the reference coordinate system and decided by the initial Euler angle $(\phi_e, \theta_e, \varphi_e)$. As is shown in Fig. 2.2, the target-local coordinate system transforms to the reference coordinate system when rotating about the z-axis for ϕ_e, the x-axis for θ_e, and the z-axis again for φ_e.

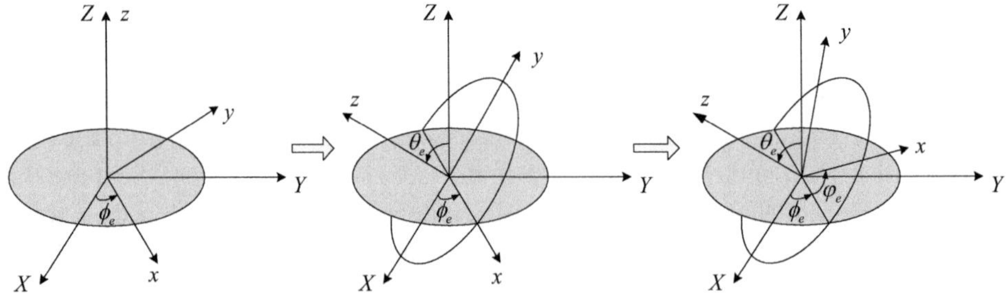

Figure 2.2 Geometry of Euler angles $(\phi_e, \theta_e, \varphi_e)$.

Suppose $\mathbf{R}_{\text{init}}$ is the Euler rotation matrix. $\mathbf{R}_{\text{init}}$ is decided by the initial Euler angle $(\phi_e, \theta_e, \varphi_e)$ and can be formulated by [1]

$$\mathbf{R}_{\text{init}} = \begin{bmatrix} \cos\varphi_e & \sin\varphi_e & 0 \\ -\sin\varphi_e & \cos\varphi_e & 0 \\ 0 & 0 & 1 \end{bmatrix} \begin{bmatrix} 1 & 0 & 0 \\ 0 & \cos\theta_e & \sin\theta_e \\ 0 & -\sin\theta_e & \cos\theta_e \end{bmatrix} \begin{bmatrix} \cos\phi_e & \sin\phi_e & 0 \\ -\sin\phi_e & \cos\phi_e & 0 \\ 0 & 0 & 1 \end{bmatrix} = \begin{bmatrix} a_{11} & a_{12} & a_{13} \\ a_{21} & a_{22} & a_{23} \\ a_{31} & a_{32} & a_{33} \end{bmatrix} \tag{2.1}$$

where

$$\begin{cases} a_{11} = \cos\phi_e \cos\varphi_e - \sin\phi_e \cos\theta_e \sin\varphi_e \\ a_{12} = -\cos\phi_e \sin\varphi_e - \sin\phi_e \cos\theta_e \cos\varphi_e \\ a_{13} = \sin\phi_e \sin\theta_e \\ a_{21} = \sin\phi_e \cos\varphi_e + \cos\phi_e \cos\theta_e \sin\varphi_e \\ a_{22} = -\sin\phi_e \sin\varphi_e + \cos\phi_e \cos\theta_e \cos\varphi_e \\ a_{23} = -\cos\phi_e \sin\theta_e \\ a_{31} = \sin\theta_e \sin\varphi_e \\ a_{32} = \sin\theta_e \cos\varphi_e \\ a_{33} = \cos\theta_e \end{cases} \tag{2.2}$$

Thus a vector represented in the target-local coordinate system will move to a new location in the reference coordinate system when multiplying by the rotation matrix $\mathbf{R}_{\text{init}}$, that is $\widehat{\boldsymbol{\omega}} = \mathbf{R}_{\text{init}}\boldsymbol{\omega}$ and $\widehat{\boldsymbol{r}}_0 = \mathbf{R}_{\text{init}}\boldsymbol{r}_0$.

The target center O at time $t = 0$ will move to O' at time t. Meanwhile, the reference coordinate system (X,Y,Z) will implement some translation accordingly, the target-local coordinate system will change to (x',y',z'), and the point scatterer P will move to P', whose corresponding vector is $\boldsymbol{r} = (r_x, r_y, r_z)^{\mathrm{T}}$ in the target-local coordinate system or $\widehat{\boldsymbol{r}} = (r_X, r_Y, r_Z)^{\mathrm{T}} = \mathbf{R}_{\text{init}}\boldsymbol{r}$ in the reference coordinate system. As shown in Fig. 2.1, the movement of point P consists of two steps: (1) translation from P to P'' with a velocity $\boldsymbol{v}$, that is, $\overrightarrow{PP''} = \boldsymbol{v}t$, and (2) rotation from P'' to P' with an angular velocity $\boldsymbol{\omega}$. And the rotation can be represented as a rotation matrix $\mathbf{R}_{\text{rotating}}$, that is, $\widehat{\boldsymbol{r}} = \mathbf{R}_{\text{rotating}}\widehat{\boldsymbol{r}}_0$. Thus the range vector from the radar to the scatterer at P' can be derived as

$$\overrightarrow{QP'} = \overrightarrow{QO} + \overrightarrow{OO'} + \overrightarrow{O'P'} = \boldsymbol{R}_0 + \boldsymbol{v}t + \mathbf{R}_{\text{rotating}}\widehat{\boldsymbol{r}}_0, \tag{2.3}$$

and the scalar range becomes

$$\begin{aligned} r(t) &= \left\| \boldsymbol{R}_0 + \boldsymbol{v}t + \mathbf{R}_{\text{rotating}}\widehat{\boldsymbol{r}}_0 \right\| \\ &= \sqrt{\left(\boldsymbol{R}_0 + \boldsymbol{v}t + \mathbf{R}_{\text{rotating}}\widehat{\boldsymbol{r}}_0\right)^{\mathrm{T}}\left(\boldsymbol{R}_0 + \boldsymbol{v}t + \mathbf{R}_{\text{rotating}}\widehat{\boldsymbol{r}}_0\right)}, \end{aligned} \tag{2.4}$$

where $\|\cdot\|$ represents the Euclidean norm.

If the radar transmits a single frequency continuous-wave signal with carrier frequency f_c as

$$p(t) = \exp(\mathrm{j}2\pi f_c t), \tag{2.5}$$

the echo signal from scatterer P' is

$$s(t) = \sigma \exp\left(\mathrm{j}2\pi f_c\left(t - \frac{2r(t)}{c}\right)\right), \tag{2.6}$$

where σ is the scattering coefficient of scatterer P'. After conducting baseband transform on $s(t)$, it can be obtained that

$$s_b(t) = \sigma \exp\left(\mathrm{j}2\pi f_c \frac{2r(t)}{c}\right) = \sigma \exp(\mathrm{j}\Phi(t)), \tag{2.7}$$

where its phase is $\Phi(t) = 2\pi f_c \frac{2r(t)}{c}$. By taking the derivative of the phase term on t, combined with Eq. (2.4), the Doppler frequency of the echo signal can be derived as

$$
\begin{aligned}
f_d &= \frac{1}{2\pi}\frac{\mathrm{d}\Phi(t)}{\mathrm{d}t} = \frac{2f_c}{c}\frac{\mathrm{d}}{\mathrm{d}t}r(t) \\
&= \frac{2f_c}{c}\frac{1}{2r(t)}\frac{\mathrm{d}}{\mathrm{d}t}\left[\left(\boldsymbol{R}_0 + \boldsymbol{v}t + \mathbf{R}_{\text{rotating}}\widehat{\boldsymbol{r}}_0\right)^{\mathrm{T}}\left(\boldsymbol{R}_0 + \boldsymbol{v}t + \mathbf{R}_{\text{rotating}}\widehat{\boldsymbol{r}}_0\right)\right] \\
&= \frac{2f_c}{c}\frac{1}{2r(t)}\left[\frac{\mathrm{d}}{\mathrm{d}t}\left(\boldsymbol{R}_0 + \boldsymbol{v}t + \mathbf{R}_{\text{rotating}}\widehat{\boldsymbol{r}}_0\right)^{\mathrm{T}}\left(\boldsymbol{R}_0 + \boldsymbol{v}t + \mathbf{R}_{\text{rotating}}\widehat{\boldsymbol{r}}_0\right)\right. \\
&\quad \left. + \left(\boldsymbol{R}_0 + \boldsymbol{v}t + \mathbf{R}_{\text{rotating}}\widehat{\boldsymbol{r}}_0\right)^{\mathrm{T}}\frac{\mathrm{d}}{\mathrm{d}t}\left(\boldsymbol{R}_0 + \boldsymbol{v}t + \mathbf{R}_{\text{rotating}}\widehat{\boldsymbol{r}}_0\right)\right] \\
&= \frac{2f_c}{c}\frac{1}{2r(t)}\left[\left(\boldsymbol{v} + \frac{\mathrm{d}}{\mathrm{d}t}\left(\mathbf{R}_{\text{rotating}}\widehat{\boldsymbol{r}}_0\right)\right)^{\mathrm{T}}\left(\boldsymbol{R}_0 + \boldsymbol{v}t + \mathbf{R}_{\text{rotating}}\widehat{\boldsymbol{r}}_0\right)\right. \\
&\quad \left. + \left(\boldsymbol{R}_0 + \boldsymbol{v}t + \mathbf{R}_{\text{rotating}}\widehat{\boldsymbol{r}}_0\right)^{\mathrm{T}}\left(\boldsymbol{v} + \frac{\mathrm{d}}{\mathrm{d}t}\left(\mathbf{R}_{\text{rotating}}\widehat{\boldsymbol{r}}_0\right)\right)\right] \\
&= \frac{2f_c}{c}\frac{1}{2r(t)}\cdot 2\left(\boldsymbol{v} + \frac{\mathrm{d}}{\mathrm{d}t}\left(\mathbf{R}_{\text{rotating}}\widehat{\boldsymbol{r}}_0\right)\right)^{\mathrm{T}}\left(\boldsymbol{R}_0 + \boldsymbol{v}t + \mathbf{R}_{\text{rotating}}\widehat{\boldsymbol{r}}_0\right) \\
&= \frac{2f_c}{c}\left[\boldsymbol{v} + \frac{\mathrm{d}}{\mathrm{d}t}\left(\mathbf{R}_{\text{rotating}}\widehat{\boldsymbol{r}}_0\right)\right]^{\mathrm{T}}\boldsymbol{n},
\end{aligned}
\tag{2.8}
$$

where $\boldsymbol{n} = \left(\boldsymbol{R}_0 + \boldsymbol{v}t + \mathbf{R}_{\text{rotating}}\widehat{\boldsymbol{r}}_0\right)/\left(\left\|\boldsymbol{R}_0 + \boldsymbol{v}t + \mathbf{R}_{\text{rotating}}\widehat{\boldsymbol{r}}_0\right\|\right)$ is the unit vector of $\overrightarrow{QP'}$. When the target is located in the far field of radar, we can get $\|\boldsymbol{R}_0\| \geq \left\|\boldsymbol{v}t + \mathbf{R}_{\text{rotating}}\widehat{\boldsymbol{r}}_0\right\|$, thus it can be approximated that $\boldsymbol{n} \approx \boldsymbol{R}_0/\|\boldsymbol{R}_0\|$, which is actually the unit vector in radar line-of-sight direction. It can be found that the first term of Eq. (2.8) is the Doppler frequency induced by target translation:

$$
f_{\text{Doppler}} = \frac{2f_c}{c}\boldsymbol{v}^{\mathrm{T}}\boldsymbol{n} = \frac{2\boldsymbol{v}^{\mathrm{T}}\boldsymbol{n}}{\lambda}, \tag{2.9}
$$

where λ is the signal wavelength. Certainly, Eq. (2.9) is consistent with Eq. (1.9). The second term is the micro-Doppler frequency induced by target rotation:

$$f_{\text{micro-Doppler}} = \frac{2f_c}{c}\left[\frac{\mathrm{d}}{\mathrm{d}t}\left(\mathbf{R}_{\text{rotating}}\,\overset{\frown}{\boldsymbol{r}}_0\right)\right]^{\mathrm{T}}\boldsymbol{n} \tag{2.10}$$

Eq. (2.10) is the general expression for micro-Doppler frequency induced by target rotation. For further derivation of micro-Doppler frequency expression, specific expression of $\mathbf{R}_{\text{rotating}}$ is needed. Here, we briefly introduce 3D rotation matrix, also called SO(3) group, special orthogonal matrix groups. SO(3) group is defined as

$$\mathrm{SO}(3) = \left\{\mathbf{R}_t \in \mathbf{R}^{3\times 3} \middle| \mathbf{R}_t\mathbf{R}_t^{\mathrm{T}} = \mathbf{I}, \det(\mathbf{R}_t) = 1\right\} \tag{2.11}$$

Taking the time derivative of both sides of $\mathbf{R}_t(t)\mathbf{R}_t^{\mathrm{T}}(t) = \mathbf{I}$, we can get

$$\mathbf{R}_t'(t)\mathbf{R}_t^{\mathrm{T}}(t) + \mathbf{R}_t(t)\mathbf{R}_t'^{\mathrm{T}}(t) = 0 \tag{2.12}$$

Thus

$$\mathbf{R}_t'(t)\mathbf{R}_t^{\mathrm{T}}(t) = -\mathbf{R}_t(t)\mathbf{R}_t'^{\mathrm{T}}(t) = -\left[\mathbf{R}_t'(t)\mathbf{R}_t^{\mathrm{T}}(t)\right]^{\mathrm{T}}, \tag{2.13}$$

which means that $\mathbf{R}_t'(t)\mathbf{R}_t^{\mathrm{T}}(t)$ is a skew symmetric matrix. Consider the following skew symmetric matrix generated by $\overset{\frown}{\boldsymbol{\omega}} = (\omega_X, \omega_Y, \omega_Z)^{\mathrm{T}}$:

$$\widehat{\boldsymbol{\omega}} = \begin{bmatrix} 0 & -\omega_Z & \omega_Y \\ \omega_Z & 0 & -\omega_X \\ -\omega_Y & \omega_X & 0 \end{bmatrix}. \tag{2.14}$$

Let $\widehat{\boldsymbol{\omega}} = \mathbf{R}_t'(t)\mathbf{R}_t^{\mathrm{T}}(t)$, and post multiply both sides by $\mathbf{R}_t(t)$, then we can get

$$\mathbf{R}_t'(t) = \widehat{\boldsymbol{\omega}}\mathbf{R}_t(t) \tag{2.15}$$

The solution of this ordinary linear differential equation is

$$\mathbf{R}_t(t) = \exp(\widehat{\boldsymbol{\omega}}t)\mathbf{R}_t(0), \tag{2.16}$$

where

$$\exp(\widehat{\boldsymbol{\omega}}t) = \mathbf{I} + \widehat{\boldsymbol{\omega}}t + \frac{(\widehat{\boldsymbol{\omega}}t)^2}{2!} + \cdots + \frac{(\widehat{\boldsymbol{\omega}}t)^n}{n!} + \cdots \tag{2.17}$$

Suppose the initial condition $\mathbf{R}_t(0) = \mathbf{I}$, then

$$\mathbf{R}_t(t) = \exp(\widehat{\boldsymbol{\omega}}t) \tag{2.18}$$

It is easy to verify that $\exp(\widehat{\boldsymbol{\omega}}t)$ is a 3D rotation matrix. Since

$$[\exp(\widehat{\boldsymbol{\omega}}t)]^{-1} = \exp(-\widehat{\boldsymbol{\omega}}t) = \exp\left(\widehat{\boldsymbol{\omega}}^{\mathrm{T}}t\right) = [\exp(\widehat{\boldsymbol{\omega}}t)]^{\mathrm{T}} \tag{2.19}$$

Thus we can get $[\exp(\widehat{\boldsymbol{\omega}}t)]^{\mathrm{T}}[\exp(\widehat{\boldsymbol{\omega}}t)] = \mathbf{I}$ and $\det[\exp(\widehat{\boldsymbol{\omega}}t)] = \pm 1$. We derive $\det[\exp(\widehat{\boldsymbol{\omega}}t)] = 1$ from

$$\det[\exp(\widehat{\boldsymbol{\omega}}t)] = \det\left[\exp\left(\frac{\widehat{\boldsymbol{\omega}}t}{2}\right)\cdot\exp\left(\frac{\widehat{\boldsymbol{\omega}}t}{2}\right)\right] = \left\{\det\left[\exp\left(\frac{\widehat{\boldsymbol{\omega}}t}{2}\right)\right]\right\}^2 \geq 0 \tag{2.20}$$

Therefore we verify that $\exp(\widehat{\boldsymbol{\omega}}t)$ is a 3D rotation matrix, which means rotating for Ωt with the angular velocity $\widehat{\boldsymbol{\omega}} = (\omega_X, \omega_Y, \omega_Z)^{\mathrm{T}}$ in physical interpretation where $\Omega = \left\|\widehat{\boldsymbol{\omega}}\right\| = \|\boldsymbol{\omega}\|$.

The unit vector of rotation angular velocity is $\widehat{\boldsymbol{\omega}}' = \widehat{\boldsymbol{\omega}}/\Omega = \left(\omega'_X, \omega'_Y, \omega'_Z\right)^{\mathrm{T}}$ in the reference coordinate system, and we construct the following skew symmetric matrix:

$$\widehat{\boldsymbol{\omega}}' = \begin{bmatrix} 0 & -\omega'_Z & \omega'_Y \\ \omega'_Z & 0 & -\omega'_X \\ -\omega'_Y & \omega'_X & 0 \end{bmatrix} \tag{2.21}$$

Therefore we can get $\mathbf{R}_{\text{rotating}} = \exp(\Omega\widehat{\boldsymbol{\omega}}'t)$. It is easy to verify $\widehat{\boldsymbol{\omega}}'^3 = -\widehat{\boldsymbol{\omega}}'$, thus

$$\begin{aligned}\mathbf{R}_{\text{rotating}} &= \exp(\Omega\widehat{\boldsymbol{\omega}}'t) = \mathbf{I} + \Omega\widehat{\boldsymbol{\omega}}'t + \frac{(\Omega\widehat{\boldsymbol{\omega}}'t)^2}{2!} + \cdots + \frac{(\Omega\widehat{\boldsymbol{\omega}}'t)^n}{n!} + \cdots \\ &= \mathbf{I} + \left(\Omega t - \frac{(\Omega t)^3}{3!} + \frac{(\Omega t)^5}{5!} - \cdots\right)\widehat{\boldsymbol{\omega}}' + \left(\frac{(\Omega t)^2}{2!} - \frac{(\Omega t)^4}{4!} + \frac{(\Omega t)^6}{6!} - \cdots\right)\widehat{\boldsymbol{\omega}}'^2 \\ &= \mathbf{I} + \widehat{\boldsymbol{\omega}}'\sin(\Omega t) + \widehat{\boldsymbol{\omega}}'^2(1 - \cos(\Omega t))\end{aligned} \tag{2.22}$$

Substitute Eq. (2.22) into Eq. (2.10), and get

$$\begin{aligned}f_{\text{micro-Doppler}} &= \frac{2f_c}{c}\left[\frac{\mathrm{d}}{\mathrm{d}t}\left(\mathbf{R}_{\text{rotating}}\widehat{\boldsymbol{r}}_0\right)\right]^{\mathrm{T}}\boldsymbol{n} = \frac{2f_c}{c}\left[\frac{\mathrm{d}}{\mathrm{d}t}\left(\exp(\Omega\widehat{\boldsymbol{\omega}}'t)\widehat{\boldsymbol{r}}_0\right)\right]^{\mathrm{T}}\boldsymbol{n} \\ &= \frac{2f_c}{c}\left[\Omega\widehat{\boldsymbol{\omega}}'\exp(\Omega\widehat{\boldsymbol{\omega}}'t)\widehat{\boldsymbol{r}}_0\right]^{\mathrm{T}}\boldsymbol{n} \\ &= \frac{2f_c}{c}\left\{\Omega\widehat{\boldsymbol{\omega}}'\left[\mathbf{I} + \widehat{\boldsymbol{\omega}}'\sin(\Omega t) + \widehat{\boldsymbol{\omega}}'^2(1 - \cos(\Omega t))\right]\mathbf{R}_{\text{init}}\boldsymbol{r}_0\right\}^{\mathrm{T}}\boldsymbol{n} \\ &= \frac{2f_c\Omega}{c}\left\{\left[\widehat{\boldsymbol{\omega}}'^2\sin(\Omega t) - \widehat{\boldsymbol{\omega}}'^3\cos(\Omega t) + \widehat{\boldsymbol{\omega}}'\left(\mathbf{I} + \widehat{\boldsymbol{\omega}}'^2\right)\right]\mathbf{R}_{\text{init}}\boldsymbol{r}_0\right\}^{\mathrm{T}}\boldsymbol{n}\end{aligned} \tag{2.23}$$

Substituting $\widehat{\boldsymbol{\omega}}'^{3} = -\widehat{\boldsymbol{\omega}}'$ into Eq. (2.23), the expression of micro-Doppler effect induced by rotation can be further written as

$$f_{\text{micro-Doppler}} = \frac{2f_c\Omega}{c}\left\{\left[\widehat{\boldsymbol{\omega}}'^{2}\sin(\Omega t) + \widehat{\boldsymbol{\omega}}'\cos(\Omega t)\right]\mathbf{R}_{\text{init}}\boldsymbol{r}_0\right\}^{\mathrm{T}}\boldsymbol{n} \qquad (2.24)$$

As is shown in Eq. (2.24), micro-Doppler frequency induced by rotation is a time-varying sine curve whose amplitude (the maximum micro-Doppler frequency) is determined by signal carrier frequency f_c, target rotation angular velocity $\boldsymbol{\omega}$, rotation vector $\boldsymbol{r}_0$, and the unit vector $\boldsymbol{n}$ in line-of-sight direction. The angular frequency of sine curve is the rotation angular frequency Ω. Therefore the period of target rotation can be extracted based on analysis of the change of echo signal micro-Doppler frequency.

Additionally, according to the following characteristic of skew symmetric matrix like Eq. (2.21):

$$\widehat{\boldsymbol{\omega}}'\widehat{\boldsymbol{r}} = \begin{bmatrix} 0 & -\omega'_Z & \omega'_Y \\ \omega'_Z & 0 & -\omega'_X \\ -\omega'_Y & \omega'_X & 0 \end{bmatrix}\begin{bmatrix} r_X \\ r_Y \\ r_Z \end{bmatrix} = \begin{bmatrix} \omega'_Y r_Z - \omega'_Z r_Y \\ \omega'_Z r_X - \omega'_X r_Z \\ \omega'_X r_Y - \omega'_Y r_X \end{bmatrix} = \widehat{\boldsymbol{\omega}}' \times \widehat{\boldsymbol{r}} \qquad (2.25)$$

Eq (2.23) can be simplified as

$$\begin{aligned} f_{\text{micro-Doppler}} &= \frac{2f_c}{c}\left[\Omega\widehat{\boldsymbol{\omega}}'\exp(\Omega\widehat{\boldsymbol{\omega}}'t)\widehat{\boldsymbol{r}}_0\right]^{\mathrm{T}}\boldsymbol{n} \\ &= \frac{2f_c}{c}\left(\Omega\widehat{\boldsymbol{\omega}}'\widehat{\boldsymbol{r}}\right)^{\mathrm{T}}\boldsymbol{n} = \frac{2f_c}{c}\left(\Omega\boldsymbol{\omega}' \times \widehat{\boldsymbol{r}}\right)^{\mathrm{T}}\boldsymbol{n} \end{aligned} \qquad (2.26)$$

The simulation: Assume that carrier frequency of the transmitted signal is $f_c = 10$ GHz and a target, located at ($U = 3$ km, $V = 4$ km, $W = 5$ km), is rotating round the x, y, and z axes with the initial Euler angles $(0,\pi/4,\pi/5)$, angular velocity $\boldsymbol{\omega} = (\pi,2\pi,\pi)^{\mathrm{T}}$ rad/s, and period of rotation $T = 2\pi/\|\boldsymbol{\omega}\| = 0.8165$ s but no translations. Suppose that the target can be modeled by three scatterers, whose coordinates are (0,0,0), (3,1.5,1.5), and (−3,−1.5,−1.5), respectively, and the unit is meter. Radar irradiates the target for 3 s. The theoretical time-varying micro-Doppler frequency curve of three scatterers is shown in Fig. 2.3A, where point (0,0,0) is located at the rotation center with no micro-Doppler frequency, corresponding to the straight line in the figure; point (3,1.5,1.5) and point (−3,−1.5,−1.5) are corresponding to two sine curves, respectively, whose periods fairly fit with the target rotation periods.

Since micro-Doppler signal is time varying, traditional Fourier analysis method is not appropriate for observing the relation between micro-Doppler frequency and time.

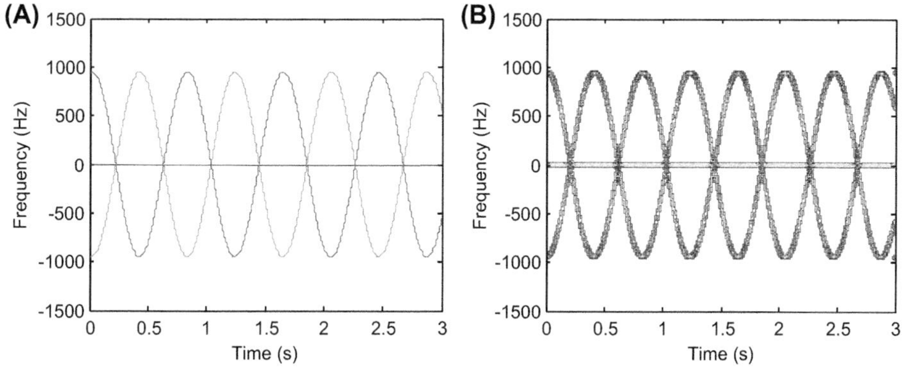

Figure 2.3 Simulation of micro-Doppler effect of target with rotation. (A) Theoretical curve of micro-Doppler frequency; (B) time-frequency analysis of micro-Doppler signal.

High-resolution time-frequency analysis methods are needed and will be introduced in Chapter 5 in detail. To verify the validity of the aforementioned micro-Doppler signal expression, the result of time-frequency analysis on echo signal (here Gabor transform is applied) is shown in Fig. 2.3B. It can be seen that the micro-Doppler frequency curve is consistent with the theoretical result.

2.2 MICRO-DOPPLER EFFECT OF TARGETS WITH VIBRATION

Fig. 2.4 shows the geometry of general targets with vibration. The radar is located at the origin Q of the radar coordinate system (U, V, W). The origin O of the reference coordinate system (X, Y, Z), which is parallel to the radar coordinate system, is located at a distance $\boldsymbol{R}_0$ from the radar with azimuth angle α and elevation angle β in the radar coordinate system. Suppose that the target can be modeled as a rigid body with a

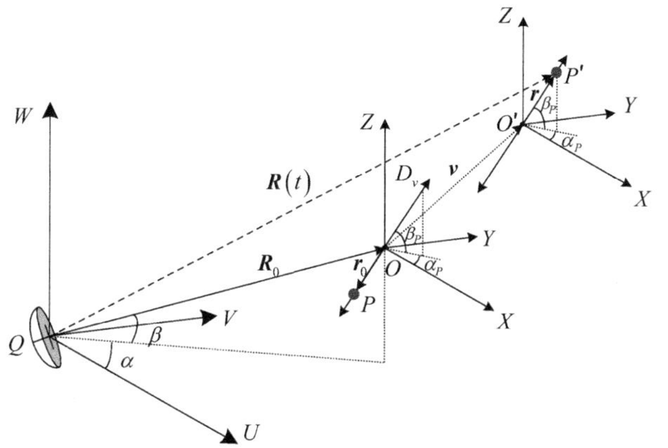

Figure 2.4 Geometry for the radar and a 3D vibrating point target.

translation velocity $\boldsymbol{v}$ with respect to the radar. While the target is in translation, the scatterer P is vibrating along some orientational axis at a frequency f_v with an amplitude D_v and the azimuth and elevation angle of the orientational axis in the reference coordinates are α_p and β_p, respectively. The corresponding vector of point P in the reference coordinate system is $\boldsymbol{r}_0 = (r_{X0}, r_{Y0}, r_{Z0})^{\mathrm{T}}$, and assume that point P is at the vibration center at the initial time. The target center O at time $t = 0$ will move to O' at time t. Meanwhile, the reference coordinate system (X,Y,Z) will implement some translation accordingly, and the point scatterer P will move to P', whose corresponding vector changes for $\boldsymbol{r} = (r_X, r_Y, r_Z)^{\mathrm{T}}$ in the reference coordinate system. The range vector from the radar to the scatterer at P' can be derived as

$$\overrightarrow{QP'} = \overrightarrow{QO} + \overrightarrow{OO'} + \overrightarrow{O'P'} = \boldsymbol{R}_0 + \boldsymbol{v}t + \boldsymbol{r} \tag{2.27}$$

and the scalar range becomes

$$\boldsymbol{r}(t) = \|\boldsymbol{R}_0 + \boldsymbol{v}t + \boldsymbol{r}\| \tag{2.28}$$

If the radar transmits a single frequency continuous-wave signal with carrier frequency f_c as

$$p(t) = \exp(\mathrm{j}2\pi f_c t) \tag{2.29}$$

the echo signal from scatterer P' is

$$s(t) = \sigma \exp\left(\mathrm{j}2\pi f_c\left(t - \frac{2r(t)}{c}\right)\right) \tag{2.30}$$

After conducting baseband transform on $s(t)$, it can be obtained that

$$s_b(t) = \sigma \exp\left(\mathrm{j}2\pi f_c \frac{2r(t)}{c}\right) = \sigma \exp(\mathrm{j}\Phi(t)), \tag{2.31}$$

where its phase is $\Phi(t) = 2\pi f_c \frac{2r(t)}{c}$. By taking the time derivative of the phase, the Doppler frequency shift of the returned signal can be derived as

$$\begin{aligned} f_d &= \frac{1}{2\pi}\frac{\mathrm{d}\Phi(t)}{\mathrm{d}t} = \frac{2f_c}{c}\frac{\mathrm{d}}{\mathrm{d}t}r(t) \\ &= \frac{2f_c}{c}\frac{1}{2r(t)}\frac{\mathrm{d}}{\mathrm{d}t}\left[(\boldsymbol{R}_0 + \boldsymbol{v}t + \boldsymbol{r})^{\mathrm{T}}(\boldsymbol{R}_0 + \boldsymbol{v}t + \boldsymbol{r})\right], \\ &= \frac{2f_c}{c}\left(\boldsymbol{v} + \frac{\mathrm{d}\boldsymbol{r}}{\mathrm{d}t}\right)^{\mathrm{T}}\boldsymbol{n} \end{aligned} \tag{2.32}$$

where $\boldsymbol{n} = (\boldsymbol{R}_0 + \boldsymbol{v}t + \boldsymbol{r})/(\|\boldsymbol{R}_0 + \boldsymbol{v}t + \boldsymbol{r}\|)$ is the unit vector of $\overrightarrow{QP'}$. If the target is located in the far field of the radar, we can get $\|\boldsymbol{R}_0\| \geq \|\boldsymbol{v}t + r\|$, thus $\boldsymbol{n}$ can be approximated as $\boldsymbol{n} \approx \boldsymbol{R}_0/\|\boldsymbol{R}_0\|$, which is the unit vector in the direction of radar sight line. And on calculation, we can get

$$\boldsymbol{n} = [\cos\alpha\cos\beta, \sin\alpha\cos\beta, \sin\beta]^{\mathrm{T}} \tag{2.33}$$

It can be found that the first term of Eq. (2.32) is the Doppler frequency induced by target translation:

$$f_{\text{Doppler}} = \frac{2f_c}{c}\boldsymbol{v}^{\mathrm{T}}\boldsymbol{n} = \frac{2\boldsymbol{v}^{\mathrm{T}}\boldsymbol{n}}{\lambda}, \tag{2.34}$$

where λ is the wavelength of transmitted signal. The second term is the Doppler shift induced by vibration of target:

$$f_{\text{micro-Doppler}} = \frac{2f_c}{c}\frac{\mathrm{d}\boldsymbol{r}^{\mathrm{T}}}{\mathrm{d}t}\boldsymbol{n} \tag{2.35}$$

At the time t, the distance from P' to O' changes for $D_v \sin(2\pi f_v t)$ and

$$\boldsymbol{r} = \begin{bmatrix} r_X \\ r_Y \\ r_Z \end{bmatrix} = D_v \sin(2\pi f_v t)\begin{bmatrix} \cos\alpha_P\cos\beta_P \\ \sin\alpha_P\cos\beta_P \\ \sin\beta_P \end{bmatrix} + \begin{bmatrix} r_{X0} \\ r_{Y0} \\ r_{Z0} \end{bmatrix} \tag{2.36}$$

Thus

$$\frac{\mathrm{d}\boldsymbol{r}^{\mathrm{T}}}{\mathrm{d}t} = 2\pi f_v D_v \cos(2\pi f_v t)(\cos\alpha_P\cos\beta_P, \sin\alpha_P\cos\beta_P, \sin\beta_P) \tag{2.37}$$

Substitute Eq. (2.33) and Eq. (2.37) into Eq. (2.35), and we can get

$$f_{\text{micro-Doppler}} = \frac{4\pi f_c f_v D_v}{c}[\cos(\alpha - \alpha_P)\cos\beta\cos\beta_P + \sin\beta\sin\beta_P]\cos(2\pi f_v t), \tag{2.38}$$

from which we can find that similarly with micro-Doppler effect induced by rotation, micro-Doppler frequency induced by vibration is also a time-varying sinusoidal curve whose amplitude (the maximum of micro-Doppler frequency) is determined by the carrier frequency f_c of the transmitted signal, the vibrating frequency f_v of the target, the amplitude D_v, azimuth angle, and elevation angle of the vibration center in the reference coordinate system and azimuth angle and elevation angle of the vibration axis in the reference coordinate system. The frequency of the sinusoidal curve is the vibrating

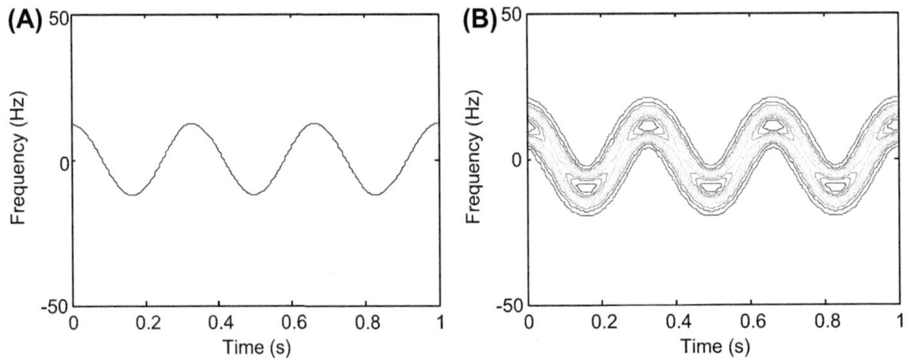

Figure 2.5 Simulation of micro-Doppler effect of target with vibration. (A) Theoretical curve of micro-Doppler frequency; (B) Time-frequency analysis of micro-Doppler signal.

frequency Ω of the target. Thus the vibrating frequency of the target can be derived based on analysis of the changing rule of the returning signal's micro-Doppler frequency and extracting the changing period. On the condition that α, β, α_P and β_P are all equal to 0, the amplitude in Eq. (2.38) reaches the maximum $4\pi f_c f_v D_v / c$.

The simulation: assume that carrier frequency of the transmitted signal is $f_c = 10$ GHz and vibration center of the target is located at ($U = 3$ km, $V = 4$ km, $W = 5$ km) with azimuth angle $\alpha = 0.9273$ rad and elevation angle $\beta = 0.7854$ rad at the initial time. Suppose the target is vibrating periodically at frequency 3 Hz with amplitude 0.01 m but no translations and the angle and elevation angle of the vibrating axis are $\alpha_P = \pi/5$ rad and $\beta_P = \pi/4$ rad, respectively. The irradiation time of the radar is 1 s. The theoretical time-varying micro-Doppler frequency curve of the vibrating scatterer points is shown in Fig. 2.5A and the result of time-frequency analysis on the returned signal is shown in Fig. 2.5B, which shows that the changing curve of micro-Doppler frequency is completely coincident with the theory.

2.3 MICRO-DOPPLER EFFECT OF TARGETS WITH PRECESSION

Precessions are usually together with spatial targets like ballistic targets. In addition to spinning around its axis of symmetry to keep stationary while going on a translation in outer space, ballistic targets will undergo a coning motion about some special orientation for a transverse force induced by the separation of missile and rocket together with the payload deployment course. The spinning and the coning make up precession, which is a significant motion dynamic to recognize the true-fake warheads in the midcourse flight of ballistic targets. So it is of great importance to analyze the micro-Doppler effects induced by precessions for recognition of true-fake warheads.

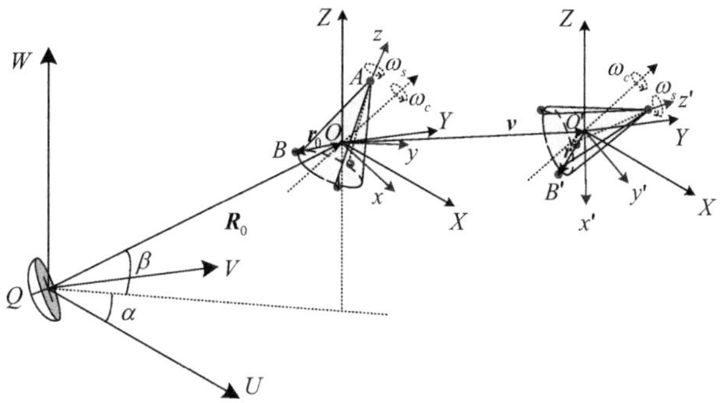

Figure 2.6 Geometry of the radar and a target with three-dimensional precession.

As shown in Fig. 2.6, coordinate system (U,V,W) is the radar coordinate system and the radar is stationary at the origin Q. The origin O of the reference coordinate system (X,Y,Z), which is parallel to the radar coordinate system, is located at a distance $\boldsymbol{R}_0$ from the radar with azimuth angle α and elevation angle β in the radar coordinate system. Suppose that the target can be modeled as a basiconic rigid body with a tail (or control surface) and it has a translation velocity $\boldsymbol{v}$ with respect to the radar. The target-local coordinate system (x,y,z) shares the same origin O with the reference coordinate system. The target is spinning around its axis of symmetry z with the angular velocity $\boldsymbol{\omega}_s$ and coning about some special orientation with the angular velocity $\boldsymbol{\omega}_c$ (the expressions of $\boldsymbol{\omega}_s$ and $\boldsymbol{\omega}_c$ are all represented in the reference coordinate system) while going on a translation. The angle between the axis of spinning and the axis of coning is called the precession angle. For a scatterer point A, which is located at the axis of spinning, the micromotion is spin with the angular velocity $\boldsymbol{\omega}_s$. And for a scatterer point B, which is located neither at the axis of spinning nor at the axis of coning, the micromotion is a combination of spin and cone, which is precession.

Suppose that the vector from O to B is represented as $\boldsymbol{r}_0 = (r_{x0}, r_{y0}, r_{z0})^{\mathrm{T}}$ in the target-local coordinate system and $\widehat{\boldsymbol{r}}_0 = \mathbf{R}_{\text{init}}\boldsymbol{r}_0$ in the reference coordinate system at the initial time. The target center O at time $t = 0$ will move to O' at time t. Meanwhile, the reference coordinate system (X,Y,Z) will implement some translation accordingly and the point scatterer B will move to B', whose corresponding vector changes for $\boldsymbol{r} = (r_x, r_y, r_z)^{\mathrm{T}}$ in the target-local coordinate system and $\widehat{\boldsymbol{r}} = \mathbf{R}_{\text{init}}\boldsymbol{r}$ in the reference coordinate system. As shown in Fig. 2.6, the movement of point B consists of three steps: (1) translation for $\boldsymbol{v}t$ with the reference coordinate system; (2) spin about the z-axis of target-local coordinate system and the vector from O to B changes for $\mathbf{R}_{\text{spinning}}\widehat{\boldsymbol{r}}_0$, where $\mathbf{R}_{\text{spinning}}$ is the rotation matrix of spinning; and (3) cone about the axis of coning and the vector from O to B changes again for $\mathbf{R}_{\text{coning}}\mathbf{R}_{\text{spinning}}\widehat{\boldsymbol{r}}_0$, where $\mathbf{R}_{\text{coning}}$ is the rotation matrix of coning. Thus the range vector from the radar to the scatterer at B' can be derived as

$$\overrightarrow{QB'} = \overrightarrow{QO} + \overrightarrow{OO'} + \overrightarrow{O'B'} = \mathbf{R}_0 + \boldsymbol{v}t + \mathbf{R}_{\text{coning}}\mathbf{R}_{\text{spinning}}\widehat{\boldsymbol{r}}_0 \quad (2.39)$$

and the scalar range becomes

$$r(t) = \left\| \boldsymbol{R}_0 + \boldsymbol{v}t + \mathbf{R}_{\text{coning}}\mathbf{R}_{\text{spinning}}\widehat{\boldsymbol{r}}_0 \right\|. \quad (2.40)$$

If the radar transmits a single frequency continuous wave with a carrier frequency nf_c, the signal returned from the point scatterer B' is

$$s_b(t) = \sigma \exp\left(\mathrm{j}2\pi f_c \frac{2r(t)}{c}\right) = \sigma \exp(\mathrm{j}\Phi(t)), \quad (2.41)$$

where the phase of the signal is $\Phi(t) = 2\pi f_c \frac{2r(t)}{c}$. By taking the time derivative of the phase, the Doppler frequency shift of the returned signal can be derived as

$$\begin{aligned} f_d &= \frac{1}{2\pi}\frac{\mathrm{d}\Phi(t)}{\mathrm{d}t} = \frac{2f_c}{c}\frac{\mathrm{d}}{\mathrm{d}t}r(t) \\ &= \frac{2f_c}{c}\frac{1}{2r(t)}\frac{\mathrm{d}}{\mathrm{d}t}\left[\left(\boldsymbol{R}_0 + \boldsymbol{v}t + \mathbf{R}_{\text{coning}}\mathbf{R}_{\text{spinning}}\widehat{\boldsymbol{r}}_0\right)^{\mathrm{T}}\left(\boldsymbol{R}_0 + \boldsymbol{v}t + \mathbf{R}_{\text{coning}}\mathbf{R}_{\text{spinning}}\widehat{\boldsymbol{r}}_0\right)\right], \\ &= \frac{2f_c}{c}\left[\boldsymbol{v} + \frac{\mathrm{d}}{\mathrm{d}t}\left(\mathbf{R}_{\text{coning}}\mathbf{R}_{\text{spinning}}\widehat{\boldsymbol{r}}_0\right)\right]^{\mathrm{T}}\boldsymbol{n}, \end{aligned} \quad (2.42)$$

where $\boldsymbol{n} = \left(\boldsymbol{R}_0 + \boldsymbol{v}t + \mathbf{R}_{\text{coning}}\mathbf{R}_{\text{spinning}}\widehat{\boldsymbol{r}}_0\right) / \left(\left\|\boldsymbol{R}_0 + \boldsymbol{v}t + \mathbf{R}_{\text{coning}}\mathbf{R}_{\text{spinning}}\widehat{\boldsymbol{r}}_0\right\|\right)$ is the unit vector of $\overrightarrow{QB'}$. If the target is located in the far field of the radar, we can get $\|\boldsymbol{R}_0\| \geq \left\|\boldsymbol{v}t + \mathbf{R}_{\text{coning}}\mathbf{R}_{\text{spinning}}\widehat{\boldsymbol{r}}_0\right\|$, thus n can be approximated as $\boldsymbol{n} \approx \boldsymbol{R}_0/\|\boldsymbol{R}_0\|$, which is the unit vector in the direction of radar sight line. The first term of Eq. (2.42) is the Doppler shift induced by transformation of target

$$f_{\text{Doppler}} = \frac{2f_c}{c}\boldsymbol{v}^{\mathrm{T}}\boldsymbol{n} = \frac{2\boldsymbol{v}^{\mathrm{T}}\boldsymbol{n}}{\lambda}, \quad (2.43)$$

where λ is the wavelength of transmitted signal. The second term is the Doppler shift induced by precessions of target:

$$
\begin{aligned}
f_{\text{micro-Doppler}} &= \frac{2f_c}{c}\left[\frac{\mathrm{d}}{\mathrm{d}t}\left(\mathbf{R}_{\text{coning}}\mathbf{R}_{\text{spinning}}\widehat{\boldsymbol{r}}_0\right)\right]^{\mathrm{T}}\boldsymbol{n} \\
&= \frac{2f_c}{c}\widehat{\boldsymbol{r}}_0^{\mathrm{T}}\left[\frac{\mathrm{d}}{\mathrm{d}t}\left(\mathbf{R}_{\text{coning}}\mathbf{R}_{\text{spinning}}\right)\right]^{\mathrm{T}}\boldsymbol{n}
\end{aligned}
\tag{2.44}
$$

Denote that $\Omega_s = \|\boldsymbol{\omega}_s\|$, $\Omega_c = \|\boldsymbol{\omega}_c\|$, $\boldsymbol{\omega}_s' = \boldsymbol{\omega}_s/\Omega_s = \left(\omega_{sX}', \omega_{sY}', \omega_{sZ}'\right)^{\mathrm{T}}$, and $\boldsymbol{\omega}_c' = \boldsymbol{\omega}_c/\Omega_c = \left(\omega_{cX}', \omega_{cY}', \omega_{cZ}'\right)^{\mathrm{T}}$, and construct the following skew symmetric matrix:

$$
\widehat{\boldsymbol{\omega}}_s' = \begin{bmatrix} 0 & -\widehat{\omega}_{sZ}' & \widehat{\omega}_{sY}' \\ \widehat{\omega}_{sZ}' & 0 & -\widehat{\omega}_{sX}' \\ -\widehat{\omega}_{sY}' & \widehat{\omega}_{sX}' & 0 \end{bmatrix}, \quad \widehat{\boldsymbol{\omega}}_c' = \begin{bmatrix} 0 & -\omega_{cZ}' & \widehat{\omega}_{cY}' \\ \widehat{\omega}_{cZ}' & 0 & -\widehat{\omega}_{cX}' \\ -\widehat{\omega}_{cY}' & \widehat{\omega}_{cX}' & 0 \end{bmatrix}, \tag{2.45}
$$

then

$$
\mathbf{R}_{\text{spinning}} = \exp\left(\Omega_s\widehat{\omega}_s't\right) = \mathbf{I} + \widehat{\omega}_s'\sin(\Omega_s t) + \widehat{\omega}_s'^2(1-\cos(\Omega_s t)) \tag{2.46}
$$

$$
\mathbf{R}_{\text{coning}} = \exp\left(\Omega_c\widehat{\boldsymbol{\omega}}_c't\right) = \mathbf{I} + \widehat{\boldsymbol{\omega}}_c'\sin(\Omega_c t) + \widehat{\boldsymbol{\omega}}_c'^2(1-\cos(\Omega_c t)) \tag{2.47}
$$

Thus we have

$$
\begin{aligned}
\frac{\mathrm{d}\mathbf{R}_{\text{spinning}}}{\mathrm{d}t} &= \Omega_s\widehat{\boldsymbol{\omega}}_s'\exp\left(\Omega_s\widehat{\boldsymbol{\omega}}_s't\right) \\
&= \Omega_s\widehat{\boldsymbol{\omega}}_s'\left[\mathbf{I} + \widehat{\boldsymbol{\omega}}_s'\sin(\Omega_s t) + \widehat{\boldsymbol{\omega}}_s'^2(1-\cos(\Omega_s t))\right] \\
&= \Omega_s\widehat{\boldsymbol{\omega}}_s'^2\sin(\Omega_s t) + \Omega_s\widehat{\boldsymbol{\omega}}_s'\cos(\Omega_s t)
\end{aligned}
\tag{2.48}
$$

and

$$
\begin{aligned}
\frac{\mathrm{d}\mathbf{R}_{\text{coning}}}{\mathrm{d}t} &= \Omega_c\widehat{\boldsymbol{\omega}}_c'\exp\left(\Omega_c\widehat{\boldsymbol{\omega}}_c't\right) \\
&= \Omega_c\widehat{\boldsymbol{\omega}}_c'\left[\mathbf{I} + \widehat{\boldsymbol{\omega}}_c'\sin(\Omega_c t) + \widehat{\boldsymbol{\omega}}_c'^2(1-\cos(\Omega_c t))\right] \\
&= \Omega_c\widehat{\boldsymbol{\omega}}_c'^2\sin(\Omega_c t) + \Omega_c\widehat{\boldsymbol{\omega}}_c'\cos(\Omega_c t),
\end{aligned}
\tag{2.49}
$$

where $\widehat{\boldsymbol{\omega}}_s'^3 = -\widehat{\boldsymbol{\omega}}_s'$ and $\widehat{\boldsymbol{\omega}}_c'^3 = -\widehat{\boldsymbol{\omega}}_c'$ are applied in the process of derivation, respectively. By inserting Eq. (2.46) to Eq. (2.49) into Eq. (2.44), the expression of $\frac{d}{dt}(\mathbf{R}_{\text{coning}}\mathbf{R}_{\text{spinning}})$ can be derived as

$$
\begin{aligned}
\frac{d}{dt}(\mathbf{R}_{\text{coning}}\mathbf{R}_{\text{spinning}}) &= \frac{d\mathbf{R}_{\text{coning}}}{dt}\mathbf{R}_{\text{spinning}} + \mathbf{R}_{\text{coning}}\frac{d\mathbf{R}_{\text{spinning}}}{dt} \\
&= \Omega_c\widehat{\boldsymbol{\omega}}_c'^2\sin(\Omega_c t) + \Omega_c\widehat{\boldsymbol{\omega}}_c'^2\widehat{\boldsymbol{\omega}}_s'\sin(\Omega_c t)\sin(\Omega_s t) + \Omega_c\widehat{\boldsymbol{\omega}}_c'^2\widehat{\boldsymbol{\omega}}_s'^2\sin(\Omega_c t) \\
&\quad -\Omega_c\widehat{\boldsymbol{\omega}}_c'^2\widehat{\boldsymbol{\omega}}_s'^2\sin(\Omega_c t)\cos(\Omega_s t) + \Omega_c\widehat{\boldsymbol{\omega}}_c'\cos(\Omega_c t) + \Omega_c\widehat{\boldsymbol{\omega}}_c'\widehat{\boldsymbol{\omega}}_s'\cos(\Omega_c t)\sin(\Omega_s t) \\
&\quad +\Omega_c\widehat{\boldsymbol{\omega}}_c'\widehat{\boldsymbol{\omega}}_s'^2\cos(\Omega_c t) - \Omega_c\widehat{\boldsymbol{\omega}}_c'\widehat{\boldsymbol{\omega}}_s'^2\cos(\Omega_c t)\cos(\Omega_s t) + \Omega_s\widehat{\boldsymbol{\omega}}_s'^2\sin(\Omega_s t) \\
&\quad +\Omega_s\widehat{\boldsymbol{\omega}}_c'\widehat{\boldsymbol{\omega}}_s'^2\sin(\Omega_s t)\sin(\Omega_c t) + \Omega_s\widehat{\boldsymbol{\omega}}_c'\widehat{\boldsymbol{\omega}}_s'^2\sin(\Omega_s t) - \Omega_s\widehat{\boldsymbol{\omega}}_c'^2\widehat{\boldsymbol{\omega}}_s'^2\sin(\Omega_s t)\cos(\Omega_c t) \\
&\quad +\Omega_s\widehat{\boldsymbol{\omega}}_s'\cos(\Omega_s t) + \Omega_s\widehat{\boldsymbol{\omega}}_c'\widehat{\boldsymbol{\omega}}_s'\cos(\Omega_s t)\sin(\Omega_c t) + \Omega_s\widehat{\boldsymbol{\omega}}_c'^2\widehat{\boldsymbol{\omega}}_s'\cos(\Omega_s t) \\
&\quad -\Omega_s\widehat{\boldsymbol{\omega}}_c'^2\widehat{\boldsymbol{\omega}}_s'\cos(\Omega_s t)\cos(\Omega_c t) \\
&= \left(\Omega_c\widehat{\boldsymbol{\omega}}_c'^2 + \Omega_c\widehat{\boldsymbol{\omega}}_c'^2\widehat{\boldsymbol{\omega}}_s'^2\right)\sin(\Omega_c t) + \left(\Omega_c\widehat{\boldsymbol{\omega}}_c' + \Omega_c\widehat{\boldsymbol{\omega}}_c'\widehat{\boldsymbol{\omega}}_s'^2\right)\cos(\Omega_c t) \\
&\quad +\left(\Omega_c\widehat{\boldsymbol{\omega}}_c'^2\widehat{\boldsymbol{\omega}}_s' + \Omega_s\widehat{\boldsymbol{\omega}}_c'\widehat{\boldsymbol{\omega}}_s'^2\right)\sin(\Omega_s t)\sin(\Omega_c t) + \left(\Omega_s\widehat{\boldsymbol{\omega}}_c'\widehat{\boldsymbol{\omega}}_s' - \Omega_c\widehat{\boldsymbol{\omega}}_c'^2\widehat{\boldsymbol{\omega}}_s'^2\right)\sin(\Omega_c t)\cos(\Omega_s t) \\
&\quad +\left(\Omega_c\widehat{\boldsymbol{\omega}}_c'\widehat{\boldsymbol{\omega}}_s' - \Omega_s\widehat{\boldsymbol{\omega}}_c'^2\widehat{\boldsymbol{\omega}}_s'^2\right)\cos(\Omega_c t)\sin(\Omega_s t) + \left(-\Omega_c\widehat{\boldsymbol{\omega}}_c'\widehat{\boldsymbol{\omega}}_s'^2 - \Omega_s\widehat{\boldsymbol{\omega}}_c'^2\widehat{\boldsymbol{\omega}}_s'\right)\cos(\Omega_c t)\cos(\Omega_s t) \\
&\quad +\left(\Omega_s\widehat{\boldsymbol{\omega}}_s'^2 + \Omega_s\widehat{\boldsymbol{\omega}}_c'^2\widehat{\boldsymbol{\omega}}_s'^2\right)\sin(\Omega_s t) + \left(\Omega_s\widehat{\boldsymbol{\omega}}_s' + \Omega_s\widehat{\boldsymbol{\omega}}_c'^2\widehat{\boldsymbol{\omega}}_s'\right)\cos(\Omega_s t)
\end{aligned}
\tag{2.50}
$$

Thus

$$
\begin{aligned}
f_{\text{micro-Doppler}} &= \frac{2f_c}{c}\widehat{\boldsymbol{r}}_0^{\mathrm{T}}\left[\frac{d}{dt}(\mathbf{R}_{\text{coning}}\mathbf{R}_{\text{spinning}})\right]^{\mathrm{T}}\boldsymbol{n} \\
&= A_1\sin(\Omega_c t) + A_2\cos(\Omega_c t) + A_3\sin(\Omega_s t)\sin(\Omega_c t) \\
&\quad +A_4\sin(\Omega_c t)\cos(\Omega_s t) + A_5\cos(\Omega_c t)\sin(\Omega_s t) \\
&\quad +A_6\cos(\Omega_c t)\cos(\Omega_s t) + A_7\sin(\Omega_s t) + A_8\cos(\Omega_s t),
\end{aligned}
\tag{2.51}
$$

where

$$
A_1 = \frac{2f_c}{c}\widehat{\boldsymbol{r}}_0^{\mathrm{T}}\left(\Omega_c\widehat{\boldsymbol{\omega}}_c'^2 + \Omega_c\widehat{\boldsymbol{\omega}}_c'^2\widehat{\boldsymbol{\omega}}_s'^2\right)^{\mathrm{T}}\boldsymbol{n}, \quad A_2 = \frac{2f_c}{c}\widehat{\boldsymbol{r}}_0^{\mathrm{T}}\left(\Omega_c\widehat{\boldsymbol{\omega}}_c' + \Omega_c\widehat{\boldsymbol{\omega}}_c'\widehat{\boldsymbol{\omega}}_s'^2\right)^{\mathrm{T}}\boldsymbol{n},
$$

$$A_3 = \frac{2f_c}{c}\widehat{\boldsymbol{r}}_0^{\mathrm{T}}\left(\Omega_c\widehat{\boldsymbol{\omega}}_c'^2\widehat{\boldsymbol{\omega}}_s' + \Omega_s\widehat{\boldsymbol{\omega}}_c'\widehat{\boldsymbol{\omega}}_s'^2\right)^{\mathrm{T}}\boldsymbol{n}, \quad A_4 = \frac{2f_c}{c}\widehat{\boldsymbol{r}}_0^{\mathrm{T}}\left(\Omega_s\widehat{\boldsymbol{\omega}}_c'\widehat{\boldsymbol{\omega}}_s' - \Omega_c\widehat{\boldsymbol{\omega}}_c'^2\widehat{\boldsymbol{\omega}}_s'^2\right)^{\mathrm{T}}\boldsymbol{n},$$

$$A_5 = \frac{2f_c}{c}\widehat{\boldsymbol{r}}_0^{\mathrm{T}}\left(\Omega_c\widehat{\boldsymbol{\omega}}_c'\widehat{\boldsymbol{\omega}}_s' - \Omega_s\widehat{\boldsymbol{\omega}}_c'^2\widehat{\boldsymbol{\omega}}_s'^2\right)^{\mathrm{T}}\boldsymbol{n}, \quad A_6 = \frac{2f_c}{c}\widehat{\boldsymbol{r}}_0^{\mathrm{T}}\left(-\Omega_c\widehat{\boldsymbol{\omega}}_c'\widehat{\boldsymbol{\omega}}_s'^2 - \Omega_s\widehat{\boldsymbol{\omega}}_c'^2\widehat{\boldsymbol{\omega}}_s'\right)^{\mathrm{T}}\boldsymbol{n},$$

$$A_7 = \frac{2f_c}{c}\widehat{\boldsymbol{r}}_0^{\mathrm{T}}\left(\Omega_s\widehat{\boldsymbol{\omega}}_s'^2 + \Omega_s\widehat{\boldsymbol{\omega}}_c'^2\widehat{\boldsymbol{\omega}}_s'^2\right)^{\mathrm{T}}\boldsymbol{n}, \quad A_8 = \frac{2f_c}{c}\widehat{\boldsymbol{r}}_0^{\mathrm{T}}\left(\Omega_s\widehat{\boldsymbol{\omega}}_s' + \Omega_s\widehat{\boldsymbol{\omega}}_c'^2\widehat{\boldsymbol{\omega}}_s'\right)^{\mathrm{T}}\boldsymbol{n},$$

Eq. (2.51) shows that compared with that by rotation or vibration, the expression of returned signal's micro-Doppler frequency induced by precession is considerably more complex, where the relation of frequency with time does not appear as a sinusoidal curve any more but as a composition of many sinusoidal components. However, the changes of frequency versus time are still periodic. And the period is the lowest common multiple of the period of coning T_c and the period of spinning T_s, that is

$$T_{pr} = k_1 T_c = k_2 T_s, \quad k_1, k_2 \in \mathbf{N}, \tag{2.52}$$

where T_{pr} is the period of precession and $\mathbf{N}$ is the set of natural numbers.

The simulation: Assume that carrier frequency of the transmitted signal is $f_c = 10$ GHz and the origin O of the target-local coordinate system is located at ($U = 3$ km, $V = 4$ km, $W = 5$ km). The initial Euler angle of the target-local coordinate system and the reference coordinate system is $(\pi/3, \pi/4, \pi/5)$ rad. On the condition of no translation, the target is spinning about the z-axis with angular velocity $\boldsymbol{\omega}_s = (19.2382, -11.1072, 22.2144)^{\mathrm{T}}$ rad/s, $\Omega_s = 10\pi$ rad/s in the reference coordinate system, and the period is $T_s = 0.2$ s. Meanwhile the angular velocity of coning is $\boldsymbol{\omega}_c = (7.6756, -2.3930, 9.6577)^{\mathrm{T}}$ rad/s, $\Omega_c = 4\pi$ rad/s in the reference coordinate system, and the period is $T_c = 0.5$ s. Suppose that there are two strong scatterer points of the target, one of which is on the top of the cone A and located at (0 m,0 m,0.5 m) in the target-local coordinate system, and the other is on the bottom of the cone B and located at (0.5 m,0 m,−0.5 m) in the target-local coordinate system. The irradiation time of the radar is 2 s.

The theoretical time-varying micro-Doppler frequency curves of the two scatterer points are shown in Fig. 2.7A, where the time-varying micro-Doppler frequency curve of the scatterer point A, which is located at the axis of spinning, is a sinusoidal curve, for the micromotion is spin merely, and that of the scatterer point B is much more complex, for the micromotion is a combination of spin and cone. And the period of curve B, as shown, is the lowest common multiple of the period of coning and the period of spinning. The result of time-frequency analysis on the returned signal is shown in Fig. 2.7B, which is completely coincident with the theoretical one.

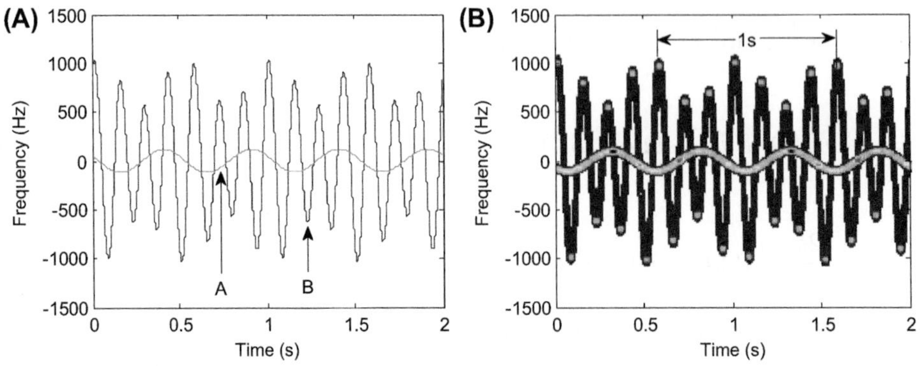

Figure 2.7 Simulation of micro-Doppler effect of target with precession. (A) Theoretical curve of micro-Doppler frequency; (B) result of time-frequency analysis on micro-Doppler signal.

2.4 INFLUENCE ON MICRO-DOPPLER EFFECT WHEN THE RADAR PLATFORM IS VIBRATING

In the practice, the spatial position of the radar receiver will change with the radar platform for mechanical vibration, which causes phase oscillation of the returned signal. The phase oscillation is also a kind of micro-Doppler modulation for the returned signal and is adverse to the analysis and abstraction on the micromotion characteristic. Now we will analyze the influence on micro-Doppler effect of returned signal when the radar platform is vibrating.

First, we take the target with rotation into account. The geometry of the radar and target shown in Fig. 2.1 can be redrawn as Fig. 2.8 when the radar platform is vibrating. The radar platform is vibrating along the direction indicated by the arrow at a frequency f_R with an amplitude D_R and the azimuth and elevation angle of the vibrating direction in the radar coordinates are α_0 and β_0, respectively.

Assume that point Q is at the vibration center at the initial time. Therefore, at time t, the range vector from point Q to the radar in the radar coordinate system can be derived as

$$\boldsymbol{r}_R = D_R \sin(2\pi f_R t)\begin{bmatrix} \cos\alpha_0 \cos\beta_0 \\ \sin\alpha_0 \cos\beta_0 \\ \sin\beta_0 \end{bmatrix}, \tag{2.53}$$

and the range vector from point P' to the radar can be derived as

$$r(t) = \left\| \boldsymbol{R}_0 + \boldsymbol{v}t + \mathbf{R}_{\text{rotating}}\widehat{\boldsymbol{r}}_0 - \boldsymbol{r}_R \right\| \tag{2.54}$$

The micro-Doppler frequency of the returned signal is

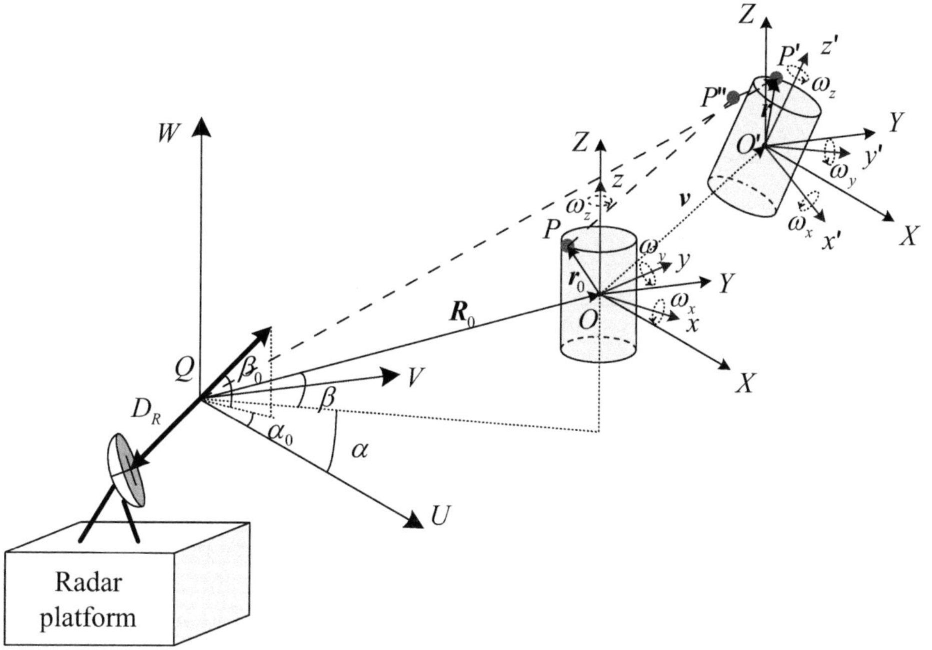

Figure 2.8 Geometry for the radar and a three-dimensional rotating target when radar platform is vibrating.

$$
\begin{aligned}
f_\mathrm{d} &= \frac{2f_\mathrm{c}}{c}\frac{\mathrm{d}}{\mathrm{d}t}r(t) \\
&= \frac{2f_\mathrm{c}}{c}\left[\boldsymbol{v}+\frac{\mathrm{d}}{\mathrm{d}t}\left(\mathbf{R}_\mathrm{rotating}\widehat{\boldsymbol{r}}_0\right)-\frac{\mathrm{d}\boldsymbol{r}_R}{\mathrm{d}t}\right]^\mathrm{T}\boldsymbol{n},
\end{aligned}
\tag{2.55}
$$

where there is an additional micro-Doppler frequency induced by radar platform vibrating compared with Eq. (2.8).

$$
\begin{aligned}
f'_\text{micro-Doppler} &= -\frac{2f_\mathrm{c}}{c}\frac{\mathrm{d}\boldsymbol{r}_R^\mathrm{T}}{\mathrm{d}t}\boldsymbol{n} \\
&= -\frac{2f_\mathrm{c}}{c}2\pi f_R D_R\cos(2\pi f_R t)\begin{bmatrix}\cos\alpha_0\cos\beta_0\\ \sin\alpha_0\cos\beta_0\\ \sin\beta_0\end{bmatrix}^\mathrm{T}\begin{bmatrix}\cos\alpha\cos\beta\\ \sin\alpha\cos\beta\\ \sin\beta\end{bmatrix} \\
&= -\frac{4\pi f_\mathrm{c} f_R D_R}{c}\left[\cos(\alpha-\alpha_0)\cos\beta\cos\beta_0+\sin\beta\sin\beta_0\right]\cos(2\pi f_R t)
\end{aligned}
\tag{2.56}
$$

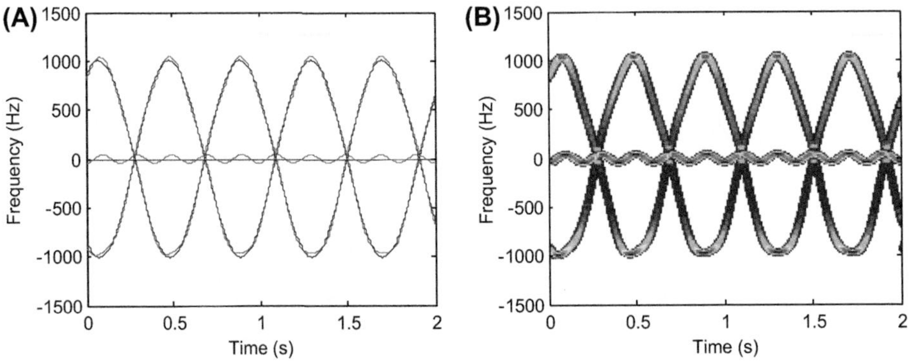

Figure 2.9 Simulation of micro-Doppler effect of target with rotation when radar platform is vibrating. (A) Theoretical curve of micro-Doppler frequency; (B) result of time-frequency analysis on micro-Doppler signal.

The existence of the additional micro-Doppler frequency will result in the curve of returned signal's micro-Doppler frequency in the time-frequency plane deviating from easy sinusoids and will bring adverse effects on the subsequent abstraction of the micromotion characteristic. The theoretical micro-Doppler frequency of target with rotation when the radar platform is vibrating and the result of time-frequency analysis are shown in Fig. 2.9. The radar platform is vibrating at 5 Hz with amplitude 0.02 m and the azimuth and elevation angle of the vibrating direction in the radar coordinates are $\alpha_0 = \pi/10$ and $\beta_0 = \pi/3$, respectively. The other simulation parameters are the same with those of Section 2.1. Fig. 2.9A shows the comparison of the changing curve (represented as full line) of returned signal's micro-Doppler frequency when radar platform is vibrating and the changing curve (represented as dotted line) of returned signal's micro-Doppler frequency when radar platform is stationary and it is clear that there is distortion on the original sinusoidal curve and the straight line corresponding to the rotation center changes for a sinusoidal curve because of radar platform vibrating. The result of time-frequency analysis on the returned signal is shown in Fig. 2.9B, which is completely coincident with the theoretical one.

For target with vibration, when radar platform is vibrating, the returned signal's micro-Doppler frequency is

$$\begin{aligned} f_{\text{micro-Doppler}} &= \frac{2f_c}{c}\left[\frac{\mathrm{d}\boldsymbol{r}^{\mathrm{T}}}{\mathrm{d}t} - \frac{\mathrm{d}\boldsymbol{r}_R^{T}}{\mathrm{d}t}\right]\boldsymbol{n} \\ &= \frac{4\pi f_c f_v D_v}{c}[\cos(\alpha - \alpha_P)\cos\beta\cos\beta_P + \sin\beta\sin\beta_P]\cos(2\pi f_v t) \\ &\quad -\frac{4\pi f_c f_R D_R}{c}[\cos(\alpha - \alpha_0)\cos\beta\cos\beta_0 + \sin\beta\sin\beta_0]\cos(2\pi f_R t), \end{aligned} \tag{2.57}$$

where the second term is the micro-Doppler modulation induced by radar platform vibrating. And while $f_v = f_R$, $\alpha_P = \alpha_0$, and $\beta_P = \beta_0$, the equation changes for

$$f_{\text{micro-Doppler}} = \frac{4\pi f_c f_v (D_v - D_R)}{c}[\cos(\alpha - \alpha_P)\cos\beta\cos\beta_P + \sin\beta\sin\beta_P]\cos(2\pi f_v t), \quad (2.58)$$

which is the same as the micro-Doppler effects induced by target vibrating with amplitude $(D_v - D_R)$ when radar platform is fixed. Usually the radar platform vibrates with an amplitude close to that of a vibrating target so it is hard to abstract the accurate vibrating characteristic of a target. While f_R, α_0, and β_0 take other values, Eq. (2.57) will be rather complex, as is shown in Fig. 2.10, where the curve of the returned signal's micro-Doppler frequency is a combination of two different sinusoidal curves.

Similarly, for target with precession, when the radar platform is vibrating, the returned signal's micro-Doppler frequency is

$$f_{\text{micro-Doppler}} = \frac{2f_c}{c}\left[\frac{\mathrm{d}}{\mathrm{d}t}\left(\mathbf{R}_{\text{coning}}\mathbf{R}_{\text{spinning}}\widehat{\boldsymbol{r}}_0\right)^{\mathrm{T}} - \frac{\mathrm{d}\boldsymbol{r}_R^{\mathrm{T}}}{\mathrm{d}t}\right]\boldsymbol{n} \quad (2.59)$$

Similarly, with the target with rotation, the vibrating radar platform will result in the curve of the returned signal's micro-Doppler frequency in the time-frequency plane deviating from a theoretical one and will bring adverse effects on the subsequent abstraction of the micromotion characteristic. As is shown in Fig. 2.11A, while the radar is vibrating at 8 Hz with amplitude 0.01 m, the changing curve (represented as full line) of returned signal's micro-Doppler frequency when radar platform is vibrating has difference with that (represented as dotted line) of when the radar platform is stationary, especially the sinusoidal curve corresponding to the scatterer point on the top of the

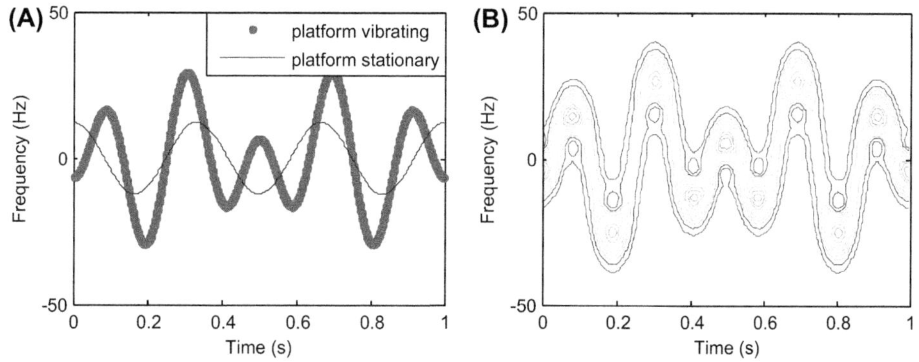

Figure 2.10 Simulation of micro-Doppler effect of target with vibration when radar platform is vibrating. (A) Theoretical curve of micro-Doppler frequency; (B) result of time-frequency analysis on micro-Doppler signal.

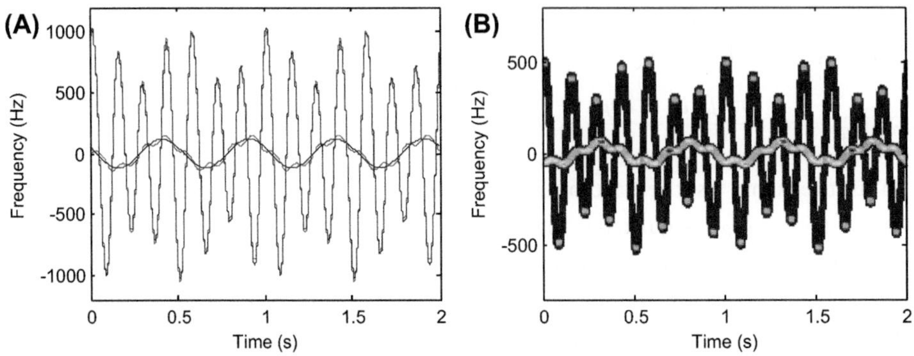

Figure 2.11 Simulation of micro-Doppler effect of target with precession when radar platform is vibrating. (A) Theoretical curve of micro-Doppler frequency; (B) result of time-frequency analysis on micro-Doppler signal.

cone, which has serious distortion. Fig. 2.11B shows the time-frequency analysis result of returned signal.

In general, when the micromotion is rotation or vibration, the maximum of micro-Doppler frequency of radar platform is usually less than the maximum of micro-Doppler frequency induced by target with rotation or precession because the amplitude of radar platform is relatively small. And while analyzing on the characteristic of returned signal's micro-Doppler frequency, there will be major relative error for the estimation of micro-Doppler frequency. But when the micromotion is vibration, induced by which the micro-Doppler shift is relatively small, radar platform vibration will produce a significant influence. Therefore it requires high stability for the radar platform so as to abstract the micro-Doppler characteristic of a target with vibration. And to eliminate the effect induced by radar platform, we can estimate the vibrating frequency of radar platform and remove its micro-Doppler component in the time-frequency plane.

In wideband radar, radar platform vibrating has a similar influence as just discussed on micro-Doppler effect of targets, thus we will not analyze it again in Chapter 3.

REFERENCE

[1] Chen VC, Li FY, Ho SS, et al. Micro-Doppler effect in radar: phenomenon, model and simulation study. IEEE Transactions on Aerospace and Electronic Systems 2006;42(1):2–21.

CHAPTER THREE

Micro-Doppler Effect in Wideband Radar

The energy of radar radiation signal (E) is a significant determinant of radar range through the radar equation. For simple rectangle pulse signal of constant carrier frequency, its energy equals the product of peak power and pulse width, that is, $E = P_t\tau$. Consequently, there are two main methods to enlarge radar range by increasing the energy of emission signal, that is, increase the peak power P_t or the pulse width τ. In general, the peak power P_t of a radar system is restricted by the maximum permissible peak power of the transmitting tube and the power capacity of the transmission line, etc. Thus, the main consideration is to increase the pulse width τ under the range of the transmitter's permissible maximum average power. However, for simple rectangle pulse signal of constant carrier frequency, its time-bandwidth product is 1. The increasing of time width means decreasing of bandwidth. According to resolution theory of radar signal, by the premise of guaranteeing certain SNR and realizing ideal process, ranging precision and range resolution depends on the frequency structure of the signal, which requires a large bandwidth. Meanwhile, velocity precision and velocity resolution depend on the time structure of the signal, which requires a large time width. Consequently, an ideal radar signal should possess a large time-bandwidth product. Obviously, the simple rectangle pulse signal of constant carrier frequency does not match such conditions.

To deal with the conflict, a complex signal of large time-bandwidth product must be adopted. The linear frequency modulated (LFM) signal, as a kind of signal of complex form, was first introduced and widely used in radar field. LFM appends linear frequency modulation to enlarge the bandwidth of signal and convert the signal into narrow pulse by pulse compression when it receives a signal. Multiple signals of large time-bandwidth product were developed later, such as nonlinear frequency-modulated signal, phase-coded signal, frequency-coded signal, etc.

From the Nyquist sampling theorem, sampling frequency must be more than two times the bandwidth of the signal to guarantee that no information is lost in A/D process. Using signal of large time-bandwidth product can take care of the radar range and generates a wideband signal at the same time. However, the increasing bandwidth of transmission signal requires an additional hardware burden of echo signal sampling. Especially in high-resolution radar, resolution of centimeter level requires the bandwidth to be up to GHz of transmission signal. To solve the conflict, a stepped-frequency radar

Micro-Doppler Characteristics of Radar Targets
ISBN 978-0-12-809861-5, http://dx.doi.org/10.1016/B978-0-12-809861-5.00003-5

signal is put forward. The characteristic of stepped-frequency radar signal is that radar transmits a group of N uniform stepped (stepped value is Δf) rectangle pulse signals. Multiple pulse signals are spectrum synthesized at a receiver to acquire a large bandwidth signal, the equivalent bandwidth of which is $N\Delta f$. The bandwidth of subpulse is small, so the sampling frequency only needs to be two times higher than the bandwidth of the subpulse, which is greatly lower than $2N\Delta f$, so that the hardware requirement of sampling velocity is lower. Stepped-frequency radar signal can acquire high-range resolution and simultaneously reduce the demand of digital signal processor (DSP) sampling bandwidth. However, long pulse repetition period, short wavelength, and high velocity of target make the signal more sensitive to Doppler effect and lead to low data utilization rate. To improve the performance of stepped-frequency radar signal, the rectangle pulse signal in stepped-frequency radar signal can be substituted with linear frequency modulated signal, which is stepped-frequency chirp signal (SFCS). SFCS possesses the advantages of both LFM signal and stepped-frequency radar signal. It can synthesize high-resolution image by DSP. SFCS is a wideband signal of high application value as well.

Other than pulse wideband signal, continuous wave signal acquires large time-bandwidth product by using signal modulation technique during the intrapulse. Among the multiple continuous wave signals, the most typical one is the LFM continuous wave (LFMCW). Compared with pulse radar, LFMCW radar transmits the signal continuously. Its energy distributes in the whole pulse repetition period, and its transmission power is relatively low; the transmission energy of pulse radar concentrates in a narrow pulse, and its peak power is relatively high, which requires more to sensors' volume, weight, and power. LFMCW radar uses a "dechirp" process method. Difference frequency signal depends on frequency modulation rate and the width of radar mapping band. In general, when the difference signal band along with the radar mapping band is low, the A/D sampling frequency will be low correspondingly. So LFMCW radar has many advantages, such as small volume, small weight, low cost, low power consumption, etc., which is applied widely in millimeter wave seeker and mini unmanned aerial vehicle.

In this chapter, we mainly introduce the micro-Doppler effect in wideband radar. Three typical kinds of wideband signals are analyzed: the LFM pulse signal, SFCS, and the LFMCW signal. LFM is the most common signal of radar, and its micro-Doppler effect represents others' well. The micro-Doppler effect of this kind of wideband signal, such as the phase-coding pulse signal and the frequency-coding pulse signal, etc. is similar to micro-Doppler effect of LFM pulse signal. The micro-Doppler effect of SFCS represents the micro-Doppler effect of such kind of wideband signal well. Conclusions drawn by SFCS fit the carrier frequency stepped rectangle signal as well. LFMCW radar is applied in antimissile missile's millimeter wave seeker. The research on the micro-Doppler effect of LFMCW radar is of great significance in detection and recognition of micromotion targets like the ballistic missile in the air-defense and antimissile combat.

3.1 WIDEBAND SIGNAL ECHO MODEL

In wideband radar signal process, radar transmits multiple pulses and makes the received echo signal get the coherent integration of pulse (even for the LFMCW radar, we can also regard each period of transmission signal as single pulse signal). Assuming that during the target observation, radar transmits M pulse signals in total, and the transmission moment of the m-th pulse is $t = mT_r$, in which T_r is pulse repetition period. For stating more convenience, radar time (total time) t is expressed as the combination of fast time (intrapulse time) t_k and slow time (interpulse) t_m, and then t becomes

$$t = t_m + t_k, \quad t_m = mT_r, \quad t_k = t - t_m \tag{3.1}$$

where t_m is the whole transmission time of each pulse and t_k is the time of each echo signal relative to each pulse transmission moment during the transmission process and the receipt process.

When the radial motion between the target and the radar exists, the echo signal will generate micro-Doppler effect. We rewrite the echo signal expression Eq. (1.7) as follows:

$$s(t) = \sigma u\left(\frac{c+v_r}{c-v_r}t - \frac{2R_0}{c-v_r}\right)\exp\left[\mathrm{j}\omega_0\left(\frac{c+v_r}{c-v_r}t - \frac{2R_0}{c-v_r}\right)\right] \tag{3.2}$$

Substituting Eq. (3.1) into Eq. (3.2), the echo signal can be expressed as

$$s(t_k, t_m) = \sigma u\left(\frac{c+v_r}{c-v_r}t_k - \frac{2R(t_m)}{c-v_r}\right)\exp\left[\mathrm{j}\omega_0\left(\frac{c+v_r}{c-v_r}t_k - \frac{2R(t_m)}{c-v_r}\right)\right] \tag{3.3}$$

Then we substitute $R(t_m) = R_0 - v_r t_m$ into Eq. (3.3) and

$$\begin{aligned} s(t_k, t_m) = {} & \sigma u\left(\frac{c+v_r}{c-v_r}t_k - \frac{2R(t_m)}{c-v_r}\right)\exp(\mathrm{j}\omega_0 t_k)\exp\left(-\mathrm{j}\omega_0\frac{2R_0}{c-v_r}\right) \\ & \cdot\exp\left(\mathrm{j}\omega_0\frac{2v_r}{c-v_r}t_k\right)\exp\left(\mathrm{j}\omega_0\frac{2v_r}{c-v_r}t_m\right) \end{aligned} \tag{3.4}$$

The echo signal can be analyzed from the previous expression. The echo signal contains four phase terms. The first one will be rejected in baseband transform. The second one is a constant term. The third one and the fourth one are both Doppler phase terms of echo signal. The Doppler angular frequency is $2v_r\omega_0/(c - v_r)$. But the meanings of the third term and the forth term are different. The third phase term is the function of fast time t_k, representing the Doppler effect of intrapulse, called intrapulse Doppler phase. It causes center deviation of echo spectrum and accordingly leads to Doppler coupling time shift of echo pulse output. The fourth term is the function of slow time t_m, representing the change of the initial phase of different pulse, which mainly affects the coherent integration between different pulses. In fact, this term is the discrete sampling of radar pulse to echo Doppler. Consequently, Doppler frequency of echoes can be estimated by discrete Fourier transform (DFT) of multiple echo pulses. This is the theoretical basis of the realization of the range Doppler algorithm in radar imaging techniques.

Eq. (3.4) expresses a wideband radar echo of a radial uniform moving target. However, analysis through Eq. (3.4) becomes quite complicated on most application conditions of radar. Accordingly, several simplified expressions are generated, among which the first-order approximate model and the "stop-go" approximate model are the most common models.

The first-order approximate model neglects the change of range induced by the target's motion during the electromagnetic wave in single-pass propagation. The delay time is approximated as

$$t_r = \frac{2(R(t_m) - v_r t_k)}{c} \tag{3.5}$$

Thus, the echo signal becomes

$$s(t_k, t_m) = \sigma u\left(t_k - \frac{2(R(t_m) - v_r t_k)}{c}\right)\exp\left[j\omega_0\left(t_k - \frac{2(R(t_m) - v_r t_k)}{c}\right)\right] \tag{3.6}$$

Substitute $R(t_m) = R_0 - v_r t_m$ into Eq. (3.6), and we can get

$$\begin{aligned} s(t_k, t_m) &= \sigma u\left(t_k - \frac{2(R(t_m) - v_r t_k)}{c}\right)\exp(j\omega_0 t_k)\exp\left(-j\omega_0\frac{2R_0}{c}\right) \\ &\cdot\exp\left(j\omega_0\frac{2v_r}{c}t_k\right)\exp\left(j\omega_0\frac{2v_r}{c}t_m\right) \end{aligned} \tag{3.7}$$

It can be found that, in this model, angular frequency of echo is $2v_r\omega_0/c$, which corresponds with Eq. (1.9). The approximate error is $2v_r\omega_0/(c - v_r) - 2v_r\omega_0/c \approx 2v_r^2\omega_0/c^2$. The model describes the echo signal quite well, so it fits to most radar application conditions.

"Stop-go" model neglects the target motion during the intrapulse based on the first-order approximate model, that is, suppose that the target is static during the intrapulse, and the range from the target to radar changes suddenly when the next pulse comes. In this case, the intrapulse Doppler frequency shift of the echo is neglected, and Eq. (3.7) is approximated further

$$s(t_k, t_m) = \sigma u\left(t_k - \frac{2R(t_m)}{c}\right)\exp\left(j\omega_0\left(t_k - \frac{2R(t_m)}{c}\right)\right) \tag{3.8}$$

"Stop-go" approximate model neglects the dilation effect of the echo envelope, the change of the range, and the intrapulse frequency deviation caused by the target motion during the electromagnetic wave in single-pass propagation. It is the most simplified model of the radar echo. Meanwhile, its approximate error is the largest. This model is used widely when describing the radar echo of narrow pulse and low target speed, such as the echoes of SAR and ISAR. When the target moves at a high speed, such as the space target or the bandwidth is relatively large such as LFMCW radar, the target motion during intrapulse cannot be neglected. The approximate error of the "stop-go" approximate model is relatively large, so the first-order approximate model should be used. In the follow-up analysis, the "stop-go" approximate model is used in the echo model of LFM signal, and the first-order approximate model is used in the echo model of LFMCW signal. SFCS is a wideband pulse signal synthesized by many subpulses. Thus, even each subpulse echo uses the "stop-go" approximate model, and the target motion between the subpulses must be considered. That is to say, the target motion during the synthesized pulse must be considered. In fact, the stepped-frequency pulse signal uses the first-order approximate model to describe the echo signal as well.

3.2 MICRO-DOPPLER EFFECT IN LINEAR FREQUENCY MODULATION SIGNAL RADAR

3.2.1 Linear Frequency Modulation Signal

The LFM signal transmitted by radar is expressed as

$$p(t) = \mathrm{rect}\left(\frac{t}{T_p}\right)\cdot\exp\left(j2\pi\left(f_c t + \frac{1}{2}\mu t^2\right)\right) \tag{3.9}$$

where

$$\mathrm{rect}\left(\frac{t}{T_p}\right) = \begin{cases} 1, & -T_p/2 \le t \le T_p/2 \\ 0, & \text{elsewhere} \end{cases} \tag{3.10}$$

where f_c is the carrier frequency, T_p is pulse width, and μ is modulation rate. The envelope of LFM signal is a rectangle pulse, whose width is T_p. But the instantaneous frequency changes with time going by and it can be expressed as:

$$f_i = f_c + \mu t \tag{3.11}$$

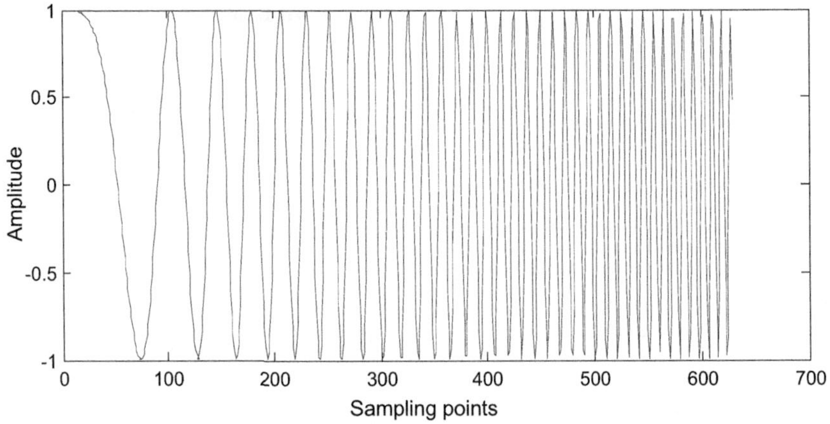

Figure 3.1 The time domain waveform of LFM signal.

From Eq. (3.11), it is obvious that the instantaneous frequency of LFM signal is proportional to time. If the time width of an LFM signal is T_p, its bandwidth will be μT_p. Thus, its time-frequency product $\mu T_p^2 \geq 1$, and the time domain waveform is demonstrated in Fig. 3.1.

3.2.2 Range Imaging

Assume that the distance from a static point to the radar is R. When radar transmits the LFM signal as the expression Eq. (3.9) expresses, the echo signal of the target received by radar is

$$s(t) = \sigma \mathrm{rect}\left(\frac{t - 2R/c}{T_p}\right) \cdot \exp\left(\mathrm{j}2\pi\left(f_c\left(t - \frac{2R}{c}\right) + \frac{1}{2}\mu\left(t - \frac{2R}{c}\right)^2\right)\right) \tag{3.12}$$

where σ is the scattering coefficient of the point target.

Similar to the entire wideband signal, range profile imaging of the target can be obtained by matched filtering, that is, taking conjugate multiplication of the echo and the transmission signal in frequency domain and taking inverse Fourier transform (IFT), by which the peak signal containing the information of the position of the target is obtained. Because of the special properties of LFM signal, a special process method called "dechirp" can be used other than the general matched filtering method.

"Dechirp" is subtracting the echo signal from the reference signal. The reference signal is a signal whose time delay is fixed and the frequency as well as the modulated frequency is same with LFM signal. In practical applications, we usually use the radar delay transmission signal, or strong scattering point on the target or on the situation as the reference signal. The "dechirp" process is described in Fig. 3.2.

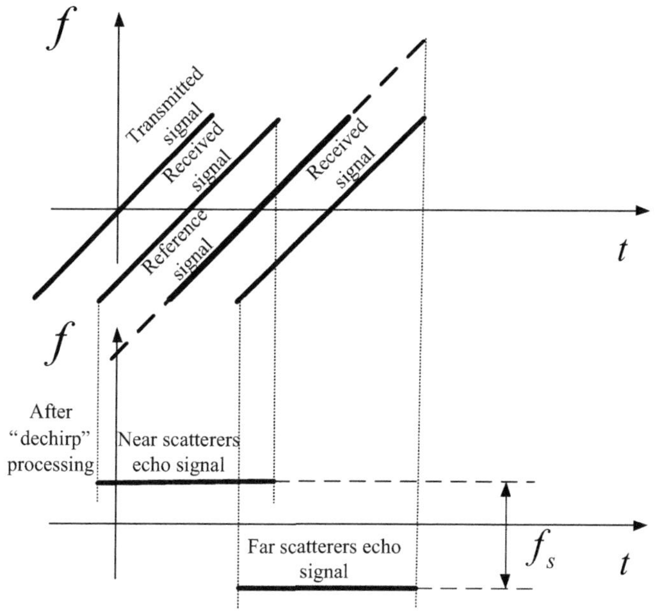

Figure 3.2 The "dechirp" processing of target echo signal.

Assume that the reference signal is R_{ref}, and R_{ref} becomes

$$s_{ref}(t) = \text{rect}\left(\frac{t-2R_{ref}/c}{T_{ref}}\right)\cdot\exp\left(\text{j}2\pi\left(f_c\left(t-\frac{2R_{ref}}{c}\right)+\frac{1}{2}\mu\left(t-\frac{2R_{ref}}{c}\right)^2\right)\right) \quad (3.13)$$

where T_{ref} is the pulse width of the reference signal. T_{ref} should be valued slightly more than T_p to avoid the width of the rectangle window T_p not changing after the conjugate multiplication of Eqs. (3.13) and (3.12). So the product of conjugate multiplication becomes

$$\begin{aligned} s_d(t) &= s(t)s^*_{ref}(t) \\ &= \sigma\text{rect}\left(\frac{t-2R/c}{T_p}\right) \\ &\quad\cdot\exp\left(\text{j}\left(-\frac{4\pi}{c}\mu\left(t-\frac{2R_{ref}}{c}\right)R_\Delta-\frac{4\pi}{c}f_cR_\Delta+\frac{4\pi\mu}{c^2}R_\Delta^2\right)\right) \end{aligned} \quad (3.14)$$

where $R_\Delta = R - R_{ref}$. Based on the time of reference point, that is $t' = t - 2R_{ref}/c$. Eq. (3.14) becomes

$$s_d(t') = \sigma\text{rect}\left(\frac{t'-2R_\Delta/c}{T_p}\right)\cdot\exp\left(\text{j}\left(-\frac{4\pi}{c}\mu R_\Delta t'-\frac{4\pi}{c}f_cR_\Delta+\frac{4\pi\mu}{c^2}R_\Delta^2\right)\right) \quad (3.15)$$

Eq. (3.15) suggests that it is a single-frequency signal after taking conjugate multiplication of the echo and the transmission signal. Its angular frequency is $-4\pi\mu R_\Delta/c$, which is proportional to the distance between the target and the reference scattering point. Consequently, LFM signal of large bandwidth converts into narrow bandwidth signal by "dechirp," so that the requirement of sampling is reduced. It is an obvious advantage compared with matched filtering. However, some phase terms not expected are introduced by "dechirp." On this condition, the compensation measures of the side lobe suppression should be taken. Eq. (3.15) after the Fourier Transform of fast time t' becomes

$$S_d(f) = \sigma T_p \text{sinc}\left(T_p\left(f + \frac{2\mu}{c}R_\Delta\right)\right)\cdot\exp\left(-j\frac{4\pi}{c}f_c R_\Delta + j\frac{4\pi\mu}{c^2}R_\Delta^2 - j\frac{4\pi f}{c}R_\Delta\right) \tag{3.16}$$

where $\text{sinc}(x) = sin(\pi x)/(\pi x)$. The right side of Eq. (3.16) contains three phase terms. The first phase term is a one degree term of R_Δ. It is a constant if the target is static. But if the target is moving, R_Δ corresponding to each echo of pulse is changing. On this condition, the first phase term comes into a changing function on R_Δ. The second phase term is called "residual video phase (RVP)." The third phase term is called the sideling term of echo envelope. Since these two terms both bring negative effects to radar imaging, they must be eliminated. Considering that the right side envelope of Eq. (3.16) is a sinc function, its peak value located in $f = -2\mu R_\Delta/c$. Thus, the last two phase terms are rewritten as

$$\frac{4\pi\mu}{c^2}R_\Delta^2 - \frac{4\pi f}{c}R_\Delta = \frac{3\pi f^2}{\mu} \tag{3.17}$$

Eq. (3.17) is obtained on the condition $f = -2\mu R_\Delta/c$. So Eq. (3.16) only needs to be multiplied by the following expression:

$$S_{\text{RVP}}(f) = \exp\left(-j\frac{3\pi f^2}{\mu}\right) \tag{3.18}$$

The RVP term and the sideling term are eliminated. So Eq. (3.16) is changed into

$$S_d(f) = \sigma T_p \text{sinc}\left(T_p\left(f + \frac{2\mu}{c}R_\Delta\right)\right)\cdot\exp\left(-j\frac{4\pi}{c}f_c R_\Delta\right) \tag{3.19}$$

The range profile is obtained from the modulus of Eq. (3.19) as

$$|S_d(f)| = \sigma T_p \text{sinc}\left(T_p\left(f + \frac{2\mu}{c}R_\Delta\right)\right) \tag{3.20}$$

The peak value is located in $f = -2\mu R_\Delta/c$. Multiplied by $-c/(2\mu)$, f is converted into the radius distance R_Δ from the point target to the reference point.

3.2.3 Micro-Doppler Effect of Targets With Rotation [1]

We still use geometric relation figure in Fig. 2.1 to analyze the micro-Doppler effect of the rotating target when the radar transmits the LFM signal. The translational velocity of the target is $\boldsymbol{v}$ relative to the radar, and meanwhile the target rotates at the angular velocity $\boldsymbol{\omega} = (\omega_x,\omega_y,\omega_z)^{\mathrm{T}}$ in space. Based on the "stop-go" approximate model as Eq. (3.8) expresses, the echo signal of target becomes

$$s(t_k, t_m) = \sigma \mathrm{rect}\left(\frac{t_k - 2r(t_m)/c}{T_p}\right) \cdot \exp\left(\mathrm{j}2\pi\left(f_c\left(t_k - \frac{2r(t_m)}{c}\right) + \frac{1}{2}\mu\left(t_k - \frac{2r(t_m)}{c}\right)^2\right)\right) \tag{3.21}$$

where

$$\begin{aligned} r(t_m) &= \left\|\boldsymbol{R}_0 + \boldsymbol{v}t_m + \mathbf{R}_{\text{rotating}}(t_m)\widehat{\boldsymbol{r}}_0\right\| \\ &= \sqrt{\left(\boldsymbol{R}_0 + \boldsymbol{v}t_m + \mathbf{R}_{\text{rotating}}(t_m)\widehat{\boldsymbol{r}}_0\right)^{\mathrm{T}}\left(\boldsymbol{R}_0 + \boldsymbol{v}t_m + \mathbf{R}_{\text{rotating}}(t_m)\widehat{\boldsymbol{r}}_0\right)} \end{aligned} \tag{3.22}$$

$$\mathbf{R}_{\text{rotating}}(t_m) = \mathbf{I} + \widehat{\omega}'\sin(\Omega t_m) + \widehat{\omega}'^2(1 - \cos(\Omega t_m)) \tag{3.23}$$

When we choose the echo signal of the local coordinate's origin as the reference signal, the reference signal is expressed as

$$\begin{aligned} s_{ref}(t_k, t_m) &= \mathrm{rect}\left(\frac{t_k - 2R_{ref}(t_m)/c}{T_{ref}}\right) \\ &\quad \cdot \exp\left(\mathrm{j}2\pi\left(f_c\left(t_k - \frac{2R_{ref}(t_m)}{c}\right) + \frac{1}{2}\mu\left(t_k - \frac{2R_{ref}(t_m)}{c}\right)^2\right)\right) \end{aligned} \tag{3.24}$$

where

$$R_{ref}(t_m) = \|\boldsymbol{R}_0 + \boldsymbol{v}t_m\| = \sqrt{(\boldsymbol{R}_0 + \boldsymbol{v}t_m)^{\mathrm{T}}(\boldsymbol{R}_0 + \boldsymbol{v}t_m)} \tag{3.25}$$

The echo signal multiplied by the reference signal comes

$$\begin{aligned} s_d(t_k, t_m) &= s(t_k, t_m)s^*_{ref}(t_k, t_m) \\ &= \sigma \mathrm{rect}\left(\frac{t_k - 2r(t_m)/c}{T_p}\right) \\ &\quad \cdot \exp\left(\mathrm{j}\left(-\frac{4\pi}{c}\mu\left(t_k - \frac{2R_{ref}(t_m)}{c}\right)R_\Delta(t_m) - \frac{4\pi}{c}f_c R_\Delta(t_m) + \frac{4\pi\mu}{c^2}R_\Delta(t_m)^2\right)\right) \end{aligned} \tag{3.26}$$

where $R_\Delta(t_m) = r(t_m) - R_{ref}(t_m)$.

Let $t' = t_k - 2R_{ref}(t_m)/c$. Take the Fourier transform (FT) of t' for Eq. (3.26) and eliminate the RVP terms and the envelop sideling term, and the expression of echo signal in fast-time domain (f_k domain) becomes

$$S_d(f_k, t_m) = \sigma T_p \mathrm{sinc}\left(T_p\left(f_k + \frac{2\mu}{c}R_\Delta(t_m)\right)\right) \cdot \exp\left(-\mathrm{j}\frac{4\pi}{c}f_c R_\Delta(t_m)\right) \tag{3.27}$$

Correspondingly, the peak value of the range profile locates at $f_k = -2\mu R_\Delta(t_m)/c$. From the previous expression we can see that the phase of $S_d(f_k,t_m)$ is modulated by $R_\Delta(t_m)$, which leads to the generation of micro-Doppler effect of echo signal in slow time (t_m) domain. In fact, it is consistent with the micro-Doppler effect induced by micro-motion in narrowband radar as Eq. (2.7) expresses. Meanwhile, from the plane of frequency—slow time ($f_k - t_m$ plane, since f_k can be converted into distance on radius by range scaling, it also can be called "range—slow time" plane), the peak value in range profile presents the curves changing with $R_\Delta(t_m)$, and meanwhile the curve reflects the motion characteristics of the micromotion points. Thus, the analysis of the micro-Doppler effect of the target can be done by the phase terms, benefiting from the high range resolution of the wideband radar. What is more, the micro-Doppler effect of the target can also be analyzed in range—slow time plane. Different from the micro-Doppler effect described by the "time-frequency image" in narrowband radar in Chapter 2, the micro-Doppler effect can be described in "range—slow time" plane in wideband radar.

The derivation of the changing curve equation of the peak value in range profile when the target is vibrating is stated as follows. The simplifying of the geometry relation between radar and the target is shown in Fig. 3.3. When the target is in the far field of radar, the following expression is derived approximately:

$$R_\Delta(t_m) = r(t_m) - R_{ref}(t_m) \approx \overrightarrow{O'H} = \left[\mathbf{R}_{\mathrm{rotating}}(t_m)\widehat{\boldsymbol{r}}_0\right]^{\mathrm{T}}\boldsymbol{n} \tag{3.28}$$

H in Fig. 3.3 is the projection point of the rotating point P' in direction of LOS.

The scalar expression of micro-Doppler curve of $R_\Delta(t_m)$ can be obtained by further derivation of Eq. (3.28), which will be beneficial to the further research on micro-Doppler feature analysis and extraction. Regardless of the translational motion of targets, a new figure of geometry relation can be described as Fig. 3.4 on the basis of Figs. 2.1 and 3.3. The plane where the circle trajectory of the rotating point P lays is

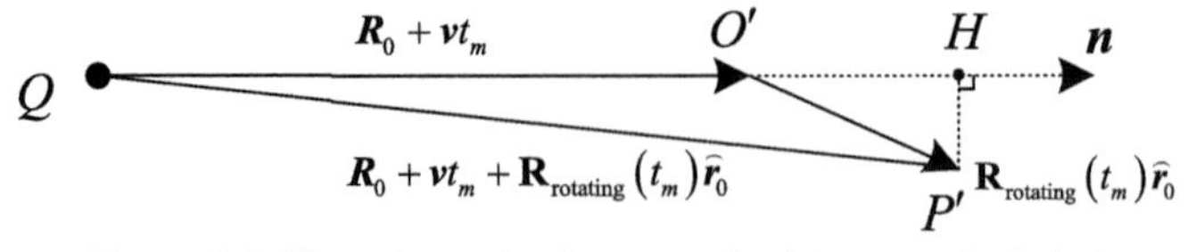

Figure 3.3 The schematic diagram of solving matrix $R_\Delta(t_m)$.

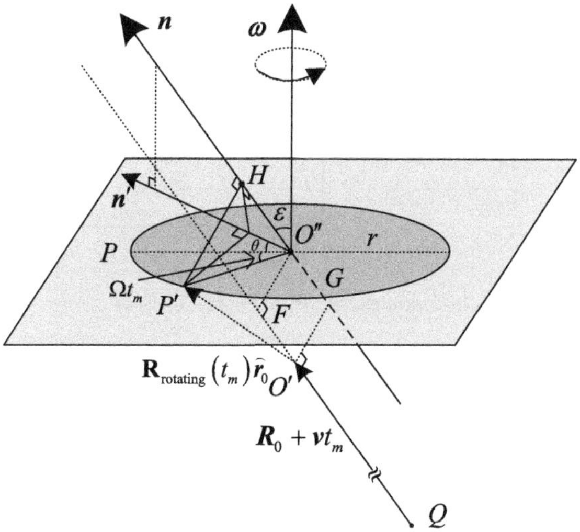

Figure 3.4 The schematic diagram of solving scalar equation $R_\Delta(t_m)$.

specially portrayed by a grey parallelogram in Fig. 3.4. O'' is the circle center of the circle trajectory and r is the radius of the circle trajectory. F is the projection of O'' in LOS. G is the projection of O' in LOS. At the moment of slow time t_m, P moves to the position of P', and the projection point of P' in LOS is H. $\boldsymbol{n}'$ is the projection vector of $\boldsymbol{n}$ in LOS on the plane. From the geometry in Fig. 3.4, we know that

$$\begin{aligned} R_\Delta(t_m) &= \left[\mathbf{R}_{\text{rotating}}(t_m)\widehat{\boldsymbol{r}}_0\right]^{\mathrm{T}}\boldsymbol{n} = \left\|\overrightarrow{GH}\right\| \\ &= \left\|\overrightarrow{GO''} + \overrightarrow{O''H}\right\| = \left\|\overrightarrow{GO''}\right\| + r\cos(\Omega t_m + \theta)\sin\varepsilon \end{aligned} \tag{3.29}$$

where θ is the initial phase and ε is the angle between $\boldsymbol{n}$ and $\boldsymbol{\omega}$. From Eq. (3.29) on the range–slow time plane, the peak value of the rotating scattering point in range profile presents such that it changes as cosine form with time going by. In this book, curves induced by the micromotion of scattering points are called "micro-Doppler feature curves," in the range–slow time profile or the time-frequency analysis. By further analysis, the angular frequency Ω of micro-Doppler curves in range–slow time profile is the rotating angular frequency of the target's scattering point. The information of the target's rotating angular frequency will be obtained by extracting the angular frequency in range–slow time profile. The amplitude of micro-Doppler curves $r\sin\varepsilon$ is the length of projection of the target's radius in LOS. If the angle ε between $\boldsymbol{n}$ and $\boldsymbol{\omega}$ is known, the information of the target's rotating radius will be obtained. However, in reality the angle ε is unknown, so it is difficult to obtain the information of the target's rotating radius. (In general, it is difficult to obtain the information of the target's rotating radius in

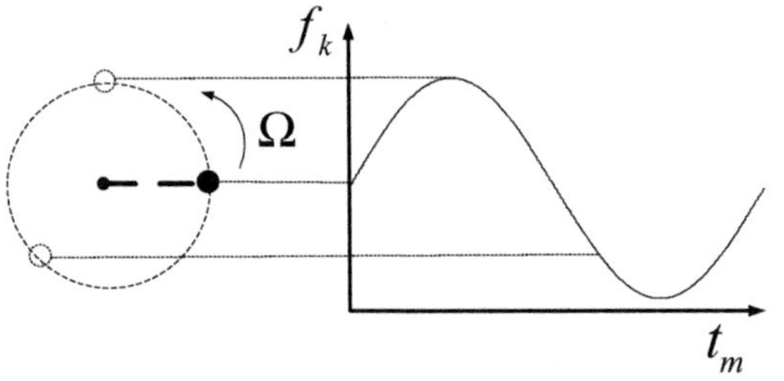

Figure 3.5 The imagery schematic diagram of micro-Doppler curves of rotating target.

single-base radar. But it can be solved in multibase radar, which will be further analyzed and stated in Chapter 6.)

It can be seen that the micro-Doppler curves of rotating targets in wideband LFM signal are the change of radius distance between the scatters on the target and the reference point at each transmission moment intrinsically. To understand more vividly, the imagery schematic diagram is drawn in Fig. 3.5. The change of the distance between the target's scatters and the reference point is corresponding to range—slow time plane ($f_k - t_m$ plane), which presents the curves changing at the cosine form.

Two groups of simulation results are given as follows. In the first group, assuming that the carrier frequency of radar is $f_c = 10$ GHz, the bandwidth is $B = 500$ MHz, and the pulse repetition frequency (PRF) = 500 Hz. The origin of local coordinate in radar coordinate is (3,4,5) km, and the initial Euler angle is $(0,\pi/4,\pi/5)$ rad. The translational velocity is 0. The rotating angular velocity is $\boldsymbol{\omega} = (\pi,2\pi,\pi)^{\mathrm{T}}$ rad/s, and the rotating period is $T = 2\pi/\|\boldsymbol{\omega}\| = 0.8165$ s. There are three scattering points on the targets, and their coordinates are as follows: (0,0,0) m, (3,1.5,1.5) m, and (−3,−1.5,−1.5) m. The radar illuminating time is 3 s. The ideal curves relative to the micro-Doppler curves, which are corresponding to the three scattering points on range—slow time plane, are shown in Fig. 3.6(A). Point (0,0,0) locates in the circle center, which does not exist in

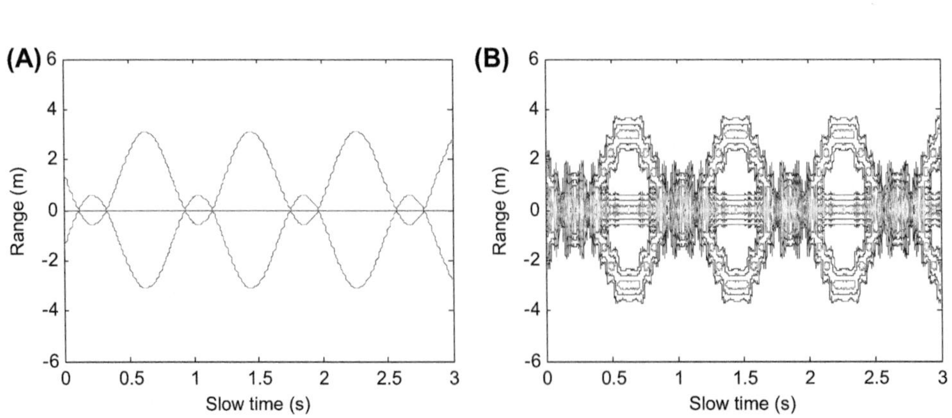

Figure 3.6 Simulation of micro-Doppler effect of the rotating target. (A) Ideal micro-Doppler curves; (B) simulated micro-Doppler curves.

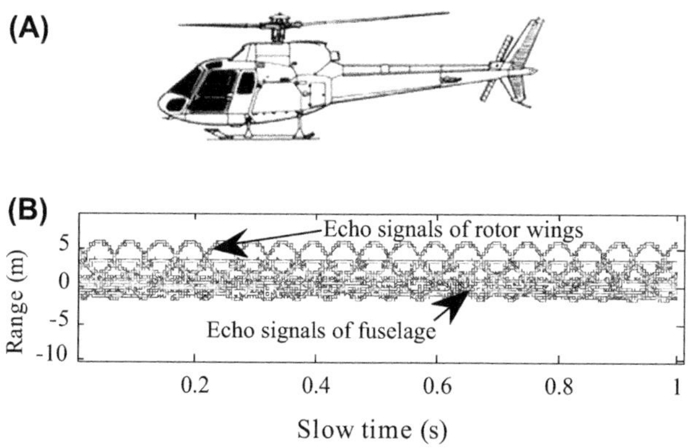

Figure 3.7 A simulation of AS350 "Squirrel" helicopter model. (A) AS350 "Squirrel" helicopter; (B) the range—slow time figure of the target when radar transmits LFM signal.

rotating motion, so it is corresponding with the straight line in the figure. Point (3,1.5,1.5) and point (−3,−1.5,−1.5) are corresponding with the two sine curves, and curves' period fits the targets' rotating period well. Compared with Fig. 2.3, because of the high-resolution of the wideband radar, when each micromotion point possesses its own rotating center, each of their micro-Doppler curves has the different baseline on range—slow time plane correspondingly. However, in narrowband radar, the sine curve corresponding with each scattering point has the same baseline.

The second group of simulation experiments based on the 1:1 model of AS350 "Squirrel" helicopter is produced by European Helicopter Company (see Fig. 3.7). AS350 "Squirrel" helicopter has three rotors, whose radius is 10.69 m and rotating speed is 394 rounds per minute, which equals to 6.5667 rounds per second. What is more, AS350 "Squirrel" helicopter has two tail wings, whose radius is 1.86 m and rotating speed is 2086 rounds per minute, which equals to 34.7667 rounds per second.

Since the empennage often is mapped by the main part in the radar observation process, the micro-Doppler effect of empennage is not considered in the simulation. It is assumed that there are two scattering points at the middle and the tail of each rotor. The pulse width of transmission signal is $T_p = 1$ μs, the carrier frequency is 10 GHz, the bandwidth is 600 MHz, the pulse repetition frequency is 1000 Hz, and the elevation of transmitting beam is 45 degrees. The velocity of target is $v = 70$ m/s. The image of range—slow time is shown as Fig. 3.7(B). Multiple sine curves with the same frequency and different phases are shown distinctively in the figure.

3.2.4 Micro-Doppler Effect of Targets With Vibration

The micro-Doppler effect of vibrating target in LFM radar is analyzed by the geometry relationship diagram of the vibrating target and radar in Fig. 2.4. Similar to the micro-Doppler effect of the rotating target, the micro-Doppler curves depend on the distance

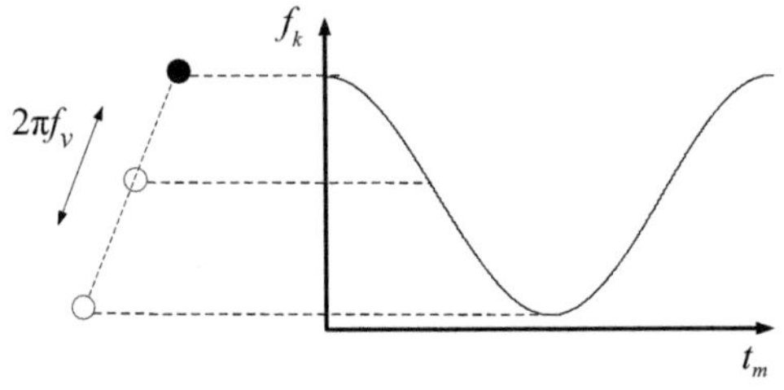

Figure 3.8 The imagery schematic diagram of the micro-Doppler curves of vibrating target.

between the vibrating point and the reference point $R_\Delta(t_m)$. When the target locates in the far field of radar, $R_\Delta(t_m)$ is expressed approximately as follows:

$$R_\Delta(t_m) = r(t_m) - R_{ref}(t_m) \approx \boldsymbol{r}^{\mathrm{T}}\boldsymbol{n} \tag{3.30}$$

Substitute Eqs. (2.33) and (2.36) into Eq. (3.30),

$$\begin{aligned} R_\Delta(t_m) &= D_v \sin(2\pi f_v t)[\cos(\alpha - \alpha_P)\cos\beta \cos\beta_P + \sin\beta \sin\beta_P] \\ &\quad + (r_{X0}\cos\alpha\cos\beta + r_{Y0}\sin\alpha\cos\beta + r_{Z0}\sin\beta) \end{aligned} \tag{3.31}$$

From Eq. (3.31), the peak value in range profile and the phase term presents the feature changing as sine form with time, in range—slow time plane, similar with the vibrating target. This is the micro-Doppler effect of wideband LFM signal. The frequency of sine curves f_v in range—slow time plane is the vibration frequency of the scattering points. The vibration information of target can be obtained through the frequency feature of curves. The amplitude of curves is the projection length of the amplitude of vibration in LOS. The imagery schematic diagram of the micro-Doppler curves generated by the vibrating target is shown in Fig. 3.8.

The micro-Doppler signature expression of the vibrating target in wideband LFM signal is deduced theoretically. However, in most real applications, the amplitude of radius vibration is quite small, some of which even smaller than the range resolution of radar. Under the circumstances, the micro-Doppler curves will not be observed in range—slow time plane. Instead, a straight line will merely be observed in the range unit where the vibrating scattering point exists.

Frequency domain extraction method is to do time-frequency analysis by extracting the signal of the range unit where the vibrating scattering points exist in range—slow time plane. When the amplitude of radius vibrating is smaller than the range resolution of radar, the micro-Doppler curve presents as a straight line in range-slow time plane approximately. Extracting the signal of the straight line is to set f_k as a fixed value f_{k0}, and then Eq. (3.27) becomes

$$
\begin{aligned}
S_d(f_{k0}, t_m) &= \sigma T_p \text{sinc}\left(T_p\left(f_{k0} + \frac{2\mu}{c}R_\Delta(t_m)\right)\right) \cdot \exp\left(-j\frac{4\pi}{c}f_c R_\Delta(t_m)\right) \\
&\approx \sigma T_p \exp\left(-j\frac{4\pi}{c}f_c R_\Delta(t_m)\right)
\end{aligned}
\tag{3.32}
$$

By taking the derivative of phase term of Eq. (3.32) with respect to slow time t_m and then dividing the result by 2π, the micro-Doppler frequency can be obtained as follows:

$$
\begin{aligned}
f_{\text{micro-Doppler}} &= \frac{2R'_\Delta(t_m)}{\lambda} \\
&= \frac{4\pi}{\lambda}f_v D_v[\cos(\alpha - \alpha_P)\cos\beta\cos\beta_P + \sin\beta\sin\beta_P]\cos(2\pi f_v t_m)
\end{aligned}
\tag{3.33}
$$

It is consistent with the expression of the vibrating target of narrowband radar. Thus, the time-frequency analysis can be used as well. The result of the signal extracted by the straight line in Fig. 3.9 by time-frequency analysis is shown in Fig. 3.10.

Since the form of $S_d(f_k, t_m)$ is a sinc function, the amplitude of the peak value differs from others' greatly, which means that researchers must try their best to extract the peak value of function "sinc" to analyze, where $f_k = f_{k0}$. In contrast, time-frequency analysis on other values will influence the power of micro-Doppler signal of the target extremely, so that the result of time-frequency analysis will be affected. However, there are some advantages of this character. When the line signal corresponding with a vibrating point is extracted, the vibration amplitude of other vibrating centers differs from the extracted one greatly. The observation and analysis is only done by the vibrating point extracted, which can avoid the interference of micro-Doppler component of other vibrating points.

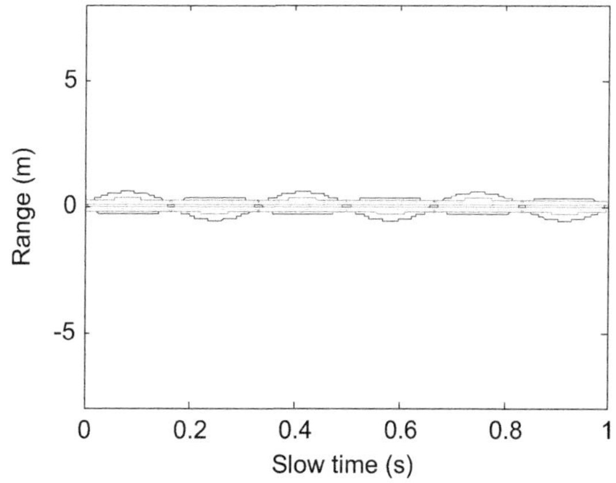

Figure 3.9 The range–slow time image when the amplitude of radius vibration is smaller than the range resolution.

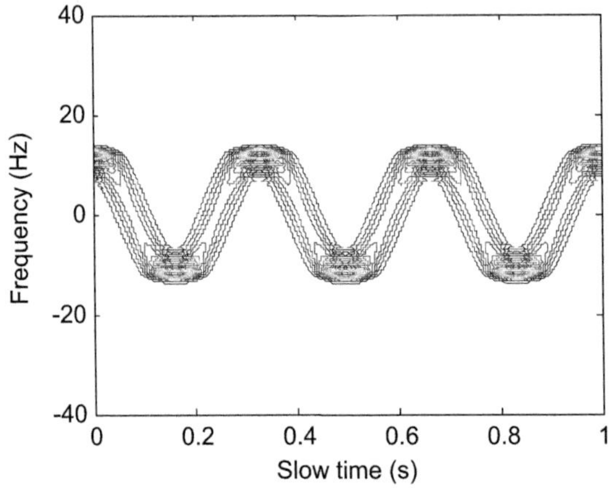

Figure 3.10 The micro-Doppler feature of vibration target by frequency domain extraction.

The time domain extraction is to extract a line signal from $S_d(t_k,t_m)$ to analyze the micro-Doppler effect. Firstly, Eq. (3.27) is transformed into fast-time t_k domain by inverse Fourier transform as

$$S_d(t_k,t_m) = \mathrm{rect}\left(\frac{t_k}{T_p}\right)\exp\left(-\mathrm{j}\frac{4\pi}{c}\mu R_\Delta(t_m)t_k\right)\exp\left(-\mathrm{j}\frac{4\pi}{c}f_c R_\Delta(t_m)\right) \quad (3.34)$$

Secondly, a line signal is extracted from $S_d(t_k,t_m)$ arbitrarily, that is $t_k = t_{k0}$. Then the phase terms are differentiated by the slow time t_m, and then divided by 2π:

$$\begin{aligned} f'_{\mathrm{micro-Doppler}} &= \left(\frac{2\mu t_{k0}}{c}+\frac{2}{\lambda}\right)2\pi f_v D_v[\cos(\alpha-\alpha_P)\cos\beta\cos\beta_P+\sin\beta\sin\beta_P] \\ &\quad \times \cos(2\pi f_v t_m) \end{aligned} \quad (3.35)$$

From the Eq. (3.35) we can see that, after the micro-Doppler frequency extracted from the time domain, its frequency bias not only depends on vibration radius, vibration frequency, and the carrier frequency but also on the value of the fast time t_{k0}. Consider that for the common radar signal, the value of $2\mu t_{k0}/c$ is a bit smaller than $2/\lambda$. For example, when the bandwidth and the fast time meet the following conditions $B < f_c/5$ and $t_{k0} \le T_p/2$, the $2\mu t_{k0}/c$ can be deduced as

$$2\mu t_{k0}/c = 2Bt_{k0}/(T_p c) \le BT_p/(T_p c) = B/c < f_c/(5c) = 1/(5\lambda) \quad (3.36)$$

The difference between it and $2/\lambda$ is one order of magnitude. Therefore, when $2\mu t_{k0}/c$ is ignorant, Eq. (3.35) is consistent with Eq. (3.33). The analysis result of time domain extraction is shown in Fig. 3.11, which is quite approximate with the one of the frequency domain extraction.

The time domain extraction method avoids being affected by the peak position of sinc function. Extracting the signal of an arbitrary position t_k in the direction of the slow

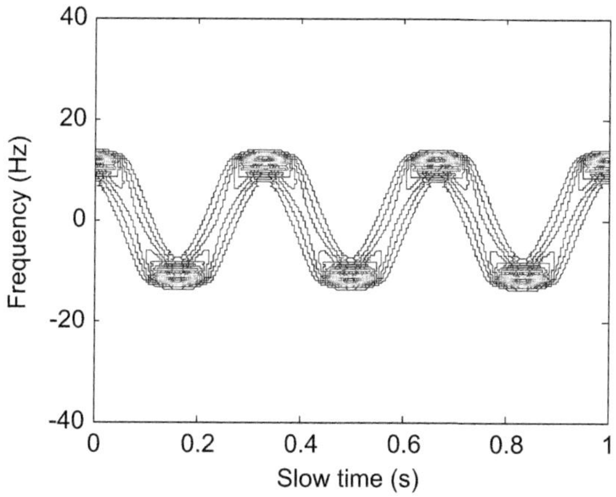

Figure 3.11 The micro-Doppler feature of vibration target by time domain extraction.

time will observe the whole micro-Doppler component. However, it also brings disadvantages to signal analysis because when a target contains many vibration centers, there will be many sinusoidal frequency modulated micro-Doppler components in echoes. On the other hand, up to now, it is difficult to get the ideal time-frequency resolution in many time-frequency analysis methods until now, which makes the estimation of the signal's parameters inaccurate. Finally, it influences the accuracy of the estimation of the target's micromotion parameters.

Based on the same theory, the frequency domain extraction fits the micro-Doppler signal analysis when the radial rotation radius is smaller than the range resolution of radar, which can be derived by readers like yourselves.

3.2.5 Micro-Doppler Effect of Targets With Procession

The geometry relation diagram of radar and the procession target is shown in Fig. 2.6. Similar with the micro-Doppler effect derivation of the rotation target and the vibration target, the micro-Doppler effect of the procession target depends on $R_\Delta(t_m)$ in wideband LFM signal. When the target is in the far field of the radar, $R_\Delta(t_m)$ can be approximated as

$$R_\Delta(t_m) = r(t_m) - R_{ref}(t_m) \approx \left(\mathbf{R}_{\text{coning}}\mathbf{R}_{\text{spinning}}\mathbf{R}_{\text{init}}\boldsymbol{r}_0\right)^{\mathrm{T}}\boldsymbol{n} \tag{3.37}$$

From Eqs. (2.46) and (2.47), we can get

$$\begin{aligned}
&\mathbf{R}_{\text{coning}}\mathbf{R}_{\text{spinning}}\\
&= \left(\mathbf{I}+\widehat{\boldsymbol{\omega}}_c'^2\right)\left(\mathbf{I}+\widehat{\boldsymbol{\omega}}_s'^2\right) + \left(\widehat{\boldsymbol{\omega}}_s' + \widehat{\boldsymbol{\omega}}_c'^2\widehat{\boldsymbol{\omega}}_s'\right)\sin(\Omega_s t_m) - \left(\widehat{\boldsymbol{\omega}}_s'^2 + \widehat{\boldsymbol{\omega}}_c'^2\widehat{\boldsymbol{\omega}}_s'^2\right)\cos(\Omega_s t_m)\\
&+ \left(\widehat{\boldsymbol{\omega}}_c' + \widehat{\boldsymbol{\omega}}_c'\widehat{\boldsymbol{\omega}}_s'^2\right)\sin(\Omega_c t_m) + \widehat{\boldsymbol{\omega}}_c'\widehat{\boldsymbol{\omega}}_s'\sin(\Omega_c t_m)\sin(\Omega_s t_m) - \widehat{\boldsymbol{\omega}}_c'\widehat{\boldsymbol{\omega}}_s'^2\sin(\Omega_c t_m)\cos(\Omega_s t_m)\\
&- \left(\widehat{\boldsymbol{\omega}}_c'^2 + \widehat{\boldsymbol{\omega}}_c'^2\widehat{\boldsymbol{\omega}}_s'^2\right)\cos(\Omega_c t_m) - \widehat{\boldsymbol{\omega}}_c'^2\widehat{\boldsymbol{\omega}}_s'\cos(\Omega_c t_m)\sin(\Omega_s t_m) + \widehat{\boldsymbol{\omega}}_c'^2\widehat{\boldsymbol{\omega}}_s'^2\cos(\Omega_c t_m)\cos(\Omega_s t_m)
\end{aligned} \tag{3.38}$$

Substitute Eq. (3.38) into Eq. (3.37), and $R_\Delta(t_m)$ becomes

$$\begin{aligned} R_\Delta(t_m) &= \boldsymbol{r}_0^{\mathrm{T}}\mathbf{R}_{\mathrm{init}}^{\mathrm{T}}(\mathbf{R}_{\mathrm{coning}}\mathbf{R}_{\mathrm{spinning}})^{\mathrm{T}}\boldsymbol{n} \\ &= A_0 + A_1\sin(\Omega_s t_m) + A_2\cos(\Omega_s t_m) + A_3\sin(\Omega_c t_m) + A_4\sin(\Omega_c t_m)\sin(\Omega_s t_m) \\ &\quad + A_5\sin(\Omega_c t_m)\cos(\Omega_s t_m) + A_6\cos(\Omega_c t_m) + A_7\cos(\Omega_c t_m)\sin(\Omega_s t_m) \\ &\quad + A_8\cos(\Omega_c t_m)\cos(\Omega_s t_m) \end{aligned} \tag{3.39}$$

where

$$\begin{aligned} A_0 &= \boldsymbol{r}_0^{\mathrm{T}}\mathbf{R}_{\mathrm{init}}^{\mathrm{T}}\left(\left(\mathbf{I}+\widehat{\boldsymbol{\omega}}_c'^2\right)\left(\mathbf{I}+\widehat{\boldsymbol{\omega}}_s'^2\right)\right)^{\mathrm{T}}\boldsymbol{n}, & A_1 &= \boldsymbol{r}_0^{\mathrm{T}}\mathbf{R}_{\mathrm{init}}^{\mathrm{T}}\left(\widehat{\boldsymbol{\omega}}_s' + \widehat{\boldsymbol{\omega}}_c'^2\widehat{\boldsymbol{\omega}}_s'\right)^{\mathrm{T}}\boldsymbol{n}, \\ A_2 &= -\boldsymbol{r}_0^{\mathrm{T}}\mathbf{R}_{\mathrm{init}}^{\mathrm{T}}\left(\widehat{\boldsymbol{\omega}}_s'^2 + \widehat{\boldsymbol{\omega}}_c'^2\widehat{\boldsymbol{\omega}}_s'^2\right)^{\mathrm{T}}\boldsymbol{n}, & A_3 &= \boldsymbol{r}_0^{\mathrm{T}}\mathbf{R}_{\mathrm{init}}^{\mathrm{T}}\left(\widehat{\boldsymbol{\omega}}_c' + \widehat{\boldsymbol{\omega}}_c'\widehat{\boldsymbol{\omega}}_s'^2\right)^{\mathrm{T}}\boldsymbol{n}, \\ A_4 &= \boldsymbol{r}_0^{\mathrm{T}}\mathbf{R}_{\mathrm{init}}^{\mathrm{T}}\left(\widehat{\boldsymbol{\omega}}_c'\widehat{\boldsymbol{\omega}}_s'\right)^{\mathrm{T}}\boldsymbol{n}, & A_5 &= -\boldsymbol{r}_0^{\mathrm{T}}\mathbf{R}_{\mathrm{init}}^{\mathrm{T}}\left(\widehat{\boldsymbol{\omega}}_c'\widehat{\boldsymbol{\omega}}_s'^2\right)^{\mathrm{T}}\boldsymbol{n}, \\ A_6 &= -\boldsymbol{r}_0^{\mathrm{T}}\mathbf{R}_{\mathrm{init}}^{\mathrm{T}}\left(\widehat{\boldsymbol{\omega}}_c'^2 + \widehat{\boldsymbol{\omega}}_c'^2\widehat{\boldsymbol{\omega}}_s'^2\right)^{\mathrm{T}}\boldsymbol{n}, & A_7 &= -\boldsymbol{r}_0^{\mathrm{T}}\mathbf{R}_{\mathrm{init}}^{\mathrm{T}}\left(\widehat{\boldsymbol{\omega}}_c'^2\widehat{\boldsymbol{\omega}}_s'\right)^{\mathrm{T}}\boldsymbol{n}, \\ A_8 &= \boldsymbol{r}_0^{\mathrm{T}}\mathbf{R}_{\mathrm{init}}^{\mathrm{T}}\left(\widehat{\boldsymbol{\omega}}_c'^2\widehat{\boldsymbol{\omega}}_s'^2\right)^{\mathrm{T}}\boldsymbol{n}. \end{aligned}$$

The right side of Eq. (3.39) is transformed to

$$\begin{aligned} A_1\sin(\Omega_s t_m) + A_2\cos(\Omega_s t_m) &= A_1'\sin(\Omega_s t_m + \phi_1), \\ A_1' &= \sqrt{A_1^2 + A_2^2}, \phi_1 = \mathrm{atan}(A_2/A_1) \end{aligned} \tag{3.40}$$

$$\begin{aligned} A_3\sin(\Omega_c t_m) + A_6\cos(\Omega_c t_m) &= A_2'\sin(\Omega_c t_m + \phi_2), \\ A_2' &= \sqrt{A_3^2 + A_6^2}, \phi_2 = \mathrm{atan}(A_6/A_3) \end{aligned} \tag{3.41}$$

$$\begin{aligned} &A_4\sin(\Omega_c t_m)\sin(\Omega_s t_m) + A_5\sin(\Omega_c t_m)\cos(\Omega_s t_m) = A_3'\sin(\Omega_c t_m)\sin(\Omega_s t_m + \phi_3) \\ &= \frac{A_3'}{2}(\cos((\Omega_c - \Omega_s)t_m - \phi_3) - \cos((\Omega_c + \Omega_s)t_m + \phi_3)), A_3' = \sqrt{A_4^2 + A_5^2}, \phi_3 \\ &= \mathrm{atan}(A_5/A_4) \end{aligned} \tag{3.42}$$

$$A_7\cos(\Omega_c t_m)\sin(\Omega_s t_m)+A_8\cos(\Omega_c t_m)\cos(\Omega_s t_m) = A_4'\cos(\Omega_c t_m)\sin(\Omega_s t_m+\phi_4)$$
$$=\frac{A_4'}{2}(\sin((\Omega_c+\Omega_s)t_m+\phi_4)-\sin((\Omega_c-\Omega_s)t_m-\phi_4)), A_4' = \sqrt{A_7^2+A_8^2}, \phi_3$$
$$=\text{atan}(A_8/A_7) \tag{3.43}$$

From the previous expressions, we can know that the micro-Doppler effect caused by procession is quite complicated. The micro-Doppler curve of the procession target is the addition of multiple sine curves in range—slow time plane, rather than a simple sine curve. Similar to the narrowband signal, the micro-Doppler curve in range—slow time plane still has its own periodicity as time going by, the period of which equals to the lowest common multiple of the conning period T_c and the rotating period is T_s.

Now a simulation is made to certify the theory. The carrier frequency of radar is $f_c = 20.5$ GHz and the pulse width is 50 μs. The bandwidth of transmission signal is 3 GHz, and the range resolution is 0.05 m. PRF equals 1000 Hz. The time length of signal equals 3 s. The origin of local coordinate in radar coordinate is (−100,−300,500) km. The translational velocity of target is 0. The target rotates around the z axis, and the vector expression of rotating angular velocity in local coordinate is $\boldsymbol{\omega}_s = (0,0,6\pi)^{\mathrm{T}}$ rad/s, $\Omega_s = 6\pi$ rad/s, and rotating period is $T_s = 0.33$ s. The vector expression of conning angular velocity in local coordinate is $\boldsymbol{\omega}_c = (5.4553,0,30.9386)^{\mathrm{T}}$ rad/s, $\Omega_c = 10\pi$ rad/s, and the conning period is $T_c = 0.2$ s. The procession angle is $\pi/18$ rad. There are three scattering points on the target. The first one is the vertex of the cone A, the coordinate of which in local coordinate is (0 m,0 m,2 m). The second and the third ones are B and C, respectively, which locate at the bottom of the cone, and they are symmetrical with the other. Their coordinates in local coordinate are (−0.6 m,0 m,−1 m) and (−0.6 m,0 m, 1 m), respectively.

The micro-Doppler curves of three scattering points in range—slow time plane are shown in Fig. 3.12. For more distinction, the curves are drawn from 0 to 1.5 s. Scattering point A is on the rotating axis, so its micromotion only contains conning. Thus its micro-Doppler curve is a simple sine curve in the slow time sequence. Differently, the micromotion of the scattering point B and C is the synthesis of conning and rotating. The micro-Doppler curves of them present as a complex form in the slow time sequence, whose period is 1 s, the least common multiple of rotating period and conning period.

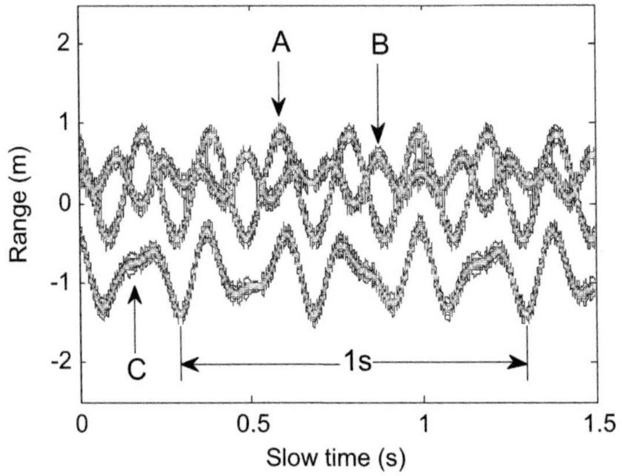

Figure 3.12 Micro-Doppler feature of procession target.

3.3 MICRO-DOPPLER EFFECT IN STEPPED-FREQUENCY CHIRP SIGNAL RADAR [2,3]

3.3.1 Range Imaging

The change of stepped-frequency chirp signal waveform in time sequence is shown in Fig. 3.13. LFM signal is the subpulse of stepped-frequency signal. Assuming that the initial time of the signal is $-T_1/2$, and the expression of subpulse signal i is

$$U_i(t_k) = u(t_k - iT_r)\exp(\mathrm{j}2\pi(f_0 + i\Delta f)(t_k - iT_r) + \mathrm{j}\theta_i), \quad 0 \le i \le N-1 \tag{3.44}$$

where $u(t) = \mathrm{rect}(t/T_1)\exp(\mathrm{j}\pi\mu t^2)$ is the LFM subpulse. The modulated gradient of subpulse is μ. The width of subpulse is T_1, and PRF of subpulse is T_r. The carrier frequency of subpulse i is $f_0 + i\Delta f$, and its initial phase is θ_i. The number of stepped-frequency chirp signals in a set of pulses is N. For point target, the echo of subpulse i is

$$\begin{aligned} s(t_k, i) &= \mathrm{rect}\left(\frac{t_k - iT_r - 2R/c}{T_1}\right) \cdot \exp\left(\mathrm{j}\pi\mu(t_k - iT_r - 2R/c)^2\right) \\ &\cdot \exp(\mathrm{j}2\pi(f_0 + i\Delta f)(t_k - iT_r - 2R/c) + \mathrm{j}\theta_i) \end{aligned} \tag{3.45}$$

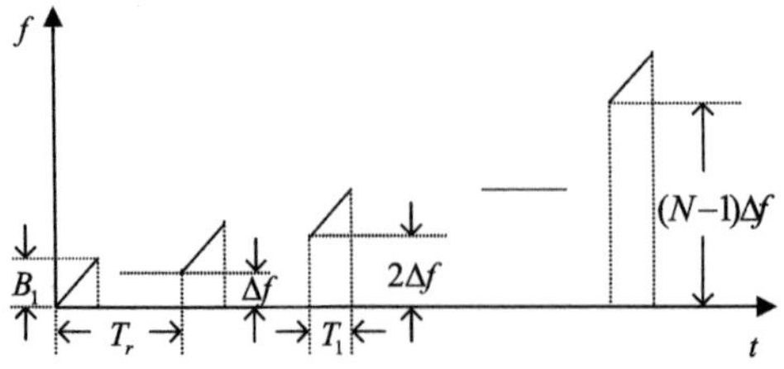

Figure 3.13 Change of the frequency of stepped-frequency chirp signals.

where R is the range from the target to the radar. The reference signal is assumed as

$$s_0(t_k,i) = \text{rect}\left(\frac{t_k - iT_r - 2R_{ref}/c}{T_{ref}}\right)\cdot\exp\left(j\pi\mu(t_k - iT_r - 2R_{ref}/c)^2\right)$$
$$\cdot\exp(j2\pi(f_0 + i\Delta f)(t_k - iT_r - 2R_{ref}/c) + j\theta_i) \tag{3.46}$$

where R_{ref} is the range from the reference point to the radar. T_{ref} is the pulse width of the signal, which is a bit larger than T_1. By "dechirp" process, it becomes

$$s_d(t_k,i) = s(t_k,i)s_0^*(t_k,i) = \text{rect}\left(\frac{t_k - iT_r - 2R/c}{T_1}\right)$$
$$\cdot\exp\left(-j\frac{4\pi\mu}{c}\left(t_k - iT_r - \frac{2R_{ref}}{c}\right)R_\Delta\right) \tag{3.47}$$
$$\cdot\exp\left(-j\frac{4\pi}{c}(f_0 + i\Delta f)R_\Delta\right)\cdot\exp\left(j\frac{4\pi\mu}{c^2}R_\Delta^2\right)$$

where $R_\Delta = R - R_{ref}$. Based on the time of reference point, that is, $t_k' = t_k - iT_r - 2R_{ref}/c$. Then Eq. (3.47) is Fourier transformed of t_k' and the RVP phase and the envelope sideling phase are removed. It becomes

$$S_d(\omega_k,i) = T_1\text{sinc}\left(T_1\left(\omega_k + \frac{4\pi\mu}{c}R_\Delta\right)\right)\cdot\exp\left(-j\frac{4\pi}{c}(f_0 + i\Delta f)R_\Delta\right) \tag{3.48}$$

From Eq. (3.48), the peak value of $|S_d(\omega_k,i)|$ locates in $\omega_k = -4\pi\mu R_\Delta/c$ that is the approximation range profile of the point target. For general micromotion targets, the radial length of radar is smaller than the range resolution of subpulse. Thus, the sampling of range unit contains whole information of all the scattering points in the target. Suppose that $\omega_k = -4\pi\mu R_\Delta/c$. Each approximation resolution of range profile is the fast Fourier transform (FFT) of i. We get

$$S(k) = C\cdot\text{sinc}\left(k + \frac{4\pi\Delta f}{c}R_\Delta\right)\cdot\exp\left(-j\frac{4\pi f_0}{c}R_\Delta\right) \tag{3.49}$$

where C is a constant. We can see that the peak value of $|S(k)|$ locates in $k = -4\pi\Delta f R_\Delta/c$, so that the fine range profile is generated.

3.3.2 Migration and Wrapping of Range Profile Induced by Micromotion

From the deductions in the previous section, the fine range profile of target is synchronized by each group of the subpulses. When the structural components have the micromotion, the motion will cause the change of the echoes' phase history of each

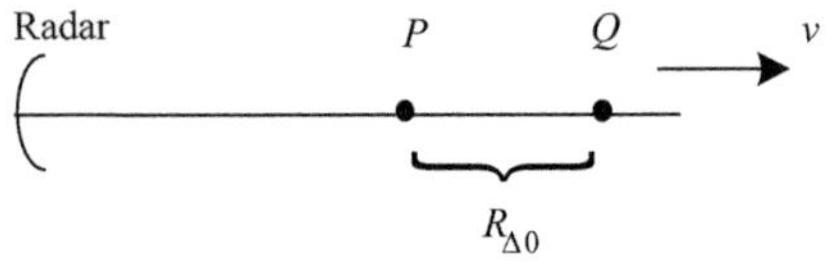

Figure 3.14 Schematic diagram of micromotion of the target.

group of the subpulses. As Fig. 3.14 shows, P is the reference point. When the micro-Doppler frequency shift of the reference point is compensated, the reference point can be regarded as a static point. Q is a micromotion point with the speed of v in the radial direction of radar relative to P. Because of the motion of Q, during a set of pulses, the range from Q to P should not be expressed as R_Δ simply in Eq. (3.47), but the following expression

$$R_\Delta = R_{\Delta 0} + iT_r v, \tag{3.50}$$

where $R_{\Delta 0}$ is the initial range from P to Q in the set of pulses. In general, during the time of the pulse train, the displacement of the micromotion point will not exceed an approximation range resolution. Thus, Eq. (3.48) can be rewritten as

$$S_d(\omega_k, i) = T_1 \text{sinc}\left(T_1\left(\omega_k + \frac{4\pi\mu}{c}R_{\Delta 0}\right)\right) \cdot \exp\left(-\mathrm{j}\frac{4\pi}{c}(f_0 + i\Delta f)(R_{\Delta 0} + iT_r v)\right) \tag{3.51}$$

In this way, the peak value of fine range profile locates in

$$k = \Phi'(i) = -\frac{4\pi}{c}\Delta f R_{\Delta 0} - \frac{4\pi}{c} f_0 T_r v - \frac{8\pi}{c}\Delta f i T_r v \tag{3.52}$$

where $\Phi(i)$ is the phase of the right-side expression of Eq. (3.51). Analyzing the three terms of the right side of Eq. (3.52), we know that the first term is the initial position term of the micromotion point, that is, the peak position of $|S(k)|$ in Eq. (3.49), which represents the real position of the micromotion point. The second term is the migration of range profile caused by the micromotion. The position migration of the micromotion in fine range profile is proportionate to its velocity. The absolute value of the third one is quite small compared with the absolute values of the first two terms, which can be neglected. For example, the parameters of radar and the target are $f_0 = 35$ GHz, $T_r = 1$ μs, $i = 1$, $\Delta f = 5$ MHz, $v = 20$ m/s, and $R_{\Delta 0} = 5$ m, respectively. The value of the first term is -1.0472, the value of the second term is -0.0293, and the value of the third term is -8.3776×10^{-6}. However, the velocity v couples with i, which makes the peak value of the micromotion point extend when Eq. (3.51) conducts FFT to synchronize the fine range profile. Especially when the velocity of the micromotion point is quite large, the influence of this term and the broadening of the fine range profile will become even worse.

The third term is neglected and Eq. (3.52) is rewritten as

$$k = -\frac{4\pi}{c}\Delta f R_{\Delta 0} - \frac{4\pi}{c} f_0 T_r v \tag{3.53}$$

This is the peak position of the micromotion point in fine range profile. The correct range can be obtained when both sides of Eq. (3.53) are divided by $-4\pi\Delta f/c$.

$$\dot{r} = R_{\Delta 0} + f_0 T_r v/\Delta f \tag{3.54}$$

The scope of fine range depends on the scope of angular frequency in fact. The fine-resolution range profile is obtained by N points DFT computation of the approximation range profile. The scope of angular frequency is $[-\pi,\pi]$. Thus, when k exceeds $[-\pi,\pi]$, the peak value of micromotion will be wrapping. When the micromotion component of the target that contains multiple scattering points moves too fast, the fine-resolution range profile will be divided into two parts, which appear in both sides of the central point. If the fine range profile is not wrapping, it must meet the following condition:

$$|k| < \pi \tag{3.55}$$

or

$$-\frac{c}{4\Delta f} < \dot{r} < \frac{c}{4\Delta f} \tag{3.56}$$

That is, the fine range profile locates in the interval of unambiguous range. Eq. (3.54) is substituted into Eq. (3.56), and the velocity must meet the condition of

$$-\frac{1}{f_0 T_r}\left(\frac{c}{4} + \Delta f R_{\Delta 0}\right) < v < \frac{1}{f_0 T_r}\left(\frac{c}{4} - \Delta f R_{\Delta 0}\right) \tag{3.57}$$

When the target's parameters of micromotion do not meet the conditions Eqs. (3.55)–(3.57), because of the wrapping effect, the position of the micromotion point depends on the following equation:

$$\hat{r} = \mathrm{mod}\left(\dot{r} + \frac{c}{4\Delta f}, \frac{c}{2\Delta f}\right) - \frac{c}{4\Delta f} \tag{3.58}$$

where mod(a,b) means complementation.

A simulation experiment is given as follows. The parameters of radar are $f_0 = 35$ GHz, $T_r = 0.78125$ μs, $\Delta f = 4.6875$ MHz, $N = 64$, and the synchronized bandwidth is $B = 300$ MHz. The range between the micromotion point and the reference point is $R_{\Delta 0} = 5$ m. Substituting the parameters into Eq. (3.57), the scope of micromotion point's velocity that meets the condition of not wrapping is $v \in (-36$ m/s, 18.8571 m/s$)$. As shown in Fig. 3.15, the initial positions of the four points all locate at the distance of $R_{\Delta 0}$ from the reference point. But their instantaneous velocity is different from each other. Accordingly, their positions caused by the range migration effect in high resolution are different from each other. Fig. 3.15A verifies the migration effect.

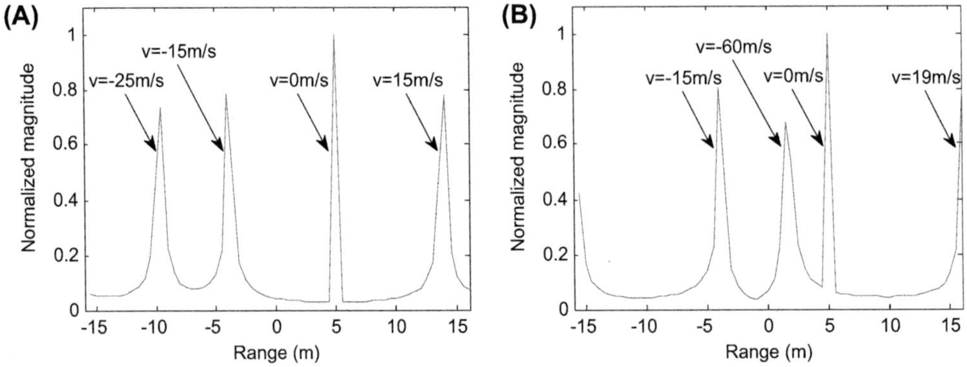

Figure 3.15 The range migration effect and the range wrapping effect of the micromotion point. (A) The range migration effect; (B) the range wrapping effect.

Fig. 3.15B analyzes the wrapping effect of the high-resolution range profile. When velocity v values 19 m/s, which approximates the critical value, the peak value splitting occurs, which appears on both sides of the range profile. When v exceeds the scope limits by Eq. (3.57), such as -60 m/s, the peak value is wrapping from the left side of the range profile to the right side, appearing at the right side of the peak value relative to $v = -15$ m/s, and its broadening of peak value is quite obvious.

3.3.3 Micro-Doppler Effect of Targets With Rotation or Vibration

To analyze the micro-Doppler effect of the target of stepped-frequency chirp signal radar further, the typical micromotion forms, that is, rotation and vibration, are discussed. In stepped-frequency chirp signal radar, when the micromotion point is rotating or vibrating and the Doppler frequency shift is compensated, the displacement expression of the micromotion point in radial direction becomes

$$R(t_m) = R_Q + \rho \cos(\Omega t_m + \theta_0) \tag{3.59}$$

where R_Q is the range from the rotating of vibrating center to the reference point, ρ is amplitude, Ω is rotating or vibrating frequency, θ_0 is initial phase, and t_m is slow time, in other words, t_m is the transmission time of each set of pulses. Its velocity is expressed as

$$v(t_m) = R'(t_m) = -\rho\Omega \sin(\Omega t_m + \theta_0) \tag{3.60}$$

Substituting Eqs. (3.59) and (3.60) into Eq. (3.53), the peak position of rotating points and vibrating points in high-resolution range profile are obtained as follows:

$$\begin{aligned} k(t_m) &= -\frac{4\pi}{c}\Delta f R(t_m) - \frac{4\pi}{c} f_0 T_r v(t_m) \\ &= \frac{-4\pi\Delta f}{c}\left(R_Q + \rho\sqrt{1 + \Omega^2 f_0^2 T_r^2/\Delta f^2}\cos(\Omega t_m + \theta_0 + \phi)\right) \end{aligned} \tag{3.61}$$

where $\phi = \arccos\left(1 \Big/ \sqrt{1+\Omega^2 f_0^2 T_r^2/\Delta f^2}\right)$. When $k(t_m)$ meets Eq. (3.55), the real position coordinate is obtained when Eq. (3.61) is divided by $-4\pi\Delta f/c$:

$$r(t_m) = R(t_m) + f_0 T_r v(t_m)/\Delta f = R_Q + \rho\sqrt{1+\Omega^2 f_0^2 T_r^2/\Delta f^2}\cos(\Omega t_m + \theta_0 + \phi) \tag{3.62}$$

When $k(t_m)$ does not meet the condition of Eq. (3.55), the high-resolution range profile will be wrapping. The position of micromotion point in high-resolution range profile becomes

$$\hat{r}(t_m) = \mathrm{mod}\left(R_Q + \rho\sqrt{1+\Omega^2 f_0^2 T_r^2/\Delta f^2}\cos(\Omega t_m + \theta_0 + \phi) + \frac{c}{4\Delta f}, \frac{c}{2\Delta f}\right) - \frac{c}{4\Delta f} \tag{3.63}$$

Obviously, Eq. (3.62) is a special situation of Eq. (3.63) on the condition of Eq. (3.55). Thus, Eq. (3.63) expresses the micro-Doppler feature of rotation and vibration in stepped-frequency chirp signal completely. Eqs. (3.62) and (3.63) can get three properties of the micro-Doppler effect of rotation and vibration in stepped-frequency chirp signal:

Property 1: When the micromotion of the rotating/vibrating point meets the condition of Eq. (3.55), the micro-Doppler feature of rotating/vibrating point presents as sine form. The period of sine curves equals the period of rotating/vibrating. But the amplitude is not the real amplitude of rotating/vibrating in the radial direction of radar. Its amplitude is magnified by $\sqrt{1+\Omega^2 f_0^2 T_r^2/\Delta f^2}$ times.

Property 2: The peak value point of the sine curve in range–slow time plane does not mean that, at this moment, the rotating/vibrating point moves to the nearest or the furthest position of the radar. Because when Eq. (3.59) takes the extreme value, Eq. (3.62) may not, and their phase difference is ϕ.

Property 3: When the micromotion of the rotating/vibrating point does not meet the condition in Eq. (3.55), the micro-Doppler feature of the rotating/vibrating point presents wrapping as the sine curve, whose period equals the period of the rotating/vibrating motion.

These three properties present the similarities and differences of micro-Doppler feature of the stepped-frequency chirp signal and the LFM signal. In LFM radar, micro-Doppler signal induced by rotating and vibrating are both modulated by sine regular. The micromotion point presents as the sine form in range–slow time plane. The period of the sine curve equals the period of rotating period and vibrating period of the scattering point, and the amplitude represents the amplitude of rotating amplitude and vibrating amplitude of the scattering point in radial direction of radar.

In stepped-frequency chirp signal radar, the parameters of curves do not represent the motion parameters directly, though the curves on range—slow time plane still contain the information of the rotating/vibrating motion. There exists the conversion relation between the curves and the motion. The main cause of the differences is the range migration effect and the range wrapping effect generated by the motion of the target in stepped-frequency chirp signal.

A simulation experiment is given to verify the previous theories. The parameters of radar signal are as follows: $f_0 = 10$ GHz, $T_r = 78.125$ μs, $\Delta f = 4.6875$ MHz, $N = 64$, the repetition frequency of the set of pulses $BRF = 200$ Hz, synchronized bandwidth $B = 300$ MHz, the range resolution $\Delta_R = 0.5$ m. Two scattering points rotate. The range from the rotating center to the reference point is $R_Q = -3$ m, and their radiuses of rotation are both 2 m. As shown in Fig. 3.16A, when the rotating frequencies of two scattering points are 2 Hz and 5 Hz, respectively, $|k|_{max} = 2.6824 < \pi$ can be obtained according to Eq. (3.55), which meet the condition of not being wrapping in Eq. (3.55). Substituting the parameters into $\sqrt{1 + \Omega^2 f_0^2 T_r^2 / \Delta f^2}$, the amplifications of the amplitude of the two sine curves relative to the two rotating points are 2.3209 and 5.3306. So the amplitude of the sine curves are 4.6418 m and 10.6613 m, respectively. Fig. 3.16A verifies the property very well.

When $t_m = 0.25$ s, the rotating point, whose rotation frequency is 5 Hz, moves to the nearest position of the radar, which is the moment at A in Fig. 3.16A. The phase shift of the curve relative to the rotating point is $\phi = 79.1875$ degrees. Thus, the trough of the sine curve appears at the moment of B. So the second property is verified.

When the frequency of rotating point adds to 8 Hz from 5 Hz, $|k|_{max} = 3.9023 > \pi$, the sine curve will occur wrapping, which is shown in Fig. 3.16B. The micro-Doppler curves of the two scattering points described in Eq. (3.63) are shown as "□" and "○" in Fig. 3.16B. We can see that the curves quite fit the curves relative to the rotating point. Property three and Eq. (3.63) are verified correctly.

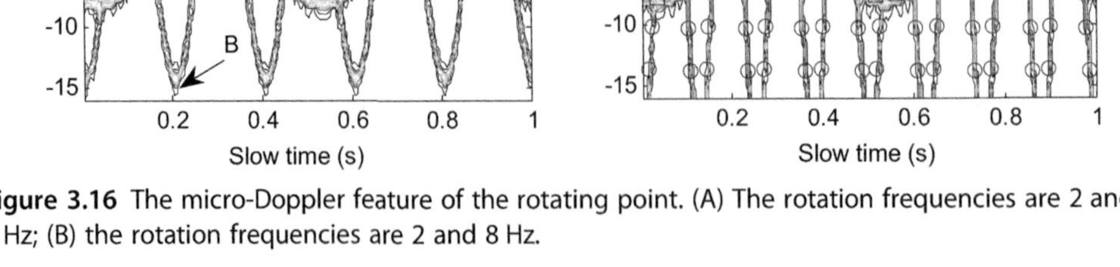

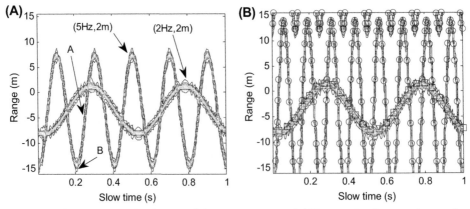

Figure 3.16 The micro-Doppler feature of the rotating point. (A) The rotation frequencies are 2 and 5 Hz; (B) the rotation frequencies are 2 and 8 Hz.

3.4 MICRO-DOPPLER EFFECT IN LINEAR FREQUENCY MODULATION CONTINUOUS WAVE SIGNAL RADAR

3.4.1 Linear Frequency Modulation Continuous Wave Signal

The transmission signal and receiving signal of pulse Doppler (PD) radar are separate on time, which accomplish the transmission process and receiving process by the converting of the transmit-receive switch. Since the transmission energy of PD radar gathers in a narrow pulse, its power of peak value is quite high. The PD radar has very high requirements for the sensor on volume, weight, and power. However, the LFMCW radar transmits the signal continuously. Consequently, the energy is distributed in the whole period. Besides, its transmission power is low, so the solid-state amplifier will meet the requirement, which leads to many advantages compared with PD radar, such as smaller volume, smaller weight, lower cost, and stronger concealment. At present, two kinds of LFMCW signal, the saw-tooth-wave frequency modulation and the triangle-wave frequency modulation, are applied popularly, which are shown in Fig. 3.17.

The expression of the saw-tooth-wave frequency modulation is

$$p(t) = \operatorname{rect}\left(\frac{t}{T_p}\right) \cdot \exp\left(j2\pi\left(f_c t + \frac{1}{2}\mu t^2\right)\right) \tag{3.64}$$

where the expression of LFMCW signal is the same as the expression of LFM signal, the difference is pulse duration T_p equals the pulse repetition interval (PRI), and pulse duration is much larger compared with LFM signal, which is often at the millisecond.

The expression of the triangle-wave frequency modulation is

$$p(t) = \begin{cases} \exp\left(j2\pi\left(f_c t + \frac{1}{2}\mu t^2\right)\right) & t \in \left[0, \frac{T_p}{2}\right) \\ \exp\left(j2\pi\left(f_c t - \frac{1}{2}\mu t^2\right)\right) & t \in \left[\frac{T_p}{2}, T_p\right) \end{cases} \tag{3.65}$$

The saw-tooth-wave frequency modulation signal is applied in imaging radar in general. But the triangle-wave frequency modulation signal is applied on the aspects of

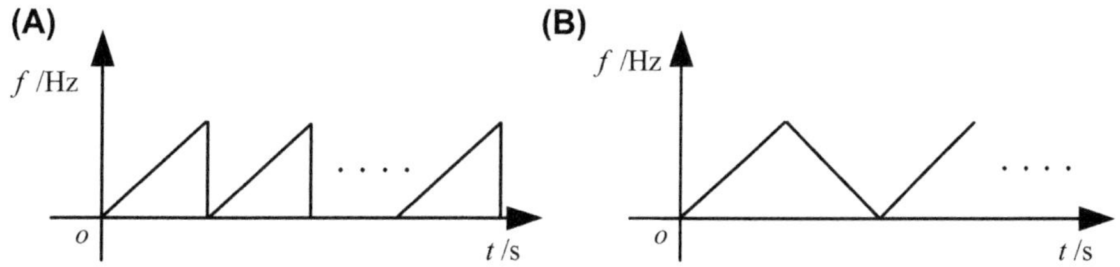

Figure 3.17 The schematic diagram of the frequency modulation form of LFMCW signal. (A) The saw-tooth-wave frequency modulation; (B) the triangle-wave frequency modulation.

target velocity estimation, motion-speed detection, etc. In this section, we mainly use the high resolution of LFMCW signal to analyze the micro-Doppler feature of the target, so the following LFMCW signal we referred to is the saw-tooth-wave frequency modulation signal.

3.4.2 Migration and Broadening of Range Profile Induced by Micromotion

Similar to the LFM signal, the target's echo signal received by LFMCW radar is expressed as

$$s(t_k) = \sigma \text{rect}\left(\frac{t_k - 2R/c}{T_p}\right) \cdot \exp\left(\text{j}2\pi\left(f_c\left(t_k - \frac{2R}{c}\right) + \frac{1}{2}\mu\left(t_k - \frac{2R}{c}\right)^2\right)\right) \tag{3.66}$$

where σ is scattering coefficient of the target, and R is the range from the static point to the radar.

LFMCW uses the "dechirp" as well. As the duration of LFMCW is quite long, when the target has radial velocity and acceleration relative to radar, it will generate Doppler modulation of the echo signal, leading to the migration and broadening of range profile. Thus, the first-order approximation model should be used to analyze, and under this condition, the range from the target and the radar is expressed as

$$R = R_0 - v_r t_k \tag{3.67}$$

where R_0 is the distance from the micromotion point to the radar at the initial moment, and v_r is the radial velocity of the target relative to the radar. The reference signal, needed in "dechirp" processing, becomes

$$s_{ref}(t_k) = \text{rect}\left(\frac{t_k - 2R_{ref}/c}{T_{ref}}\right) \cdot \exp\left(\text{j}2\pi\left(f_c\left(t_k - \frac{2R_{ref}}{c}\right) + \frac{1}{2}\mu\left(t_k - \frac{2R_{ref}}{c}\right)^2\right)\right) \tag{3.68}$$

After conducting the "dechirp" process, the echo signal of the single scattering point and the reference signal contain positive frequency and negative frequency as two parts. When the time delay of the echo signal is smaller than the time delay of the reference, the time length of negative frequency is much smaller than that of positive frequency. In contrast, when the time delay of the echo signal is larger than the time delay of the reference, the time length of positive frequency is much smaller than that of negative frequency. And the absolute value of positive frequency is smaller than that of negative frequency. Thus, when the signal after "dechirp" process gets low-pass filtered, the spectrum mutation of the echo signal of the single scattering point will be removed. In general, the range of LFMCW radar is small. When some point on the target is regarded as the reference point, the difference between the time delay of the echo signal and the time delay of the reference signal is quite small, so the energy consumption caused by low-pass filtering can be neglected. Thus, the echo signal and the reference signal processed by "dechirp" becomes

$$\begin{aligned} s_d(t_k) &= s(t_k)s^*_{ref}(t_k) \\ &= \sigma \mathrm{rect}\left(\frac{t_k - 2R/c}{T_p}\right) \cdot \exp\left(\mathrm{j}\left(-\frac{4\pi}{c}\mu\left(t_k - \frac{2R_{ref}}{c}\right)R_\Delta - \frac{4\pi}{c}f_c R_\Delta + \frac{4\pi\mu}{c^2}R_\Delta^2\right)\right) \\ &\quad \cdot \exp\left(\mathrm{j}\left(-\frac{4\pi\mu v_r t_k^2}{c} + \frac{4\pi f_c v_r t_k}{c}\right)\right) \exp\left(\mathrm{j}\left(-\frac{8\pi\mu R_0 v_r t_k}{c^2} + \frac{4\pi\mu v_r^2 t_k^2}{c^2}\right)\right) \end{aligned} \tag{3.69}$$

where $R_\Delta = R_0 - R_{ref}$. The last phase is quite small, which is often neglected. Calculating the derivative of the right-side phase term and dividing the derivative by 2π, the instantaneous frequency of micro-Doppler signal is

$$f = -\frac{2\mu}{c}R_\Delta + \frac{2v_r}{\lambda} + \frac{4\mu}{c}v_r t_k \tag{3.70}$$

By analyzing the three terms on the right side, we know that the first term is the initial position of the micromotion; the second term is the migration of the range profile induced by the motion during the internal pulse, and the amount of migration is proportional to its radial velocity; the absolute value of the third term is relatively small compared with the absolute value of the first two terms, which can be neglected. Because of the coupling of v_r and t_k, when Eq. (3.69) is processed by FFT to obtain the range profile, the third term will make the broadening of the peak value relative to the micromotion. The more the velocity of the micromotion is, the higher will be the degree of broadening.

Neglecting the third term, Eq. (3.70) is rewritten as

$$f = -\frac{2\mu}{c}R_\Delta + \frac{2v_r}{\lambda} \tag{3.71}$$

This is the peak position of micromotion point in range profile. Eq. (3.71) is divided by $-2\mu/c$, and the range scaling is:

$$\dot{r} = R_\Delta - \frac{v_r c}{\lambda\mu} \tag{3.72}$$

We can see that the quantity of migration in range profile induced by internal pulse motion of the target is $v_r c/(\lambda\mu)$.

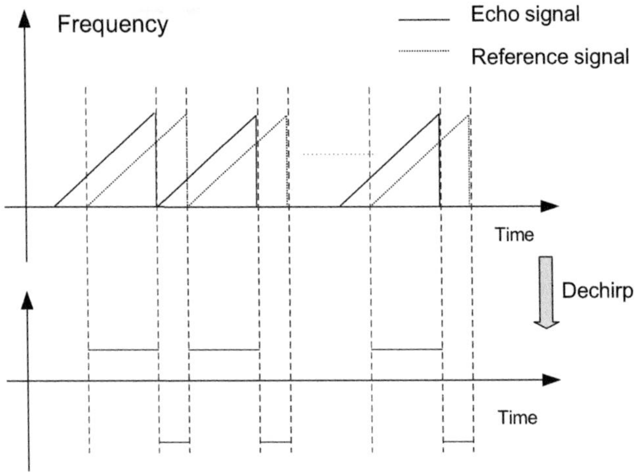

Figure 3.18 Schematic diagram of LFMCW signal processed by "dechirp."

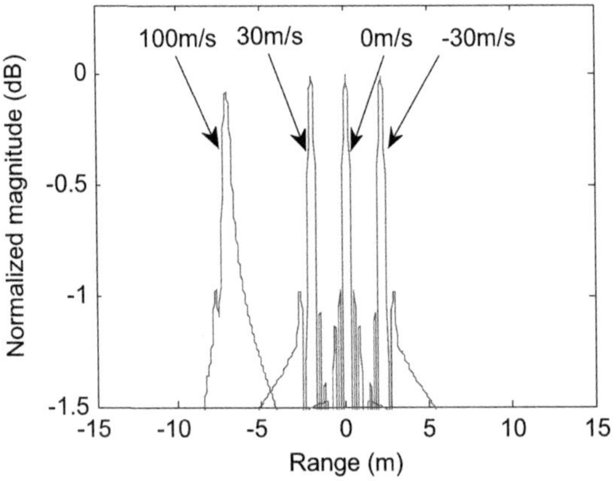

Figure 3.19 Migration and broadening of range profile induced by micromotion.

A simulation experiment is given for verification. The parameters are set as follows. The carrier frequency $f_c = 35$ GHz, the pulse width $T_p = 1$ ms, the bandwidth $B = 500$ MHz, and $R_\Delta = 5$ m. The range profiles are shown in Fig. 3.19 whose velocity of the micromotion points are 0, -30, 30, and 100 m/s, respectively. We can see that with the change of targets' velocity, the positions of the range profile vary as well. And the broadening of the peak value of range profile is more obvious with the addition of the velocity.

3.4.3 Micro-Doppler Effect of Targets With Rotation or Vibration

When the micromotion point is rotating or vibrating and the radar is transmitting the m-th LFM pulse, after the translational motion compensation, the displacement expression of the micromotion point in the radial direction of the radar is

$$R(t_m, t_k) = R_Q + \rho \cos(\Omega(mT_p + t_k) + \theta_0) \tag{3.73}$$

where R_Q is the range from the rotation center or the vibration center to the reference point, ρ is the amplitude, Ω is the angular frequency of rotation or vibration, θ_0 is the initial phase, and t_m is slow time, which is also called the transmission time of each pulse train. The velocity is expressed as

$$v_r(t_m, t_k) = R'(t_m, t_k) = -\rho\Omega \sin(\Omega(mT_p + t_k) + \theta_0) \tag{3.74}$$

Because the bandwidth of LFMCW signal is millisecond in general, the motion during the internal pulse can be regarded as the translational motion. Thus, Eq. (3.74) can be rewritten as

$$v_r(t_m) = -\rho\Omega \sin(\Omega t_m + \theta_0). \tag{3.75}$$

Eq. (3.73) can be rewritten as

$$\begin{aligned} R(t_m, t_k) &= R_Q + \rho \cos(\Omega t_m + \theta_0) + v(t_m)t_k \\ &= R_\Delta(t_m) + v(t_m)t_k \end{aligned} \tag{3.76}$$

where $R_\Delta(t_m) = R_Q + \rho\cos(\Omega t_m + \theta_0)$. Substitute R_Δ and v_r in Eq. (3.72) with $R_\Delta(t_m)$ and $v_r(t_m)$, then

$$\dot{r}(t_m) = R_\Delta(t_m) - \frac{c}{\lambda\mu} v_r(t_m) \tag{3.77}$$

Substitute the expression of $R_\Delta(t_m)$ and $v_r(t_m)$ into Eq. (3.77), then

$$\dot{r}(t_m) = R_Q + \frac{\rho}{\lambda\mu}\sqrt{c^2\Omega^2 + \lambda^2\mu^2}\cos(\Omega t_m + \theta_0 + \phi) \tag{3.78}$$

where $\phi = \arctan(\lambda\mu/(c\Omega)) - \pi/2$.

Eq. (3.78) describes the micro-Doppler effect induced by rotation or vibration of the target in LFMCW signal. From Eq. (3.78) we can conclude two properties:

Property 1: In range—slow time plane, the micro-Doppler feature of the rotating/vibrating point presents as cosine curves, and the period of the cosine curves equals to the period of rotating/vibrating motion. But the amplitude is not the real radius (amplitude) of the rotating/vibrating point in the radial direction of the radar, which is amplified by $\sqrt{c^2\Omega^2 + \lambda^2\mu^2}\Big/(\lambda\mu)$ times.

Property 2: The peak value point of the sine curve in range—slow time plane does not mean that, at this moment, the rotating/vibrating point moves to the nearest or the furthest position of the radar. Because when Eq. (3.73) takes the extreme value Eq. (3.78) cannot, and their phase difference is ϕ.

We can see that there are some similarities between these two properties and the micro-Doppler effect of the target in SFCS signal. However, in LFMCW signal, the wrapping effect will not appear in the micro-Doppler spectrum. When receiving the LFMCW echo signal, the sampling rate during the internal pulse is quite high, which avoids the wrapping effect in Doppler spectrum. But the equivalent sampling rate in the pulse train of SFCS signal depends on the number of subpulses and the time length of the pulse train, which is difficult to obtain to be large enough, so that the wrapping of Doppler spectrum is easy to emerge.

A simulation experiment is given to verify the previous properties. The parameters of radar signal are set as follows. The carrier frequency $f_c = 35$ GHz, the pulse width $T_p = 1$ ms, the bandwidth $B = 500$ MHz, $R_Q = 3$ m, $\rho = 6$ m, $\Omega = 8\pi$ rad/s, and $\theta_0 = 0$. According to Eq. (3.78), the amplifier time of the amplitude is 2.0236, the amplitude of the curve is 12.1418 m, and the phase shift $\phi = -60.3856$ degrees. The range—slow time image is shown in Fig. 3.20. From Fig. 3.20 we can see that the period of the curve is 0.25 s; the difference between the maximum value and the minimum value is 24.3 m, that is to say, the amplitude is 12.15 m; the first peak value locates at 0.208 s, and the phase shift is 0.208 s/0.25 s × 360 degrees = 299.52 degrees, which is equivalent to 299.52 degrees − 360 degrees = −60.48 degrees. The value is very close to the theoretical values.

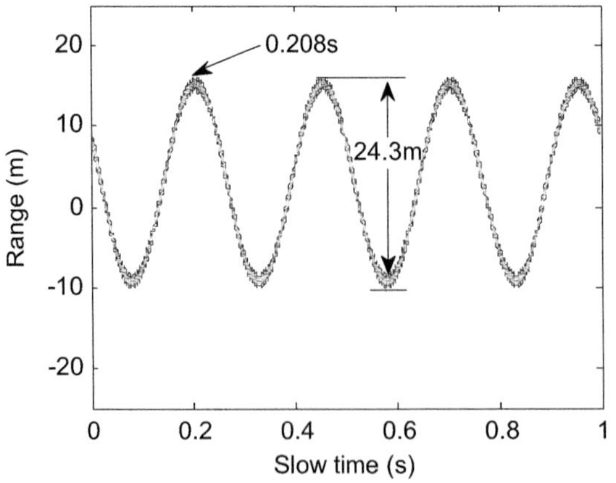

Figure 3.20 The micro-Doppler effect of rotation and vibration in LFMCW signal.

REFERENCES

[1] Zhang Q, Yeo TS, Tan HS, et al. Imaging of a moving target with rotating parts based on the Hough transform. IEEE Transactions on Geoscience and Remote Sensing 2008;46(1):291–9.
[2] Luo Y, Zhang Q, Qiu C-W, Liang X-J, Li K-M. Micro-doppler effect analysis and feature extraction in ISAR imaging with stepped-frequency chirp signals. IEEE Transactions on Geoscience and Remote Sensing 2010;48(4):2087–98.
[3] Luo Y, Zhang Q, Bai Y-Q, Zhu F. Analysis of micro-doppler effect and feature extraction of target in frequency-stepped chirp signal radar. ACTA Electronica Sinica 2009;37(12):2741–6.

CHAPTER FOUR

Micro-Doppler Effect in Bistatic Radar

In the previous sections we have introduced the micro-Doppler effect in narrowband radar and wideband radar under monostatic mode. With the rapid development of stealth targets, integrated electronic jamming, over-low altitude penetration technologies, and antiradiation missiles, monostatic radar is facing an increasingly serious survival crisis. Compared with the monostatic radar, bistatic radar has many special characteristics due to its spatial bistatic structures. In this section, we mainly take the rotation targets for examples to derive the micro-Doppler effect in narrowband radar and wideband radar under bistatic mode.

4.1 BISTATIC RADAR

The name of bistatic radar was first put forward by K.M. Siegel and R.E. Machol in 1952. In IEEE standard, bistatic radar is defined as a kind of radar whereby the transmitter and the receiver are mounted on two separate platforms. Long before this standard, people had already used this kind of radar mode. In the early development of radar, due to the lack of effective isolation method, the test radar could only use the bistatic mode. Early radar defense systems in the United States, Britain, France, Germany, Russia, and other countries were all bistatic mode. For example, in 1922 the US Navy lab performed experiments to detect battleships using bistatic continuous wave radar. In 1936, the high-power magnetron was invented, and after that the monostatic mode radar became main stream due to its simple structure. Since then, interest was lost in using the bistatic radar system.

From the beginning of the 1950s, the rapid development of science put military technologies such as stealth targets, integrated electronic jamming, over-low altitude penetration technologies, and antiradiation missiles at a new level, putting monostatic radar into serious survival crisis. Compared with the monostatic radar, bistatic radar has many advantages: (1) It has much longer working distance, as the bistatic radar has overcome the R^4 attenuation weakness in monostatic radar [1]; (2) it has stronger anti-interference ability: whether it is an active or passive jamming interference, the probability of bistatic radar interference is greatly reduced [2]; (3) it has good anti-destroy ability: due to the postposition of the transmitter and the silent working mode of the receiver, bistatic radar has obvious advantages in the fight against antiradiation missiles, precision-guided weapons, and combating directed energy weapons [3]; (4) it has antistealth capability: due to the side/forward scattering enhancements and the Doppler

Micro-Doppler Characteristics of Radar Targets
ISBN 978-0-12-809861-5, http://dx.doi.org/10.1016/B978-0-12-809861-5.00004-7

beat frequency effect on the stealth aircraft, the bistatic radar system has strong antistealth capability, so the bistatic radar is also known as the "bane" of stealth targets; and (5) it has antialtitude/low altitude penetration capability: benefiting from its spatial diversity characteristics, bistatic radar has strong low-altitude target detection capability. Based on these advantages, the bistatic radar system regained the attention of radar industry. Since the 1970s, many foreign countries started carrying out research and experimental work on bistatic/multistatic radar. By the 1980s, bistatic radar systems had been further developed and many countries established their automation radar defense networks using it. The United States Rome Lab and its partners began to study and test the man-portable passive bistatic radar system beginning in the 1980s. This radar system has advantages such as small size, low weight, low power, less cost, etc. [4] In the 1980s, China also began researching various systems of bistatic radar for remote alert, medium-range guidance, space-based and airborne.

Due to its special spatial composition of bistatic radar, the target micro-Doppler effect demonstrates some new features. By studying the bistatic radar target micro-Doppler effect, we may extract the structural features and motion features of the target more precisely and further more provide strong information support for the bistatic radar target identification and classification.

4.2 MICRO-DOPPLER EFFECT IN NARROWBAND BISTATIC RADAR

We take the micromotion induced by rotation of a target, for example, to analyze the micro-Doppler effect in narrowband bistatic radar. Bistatic radar system is shown in Fig. 4.1. We define a global coordinate system (U, V, W). The transmitter is located at O_T

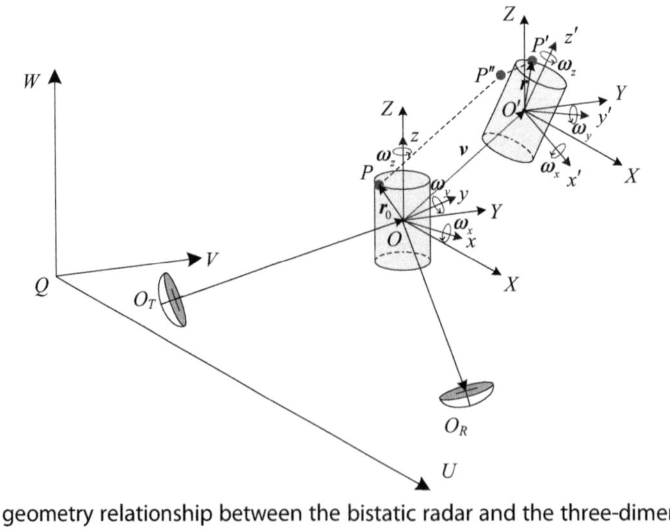

Figure 4.1 The geometry relationship between the bistatic radar and the three-dimensional rotation target.

and the receiver located at O_R. The reference coordinate system (X,Y,Z) is parallel to the radar coordinate system, originated at O. The target-local coordinate system (x,y,z) shares the same origin O with the reference coordinate system. Suppose that the target can be modeled as a rigid body with a translation velocity $\boldsymbol{v}$ with respect to the radar. The target rotates about the x-axis, y-axis, and z-axis with the angular rotation velocity ω_x, ω_y, and ω_z, respectively, when undergoing a translation, which can be either represented in the target-local coordinate system as $\boldsymbol{\omega} = (\omega_x,\omega_y,\omega_z)^{\mathrm{T}}$ (the superscript "T" means transposition) or represented in the reference coordinate system as $\widehat{\boldsymbol{\omega}} = (\omega_X, \omega_Y, \omega_Z)^{\mathrm{T}}$. A scatterer point of the target is located at $\boldsymbol{r}_0 = (r_{x0},r_{y0},r_{z0})^{\mathrm{T}}$ in the target-local coordinate system and $\widehat{\boldsymbol{r}}_0 = (r_{X0}, r_{Y0}, r_{Z0})^{\mathrm{T}}$ in the reference coordinate system at the initial time.

After time t, the target center will move from O to O'. The reference coordinate system (X,Y,Z) will implement some translation accordingly, the target-local coordinate system will change to (x',y',z'), and the scattering point P will move to P', whose corresponding vector is $\boldsymbol{r} = (r_x,r_y,r_z)^{\mathrm{T}}$ in the target-local coordinate system or $\widehat{\boldsymbol{r}} = (r_X, r_Y, r_Z)^{\mathrm{T}} = \mathbf{R}_{\text{init}}\boldsymbol{r}$ in the reference coordinate system. At time t, the scalar range becomes:

$$r(t) = \left\| \overrightarrow{O_T O} + \boldsymbol{v}t + \mathbf{R}_{\text{rotating}}\widehat{\boldsymbol{r}}_0 \right\| + \left\| \overrightarrow{O_R O} + \boldsymbol{v}t + \mathbf{R}_{\text{rotating}}\widehat{\boldsymbol{r}}_0 \right\|, \tag{4.1}$$

where the expression of $\mathbf{R}_{\text{rotating}}$ is the same as in Eq. (2.22).

If the radar transmits a single frequency continuous wave with a carrier frequency f_c:

$$p(t) = \exp(\mathrm{j}2\pi f_c t) \tag{4.2}$$

The echo signal from the scattering point P' can be written as:

$$s(t) = \sigma \exp\left(\mathrm{j}2\pi f_c\left(t - \frac{r(t)}{c}\right)\right) \tag{4.3}$$

After baseband transformation of $s(t)$, we can get:

$$s_b(t) = \sigma \exp\left(\mathrm{j}2\pi f_c\frac{r(t)}{c}\right) = \sigma \exp(\mathrm{j}\Phi(t)) \tag{4.4}$$

where the phase of the baseband signal is $\Phi(t) = 2\pi f_c\frac{r(t)}{c}$. By taking the time derivative of the phase, the Doppler frequency shift of the echo signal can be derived as:

$$\begin{aligned} f_{\mathrm{d}} &= \frac{1}{2\pi}\frac{\mathrm{d}\Phi(t)}{\mathrm{d}t} = \frac{f_c}{c}\frac{\mathrm{d}}{\mathrm{d}t}r(t) \\ &= \frac{f_c}{c}\left[\frac{\mathrm{d}}{\mathrm{d}t}\left(\left\|\overrightarrow{O_T O} + \boldsymbol{v}t + \mathbf{R}_{\text{rotating}}\widehat{\boldsymbol{r}}_0\right\|\right) + \frac{\mathrm{d}}{\mathrm{d}t}\left(\left\|\overrightarrow{O_R O} + \boldsymbol{v}t + \mathbf{R}_{\text{rotating}}\widehat{\boldsymbol{r}}_0\right\|\right)\right] \end{aligned} \tag{4.5}$$

The first term can be written as,

$$
\begin{aligned}
&\frac{\mathrm{d}}{\mathrm{d}t}\left(\left\|\overrightarrow{O_T O}+\boldsymbol{v}t+\mathbf{R}_{\text{rotating}}\widehat{\boldsymbol{r}}_0\right\|\right)\\
&=\frac{1}{2\left\|\overrightarrow{O_T O}+\boldsymbol{v}t+\mathbf{R}_{\text{rotating}}\widehat{\boldsymbol{r}}_0\right\|}\cdot\frac{\mathrm{d}}{\mathrm{d}t}\left[\left(\overrightarrow{O_T O}+\boldsymbol{v}t+\mathbf{R}_{\text{rotating}}\widehat{\boldsymbol{r}}_0\right)^{\mathrm{T}}\left(\overrightarrow{O_T O}+\boldsymbol{v}t+\mathbf{R}_{\text{rotating}}\widehat{\boldsymbol{r}}_0\right)\right]\\
&=\frac{1}{2\left\|\overrightarrow{O_T O}+\boldsymbol{v}t+\mathbf{R}_{\text{rotating}}\widehat{\boldsymbol{r}}_0\right\|}\cdot\left[\frac{\mathrm{d}}{\mathrm{d}t}\left(\overrightarrow{O_T O}+\boldsymbol{v}t+\mathbf{R}_{\text{rotating}}\widehat{\boldsymbol{r}}_0\right)^{\mathrm{T}}\left(\overrightarrow{O_T O}+\boldsymbol{v}t+\mathbf{R}_{\text{rotating}}\widehat{\boldsymbol{r}}_0\right)\right.\\
&\quad\left.+\left(\overrightarrow{O_T O}+\boldsymbol{v}t+\mathbf{R}_{\text{rotating}}\widehat{\boldsymbol{r}}_0\right)^{\mathrm{T}}\frac{\mathrm{d}}{\mathrm{d}t}\left(\overrightarrow{O_T O}+\boldsymbol{v}t+\mathbf{R}_{\text{rotating}}\widehat{\boldsymbol{r}}_0\right)\right]\\
&=\frac{1}{2\left\|\overrightarrow{O_T O}+\boldsymbol{v}t+\mathbf{R}_{\text{rotating}}\widehat{\boldsymbol{r}}_0\right\|}\cdot\left\{\left[\boldsymbol{v}+\frac{\mathrm{d}}{\mathrm{d}t}\left(\mathbf{R}_{\text{rotating}}\widehat{\boldsymbol{r}}_0\right)^{\mathrm{T}}\right]\left(\overrightarrow{O_T O}+\boldsymbol{v}t+\mathbf{R}_{\text{rotating}}\widehat{\boldsymbol{r}}_0\right)\right.\\
&\quad\left.+\left(\overrightarrow{O_T O}+\boldsymbol{v}t+\mathbf{R}_{\text{rotating}}\widehat{\boldsymbol{r}}_0\right)^{\mathrm{T}}\left[\boldsymbol{v}+\frac{\mathrm{d}}{\mathrm{d}t}\left(\mathbf{R}_{\text{rotating}}\widehat{\boldsymbol{r}}_0\right)\right]\right\}\\
&=\frac{1}{2\left\|\overrightarrow{O_T O}+\boldsymbol{v}t+\mathbf{R}_{\text{rotating}}\widehat{\boldsymbol{r}}_0\right\|}\cdot 2\left[\boldsymbol{v}+\frac{\mathrm{d}}{\mathrm{d}t}\left(\mathbf{R}_{\text{rotating}}\widehat{\boldsymbol{r}}_0\right)^{\mathrm{T}}\right]\left(\overrightarrow{O_T O}+\boldsymbol{v}t+\mathbf{R}_{\text{rotating}}\widehat{\boldsymbol{r}}_0\right)\\
&=\left[\boldsymbol{v}+\frac{\mathrm{d}}{\mathrm{d}t}\left(\mathbf{R}_{\text{rotating}}\widehat{\boldsymbol{r}}_0\right)^{\mathrm{T}}\right]\frac{\left(\overrightarrow{O_T O}+\boldsymbol{v}t+\mathbf{R}_{\text{rotating}}\widehat{\boldsymbol{r}}_0\right)}{\left\|\overrightarrow{O_T O}+\boldsymbol{v}t+\mathbf{R}_{\text{rotating}}\widehat{\boldsymbol{r}}_0\right\|}
\end{aligned}
\tag{4.6}
$$

So is the second term is:

$$
\frac{\mathrm{d}}{\mathrm{d}t}\left(\left\|\overrightarrow{O_R O}+\boldsymbol{v}t+\mathbf{R}_{\text{rotating}}\widehat{\boldsymbol{r}}_0\right\|\right)=\left[\boldsymbol{v}+\frac{\mathrm{d}}{\mathrm{d}t}\left(\mathbf{R}_{\text{rotating}}\widehat{\boldsymbol{r}}_0\right)^{\mathrm{T}}\right]\frac{\left(\overrightarrow{O_R O}+\boldsymbol{v}t+\mathbf{R}_{\text{rotating}}\widehat{\boldsymbol{r}}_0\right)}{\left\|\overrightarrow{O_R O}+\boldsymbol{v}t+\mathbf{R}_{\text{rotating}}\widehat{\boldsymbol{r}}_0\right\|}
\tag{4.7}
$$

Therefore,

$$f_{\mathrm{d}}=\frac{f_{\mathrm{c}}}{c}\left[\boldsymbol{v}+\frac{\mathrm{d}}{\mathrm{d}t}\left(\mathbf{R}_{\text{rotating}}\widehat{\boldsymbol{r}}_{0}\right)\right]^{\mathrm{T}}\left(\frac{\overrightarrow{O_{T}O}+\boldsymbol{v}t+\mathbf{R}_{\text{rotating}}\widehat{\boldsymbol{r}}_{0}}{\left\|\overrightarrow{O_{T}O}+\boldsymbol{v}t+\mathbf{R}_{\text{rotating}}\widehat{\boldsymbol{r}}_{0}\right\|}+\frac{\overrightarrow{O_{R}O}+\boldsymbol{v}t+\mathbf{R}_{\text{rotating}}\widehat{\boldsymbol{r}}_{0}}{\left\|\overrightarrow{O_{R}O}+\boldsymbol{v}t+\mathbf{R}_{\text{rotating}}\widehat{\boldsymbol{r}}_{0}\right\|}\right) \tag{4.8}$$

If the target is located in the far field of the radar, we can get $\left\|\overrightarrow{O_{T}O}+\boldsymbol{v}t\right\| \geq \left\|\mathbf{R}_{\text{rotating}}\widehat{\boldsymbol{r}}_{0}\right\|$, $\left\|\overrightarrow{O_{R}O}+\boldsymbol{v}t\right\| \geq \left\|\mathbf{R}_{\text{rotating}}\widehat{\boldsymbol{r}}_{0}\right\|$, thus:

$$\begin{aligned}\frac{\overrightarrow{O_{T}O}+\boldsymbol{v}t+\mathbf{R}_{\text{rotating}}\widehat{\boldsymbol{r}}_{0}}{\left\|\overrightarrow{O_{T}O}+\boldsymbol{v}t+\mathbf{R}_{\text{rotating}}\widehat{\boldsymbol{r}}_{0}\right\|}&\approx\frac{\overrightarrow{O_{T}O}+\boldsymbol{v}t}{\left\|\overrightarrow{O_{T}O}+\boldsymbol{v}t\right\|},\\ \frac{\overrightarrow{O_{R}O}+\boldsymbol{v}t+\mathbf{R}_{\text{rotating}}\widehat{\boldsymbol{r}}_{0}}{\left\|\overrightarrow{O_{R}O}+\boldsymbol{v}t+\mathbf{R}_{\text{rotating}}\widehat{\boldsymbol{r}}_{0}\right\|}&\approx\frac{\overrightarrow{O_{R}O}+\boldsymbol{v}t}{\left\|\overrightarrow{O_{R}O}+\boldsymbol{v}t\right\|}\end{aligned} \tag{4.9}$$

The expression (4.8) can be simplified as:

$$\begin{aligned}f_{\mathrm{d}}&=\frac{f_{\mathrm{c}}}{c}\left[\boldsymbol{v}+\frac{\mathrm{d}}{\mathrm{d}t}\left(\mathbf{R}_{\text{rotating}}\widehat{\boldsymbol{r}}_{0}\right)\right]^{\mathrm{T}}\left(\frac{\overrightarrow{O_{T}O}+\boldsymbol{v}t_{0}}{\left\|\overrightarrow{O_{T}O}+\boldsymbol{v}t\right\|}+\frac{\overrightarrow{O_{R}O}+\boldsymbol{v}t}{\left\|\overrightarrow{O_{R}O}+\boldsymbol{v}t\right\|}\right)\\&=\frac{f_{\mathrm{c}}}{c}\left[\boldsymbol{v}+\frac{\mathrm{d}}{\mathrm{d}t}\left(\mathbf{R}_{\text{rotating}}\widehat{\boldsymbol{r}}_{0}\right)\right]^{\mathrm{T}}\boldsymbol{n}_{b}(t)\end{aligned} \tag{4.10}$$

In fact, the $\boldsymbol{n}_b(t)$ above is a vector, which is opposite to the angular bisector vector of target relative transmit-receive array element at t. The first term of (4.10) is the Doppler frequency shift induced by the movement of the main target. The second term is the micro-Doppler shift induced by rotation of the target:

$$f_{\text{micro-Doppler}}=\frac{f_{\mathrm{c}}}{c}\left[\frac{\mathrm{d}}{\mathrm{d}t}\left(\mathbf{R}_{\text{rotating}}\widehat{\boldsymbol{r}}_{0}\right)\right]^{\mathrm{T}}\boldsymbol{n}_{b}(t) \tag{4.11}$$

Put expressions (2.21) and (2.22) into (4.11), we will get:

$$\begin{aligned}
f_{\text{micro-Doppler}} &= \frac{f_c}{c}\left[\frac{\mathrm{d}}{\mathrm{d}t}\left(e^{\widehat{\omega}t}\widehat{\boldsymbol{r}}_0\right)\right]^{\mathrm{T}}\boldsymbol{n}_b(t) \\
&= \frac{f_c}{c}\left[\widehat{\omega}e^{\widehat{\omega}t}\widehat{\boldsymbol{r}}_0\right]^{\mathrm{T}}\boldsymbol{n}_b(t) \\
&= \frac{\Omega f_c}{c}\left[\widehat{\omega}'(\mathbf{I}+\widehat{\omega}'\sin\Omega t+\widehat{\omega}'^2(1-\cos\Omega t))\widehat{\boldsymbol{r}}_0\right]^{\mathrm{T}}\boldsymbol{n}_b(t) \\
&= \frac{\Omega f_c}{c}\left[\widehat{\omega}'(\widehat{\omega}'\sin\Omega t+\mathbf{I}\cos\Omega t)\widehat{\boldsymbol{r}}_0\right]^{\mathrm{T}}\boldsymbol{n}_b(t),
\end{aligned} \tag{4.12}$$

where $\Omega = \left\|\widehat{\omega}\right\| = \|\boldsymbol{\omega}\|$. In (4.12), we usually ignore the micro-Doppler frequency error caused by $\boldsymbol{n}_b(t)$ that is $\boldsymbol{n}_b(t) \approx \boldsymbol{n}_b(0) = \boldsymbol{n}_{b0}$. Assume that carrier frequency of the transmitted signal is $f_c = 6$ GHz and a transmitter located at (0,0,0) m, a receiver located at (7000,0,0) m. The target center is at (3,4,5) km, $\boldsymbol{v} = (150,150,150)^{\mathrm{T}}$ m/s, $\boldsymbol{\omega} = (5\pi,10\pi,5\pi)^{\mathrm{T}}$ rad/s, $\widehat{\boldsymbol{r}}_0 = (0.2, 0.2, -0.6)^{\mathrm{T}}$. The micro-Doppler frequency error caused by $\boldsymbol{n}_b(t) \approx \boldsymbol{n}_b(0) = \boldsymbol{n}_{b0}$ is only 3 Hz at time $t = 0.25$ s. The time-varying and no time-varying micro-Doppler frequency curves are shown in Fig. 4.2. We can see that they are barely the same, thus (4.12) can be approximated to:

$$f_{\text{micro-Doppler}} = \frac{\Omega f_c}{c}\left[\widehat{\omega}'(\widehat{\omega}'\sin\Omega t+\mathbf{I}\cos\Omega t)\widehat{\boldsymbol{r}}_0\right]^{\mathrm{T}}\boldsymbol{n}_{b0} \tag{4.13}$$

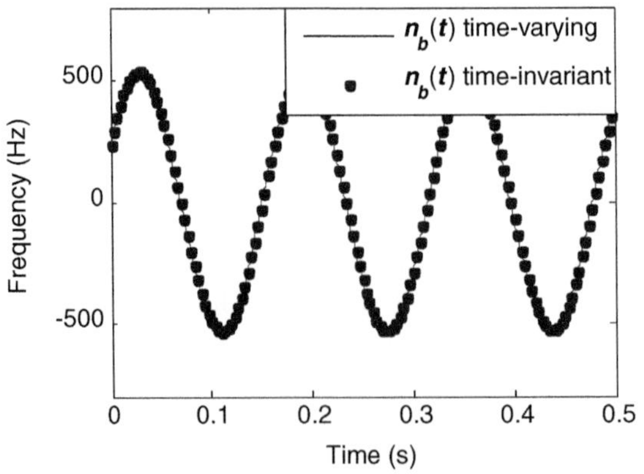

Figure 4.2 The time-varying and no time-varying micro-Doppler frequency curves.

Assume that $\boldsymbol{n}_{b0} = (n_{b0x}, n_{b0y}, n_{b0z})^{\mathrm{T}}$, when we put $\widehat{\boldsymbol{r}}_0 = (r_{X0}, r_{Y0}, r_{Z0})^{\mathrm{T}}$ and (2.21) into (4.13), we can get:

$$f_{\text{micro-Doppler}} = \frac{\Omega f_c}{c}(A \sin \Omega t + B \cos \Omega t) = \frac{\Omega f_c}{c}\sqrt{A^2 + B^2}\sin(\Omega t + \phi) \tag{4.14}$$

In (4.14),

$$\begin{aligned} A = & -n_{b0x} r_{X0} \omega_Z'^2 - n_{b0x} r_{X0} \omega_Y'^2 + n_{b0x} r_{Y0} \omega_X' \omega_Y' + n_{b0x} r_{Z0} \omega_X' \omega_Z' + n_{b0y} r_{X0} \omega_X' \omega_Y' \\ & - n_{b0y} r_{Y0} \omega_Z'^2 - n_{b0y} r_{Y0} \omega_X'^2 + n_{b0y} r_{Z0} \omega_Y' \omega_Z' + n_{b0z} r_{X0} \omega_X' \omega_Z' + n_{b0z} r_{Y0} \omega_Y' \omega_Z' \\ & - n_{b0z} r_{Z0} \omega_Y'^2 - n_{b0z} r_{Z0} \omega_X'^2 \end{aligned}$$

$$B = -n_{b0x} r_{Y0} \omega_Z' + n_{b0x} r_{Z0} \omega_Y' + n_{b0y} r_{X0} \omega_Z' - n_{b0y} r_{Z0} \omega_X' - n_{b0z} r_{X0} \omega_Y' + n_{b0z} r_{Y0} \omega_X'$$

$$\phi = \operatorname{atan}(B/A)$$

From (4.14) we can see that the micro-Doppler frequency induced by rotation is a time-varying sinusoidal curve whose amplitude (the maximum of micro-Doppler frequency) M_a is determined by the carrier frequency f_c of the transmitted signal, the angular rotation velocity $\boldsymbol{\omega}$ of the target, the rotation vector $\boldsymbol{r}_0$, and the unit vector in the direction of radar sight line $\boldsymbol{n}$.

$$M_a = \Omega f_c \sqrt{A^2 + B^2} \Big/ c, \tag{4.15}$$

where the angular frequency of the sinusoidal curve is the angular rotation frequency Ω. From (4.15) we can see that the expression of the sinusoidal curve in bistatic radar is more complicated than those in monostatic radar.

Now a simulation is made to certify the theory: Assume that there is a transmitter, located at (0,0,0) m, a receiver located at (7000,0,0) m. The target center is at (3,4,5) km. The carrier frequency of the transmitted signal is 10 GHz. Assume that there are three strong scattering points of the target, and their locations are shown in Fig. 4.3. Three

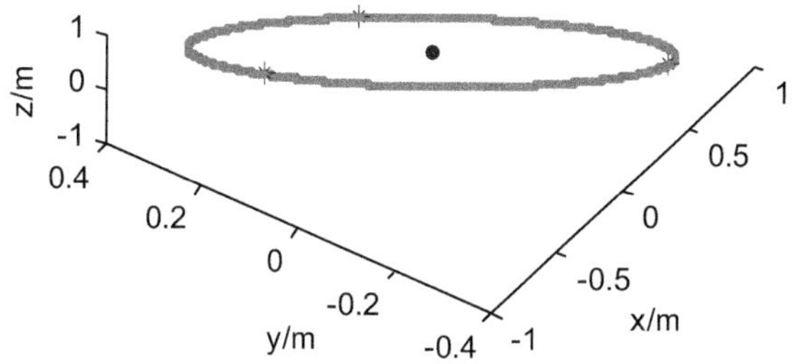

Figure 4.3 Target scatter point model.

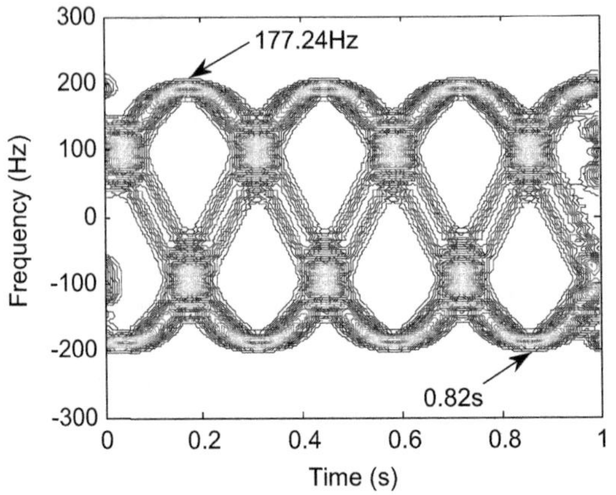

Figure 4.4 Time-frequency analysis of micro-Doppler signal.

scattering points are evenly spaced on the rotation circle to do the uniform circular motion, with same angular rotation velocity $\widehat{\boldsymbol{\omega}} = (\pi, 2\pi, \pi)^{\mathrm{T}}$rad/s and same radius $\left\|\widehat{\boldsymbol{r}}_0\right\| = 0.6633$ m. The original rotation radius vectors of these three scatters are $(0.2,0.2,-0.6)^{\mathrm{T}}$ m, $(-0.5950,0.1828,0.2293)^{\mathrm{T}}$ m, and $(0.3950,-0.3828,0.3707)^{\mathrm{T}}$ m. From the previous parameters we can get $\Omega = 7.6953$ rad/s, thus the rotation period is 0.8165 s and $M_a = 178.2$ Hz. The sampling time is 1 s and the sampling rate is 3 kHz. Fig. 4.4 gives the outcome of a micro-Doppler signal after Gabor transformation. We can see that the period and the amplitude of the sinusoidal curve is very close to the theoretical value.

4.3 MICRO-DOPPLER EFFECT IN WIDEBAND BISTATIC RADAR

We use the similar analysis method in Chapter 3 to deduce the micro-Doppler effect in wideband bistatic radar. In Chapter 3, we have proven that in wideband radar, the micro-Doppler effect can be analyzed not only by the echo phase but also by the range—slow time image. When we use range—slow time image to analyze micro-Doppler effect, we can observe the micro-Doppler characteristic curve if the radial micromotion amplitude is bigger than the range resolution cell of the radar. If the radial micromotion amplitude is equal to or less than the range resolution cell, we can only observe a straight line. Therefore, the range resolution capability of the radar may influence the observability of target micro-Doppler effect. Due to transceiver characteristics, the bistatic radar range resolution is different from the resolution of single radar.

4.3.1 Range Resolution of Bistatic Radar

Range resolution is defined as the minimum distance that the radar system can distinguish between two scatters on the target [5]. The range resolution of monostatic radar is determined by the bandwidth of the radar and is a system parameter that is completely unrelated to the target state and the geometric relationship between the target and the radar. The range resolution of monostatic radar can be written as

$$\Delta R = c/(2B) \tag{4.16}$$

Thus, the isorange contours in monostatic radar system are a series of concentric circles, whose center is radar and difference in range direction is range resolution. The width of the range bin in Mono-ISAR is the space between the two concentric isorange contours.

Compared with the range resolution of monostatic radar, the range resolution of bistatic radar has many differences due to its spatial bistatic structures. We assume the distance between the transmitter and the target is R_T, and R_R is the distance between the receiver and the target. $R_T + R_R$ is a very important parameter in bistatic radar because it determines the delay time τ of echo signal:

$$\tau = (R_T + R_R)/c. \tag{4.17}$$

Therefore, the echo signal of the bistatic radar is a function of $R_T + R_R$. In three-dimensional space, the isorange contours in bistatic radar system are a series of ellipses, whose two foci are the transmitter and receiver sites. Due to that the bistatic angle bisector is perpendicular to the tangent line at the intersection point of the ellipse and the bistatic angle bisector, the range resolution in bistatic system is usually defined as the range resolution in the bistatic angle bisector. Fig. 4.5 is drawn to convey the basic meaning of the range resolution in bistatic system for clarity: The two curves are the isoranges whose intersection points with the bistatic angle bisector are the location of target 1 and target 2, and the space between the two intersection points of the bistatic

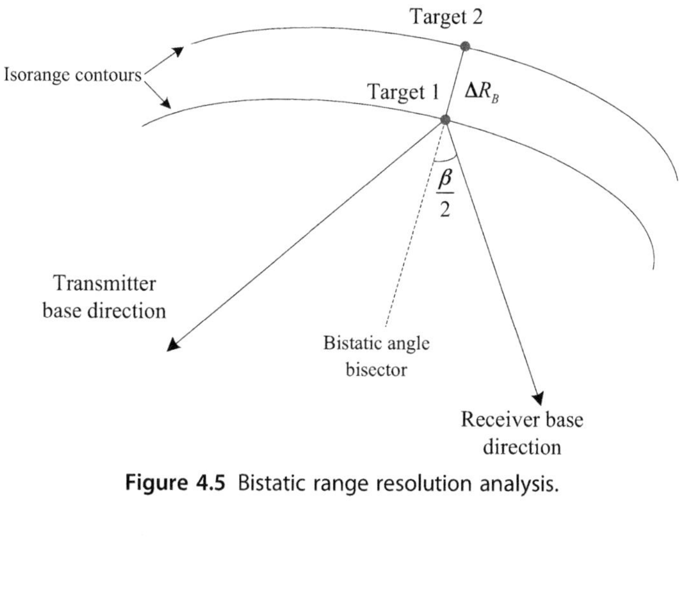

Figure 4.5 Bistatic range resolution analysis.

angle bisector and the adjacent two ellipses is just the range bin of bistatic system, which is equivalent to the range resolution. Due to the bandwidth of system B, the difference of range sum, ΔR_B, can be expressed approximately by

$$\Delta R_B \approx \frac{c}{2B\cos(\beta/2)}, \tag{4.18}$$

where β is the bistatic angle. According to the definition of range resolution, the range resolution in bistatic radar is ΔR_B.

From the previous analysis it can be seen that the range bins are partitioned according to those ellipses, and the range resolution is related to not only the bandwidth but also the target location in the bistatic system, that is, the range resolution in bistatic system is inversely proportional to the cosine of half of the bistatic angle. It is also the reflection of the space-variant characteristic of the range resolution in bistatic radar system. In particular, if $\beta = 0$, the expression in (4.18) will convert into (4.16), which means that the monostatic system can be regarded as a special example of the bistatic system.

From (4.18) we can see that the bistatic angle β can greatly influence the range resolution. From a strict mathematics perspective, if the bistatic angles of scattering points on the same target are different, the corresponding range resolutions are different. However, in real bistatic radar systems, the size of the target is much less than the distance between the target and the radar transmitters or receivers, so β of scattering points on the same target are barely the same. For example, there is a square-shaped target, the length of the side is 20 m, the distance between the target center and the radar transmitter is $R_T = 21.3$ km, the distance between the target center and the receiver is $R_R = 17.1$ km. From these parameters we can calculate that the bistatic angle of the target center is $\beta = 61.74°$, the max bistatic angle and the minimum bistatic angle of the target are $\beta_{max} = 61.82°$ and $\beta_{min} = 61.66°$. When we use the bistatic angle of the target center to replace other bistatic angles, the error can be less than $0.1°$, so usually we can consider that all the bistatic angles of the target are the same. In other words, we can partition the range unit according to the same range width.

4.3.2 Micro-Doppler Effect Analysis

We still use the geometry model shown in Fig. 4.1 to analyze the micro-Doppler effect induced by rotation of a target. When radar transmits a linear frequency-modulated signal, the received echo signal at slow time t_m from a scattering point P is

$$s(t_k, t_m) = \sigma \mathrm{rect}\left(\frac{t_k - r(t_m)/c}{T_p}\right) \cdot \exp\left(\mathrm{j}2\pi\left(f_c\left(t_k - \frac{r(t_m)}{c}\right) + \frac{1}{2}\mu\left(t_k - \frac{r(t_m)}{c}\right)^2\right)\right), \tag{4.19}$$

where $r(t_m)$ is given by:

$$r(t_m) = \left\| \overrightarrow{O_T O} + \boldsymbol{v} t_m + \mathbf{R}_{\text{rotating}}(t_m) \widehat{\boldsymbol{r}}_0 \right\| + \left\| \overrightarrow{O_R O} + \boldsymbol{v} t_m + \mathbf{R}_{\text{rotating}}(t_m) \widehat{\boldsymbol{r}}_0 \right\| \quad (4.20)$$

The expression of $\mathbf{R}_{\text{rotating}}(t_m)$ is the same in (3.23).

Taking the received signal of target center O as the reference signal $s_{ref}(t_k,t_m)$, then we have:

$$\begin{aligned} s_{ref}(t_k, t_m) = {} & \text{rect}\left(\frac{t_k - R_{ref}(t_m)/c}{T_{ref}}\right) \cdot \exp\left(j2\pi\left(f_c\left(t_k - \frac{R_{ref}(t_m)}{c}\right)\right.\right. \\ & \left.\left. + \frac{1}{2}\mu\left(t_k - \frac{R_{ref}(t_m)}{c}\right)^2\right)\right), \end{aligned} \quad (4.21)$$

where $R_{ref}(t_m)$ can be written as:

$$R_{ref}(t_m) = \left\| \overrightarrow{O_T O} + \boldsymbol{v} t_m \right\| + \left\| \overrightarrow{O_R O} + \boldsymbol{v} t_m \right\| \quad (4.22)$$

After the "dechirp" process, the output signal can be written as:

$$\begin{aligned} s_d(t_k, t_m) = {} & s(t_k, t_m) s_{ref}^*(t_k, t_m) \\ = {} & \sigma \text{rect}\left(\frac{t_k - r(t_m)/c}{T_p}\right) \exp\left(j2\pi\left(-\frac{\mu}{c}\left(t_k - \frac{R_{ref}(t_m)}{c}\right) R_\Delta(t_m)\right.\right. \\ & \left.\left. - \frac{f_c}{c} R_\Delta(t_m) + \frac{\mu}{c^2} R_\Delta(t_m)^2\right)\right) \end{aligned} \quad (4.23)$$

where $R_\Delta(t_m) = r(t_m) - R_{ref}(t_m)$.

Let $t' = t_k - R_{ref}(t_m/c)$, after taking the Fourier transform of (4.23) with respect to the fast time t' and compensating the (RVP) error and envelope oblique item, we will obtain the expression of echo signal in fast time-frequency domain (f_k domain) as follows:

$$S_d(f_k, t_m) = \sigma T_p \text{sinc}\left(T_p\left(f_k + \frac{\mu}{c} R_\Delta(t_m)\right)\right) \cdot \exp\left(-j\frac{2\pi}{c} f_c R_\Delta(t_m)\right) \quad (4.24)$$

It can be seen that the value of the range profiles will peak at $f_k = -\mu R_\Delta(t_m)/c$.

From (4.24) we can find that the phase of $S_d(f_k,t_m)$ is modulated by $R_\Delta(t_m)$ and causes the micro-Doppler effect in slow time domain. This is essentially the same reason the micro-Doppler effect is caused by micromotion in narrowband bistatic radar. Similar to the monocratic radar system, the pick range of the $|S_d(f_k,t_m)|$ is a curve vary from $R_\Delta(t_m)$

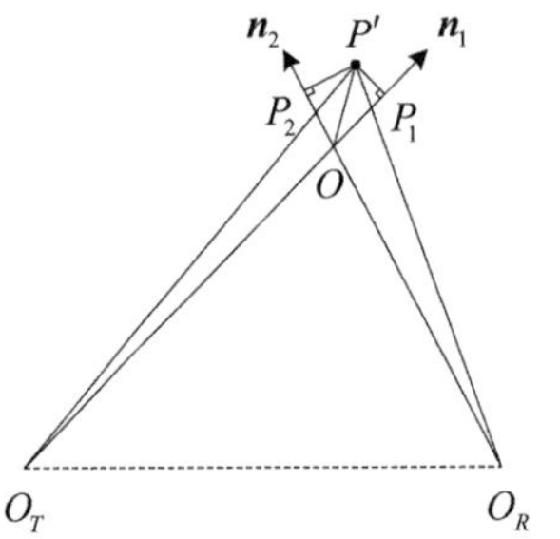

Figure 4.6 The geometry model of bistatic radar and rotation micromotion target.

in the range—slow time plane. From this curve we can extract micro-Doppler features of the moving target.

In order to account for $R_\Delta(t_m)$ conveniently, the geometry relationship model between the target and the bistatic radar at time t_m is shown in Fig. 4.6, where $\boldsymbol{n}_1$ and $\boldsymbol{n}_2$ denote the unit vector of transmitter $O_T$and receiver O_R, respectively. P_1, P_2 denote the projection of P' on $\boldsymbol{n}_1$ and $\boldsymbol{n}_2$, respectively. $\boldsymbol{n}_1$ and $\boldsymbol{n}_2$ remain unchanged in a short period of time. For $\left\|\overrightarrow{OP'}\right\| \leq \left\|\overrightarrow{O_TO}\right\|$ and $\left\|\overrightarrow{OP'}\right\| \leq \left\|\overrightarrow{O_RO}\right\|$, we can get $\left\|\overrightarrow{O_TP'}\right\| \approx \left\|\overrightarrow{O_TP_1}\right\|$ and $\left\|\overrightarrow{O_RP'}\right\| \approx \left\|\overrightarrow{O_RP_1}\right\|$. Then $R_\Delta(t_m)$ can be written as:

$$R_\Delta(t_m) \approx \left\|\overrightarrow{OP_1}\right\| + \left\|\overrightarrow{OP_2}\right\| \tag{4.25}$$

When the scattering point P rotates around the center O, the geometry of the target at t_m is drawn in Fig. 4.7. The surface in the figure is the rotation moving surface of P. $\overrightarrow{OP'}$ is the vector of the radius with length of r, $\boldsymbol{n}_1'$ and $\boldsymbol{n}_2'$ are the projections of $\boldsymbol{n}_1$ and $\boldsymbol{n}_2$

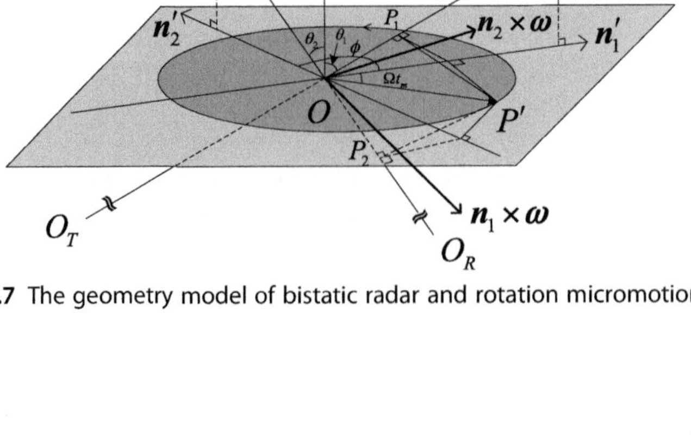

Figure 4.7 The geometry model of bistatic radar and rotation micromotion target.

on the plane spanned by the trajectory of scattering point P', respectively, P_1 and P_2 are the projections of P' on $\boldsymbol{n}_1$ and $\boldsymbol{n}_2$, respectively. Assume the included angle between $\overrightarrow{OP'}$ and $\boldsymbol{n}_1'$ at t_m is Ωt_m, the lengths of $\overrightarrow{OP_1}$ and $\overrightarrow{OP_2}$ are positive when $\overrightarrow{OP_1}$ and $\overrightarrow{OP_2}$ have the same directions with $\boldsymbol{n}_1$ and $\boldsymbol{n}_2$, respectively, and vice versa. Then we have:

$$\left\|\overrightarrow{OP_1}\right\| = \left\|\overrightarrow{OP'}\right\|\cos(\Omega t_m)\sin\theta_1, \quad \left\|\overrightarrow{OP_2}\right\| = \left\|\overrightarrow{OP'}\right\|\cos(\Omega t_m + \phi)\sin\theta_2 \tag{4.26}$$

where θ_1 is the included angle between $\boldsymbol{\omega}$ and $\boldsymbol{n}_1$, θ_2 is the included angle between $\boldsymbol{\omega}$ and $\boldsymbol{n}_2$, and ϕ is the included angle between $\boldsymbol{n}_1'$ and $\boldsymbol{n}_2'$.

Substitute (4.26) to (4.25), and it yields:

$$\begin{aligned} R_\Delta(t_m) &= r\cos(\Omega t_m)\sin\theta_1 + r\cos(\Omega t_m + \phi)\sin\theta_2 \\ &= (r\sin\theta_1 + r\sin\theta_2\cos\phi)\cos(\Omega t_m) - r\sin\theta_2\sin\phi\sin(\Omega t_m) \\ &= r\sqrt{\sin^2\theta_1 + \sin^2\theta_2 + 2\sin\theta_1\sin\theta_2\cos\phi}\cdot\sin(\Omega t_m - \varphi) \end{aligned} \tag{4.27}$$

where

$$\varphi = \arctan\left(\frac{\sin\theta_1 + \sin\theta_2\cos\phi}{\sin\theta_2\sin\phi}\right)$$

From (4.27), it can be found that: on the $f - t_m$ plane determined by (4.24), after the range scaling by $f = -\mu\Delta R(t_m)/c$, the locations of peaks in a scattering point's profile appear as a sinusoidal curve with respect to t_m, and the period of the curve is equivalent to the rotation period of P. For a given value of f_k, taking the derivative of the phase term on the right side of (4.24) in terms of t_m, then the instantaneous frequency of P is obtained as:

$$f_{\text{micro-Doppler}}(t_m) = -\frac{f_c}{c}\Omega r\sqrt{\sin^2\theta_1 + \sin^2\theta_2 + 2\sin\theta_1\sin\theta_2\cos\phi}\cdot\cos(\Omega t_m - \varphi) \tag{4.28}$$

From the geometry of Fig. 4.7, we can also get:

$$\theta_1 = \arccos\left(\frac{\boldsymbol{\omega}\cdot\boldsymbol{n}_1}{\|\boldsymbol{\omega}\|\cdot\|\boldsymbol{n}_1\|}\right), \quad \theta_2 = \arccos\left(\frac{\boldsymbol{\omega}\cdot\boldsymbol{n}_2}{\|\boldsymbol{\omega}\|\cdot\|\boldsymbol{n}_2\|}\right), \quad 0 \le \theta_1, \theta_2 \le \pi \tag{4.29}$$

When $0 \le \phi \le \pi$, ϕ is equal to the angle between $\boldsymbol{n}_1 \times \boldsymbol{\omega}$ and $\boldsymbol{n}_2 \times \boldsymbol{\omega}$, ie, $\phi = (\boldsymbol{n}_1 \times \boldsymbol{\omega}, \boldsymbol{n}_2 \times \boldsymbol{\omega})$; when $\pi \le \phi \le 2\pi$, $\phi = 2\pi - (\boldsymbol{n}_1 \times \boldsymbol{\omega}, \boldsymbol{n}_2 \times \boldsymbol{\omega})$. Therefore,

$$\cos\phi = \frac{(\boldsymbol{n}_1 \times \boldsymbol{\omega})\cdot(\boldsymbol{n}_2 \times \boldsymbol{\omega})}{\|\boldsymbol{n}_1 \times \boldsymbol{\omega}\|\cdot\|\boldsymbol{n}_2 \times \boldsymbol{\omega}\|} \tag{4.30}$$

From the previous discussion, compared with the micro-Doppler effect in monostatic radar system some conclusions about the micro-Doppler effect in bistatic radar system can be obtained:

1. The micro-Doppler effect induced by the rotation appears as a sinusoidal curve on the range–slow time plane, and the period of the curve is equivalent to the rotation period Ω, which is similar to that in the monostatic radar system.
2. The amplitudes of the micro-Doppler curves are determined by r, θ_1, θ_2, and ϕ, which is different from the amplitudes of the micro-Doppler curves in the monostatic radar system, whose amplitudes are determined by the projections of rotation radius in radar view sight.
3. The phase of the micro-Doppler curves does not coincide with that of the motion equation of the rotating scattering point; it has a phase difference φ determined by θ_1, θ_2, and ϕ. However, in a monostatic radar system, the phase of the micro-Doppler curve coincides with that of the motion equation of the rotating scattering point. So in bistatic radar we must consider the geometry relationship between the bistatic radar and the rotation micromotion target to ensure the instantaneous motion status of the rotating scattering point.

Fig. 4.8 gives a comparison of the micro-Doppler curve obtained by a monostatic radar and a bistatic radar. In the simulation, transmitter O_T is located at the center of the overall coordinate system. Two transmitters O_{R1} and O_{R2} are located at the center of the overall coordinate system and (7000,0,0) m, respectively. Transmitter O_T and the receiver O_{R1} actually simulate a monostatic radar. The carrier frequency is 10 GHz and the bandwidth of the transmitted signals are 500 MHz with $T_p = 4$ μs. The original location of the target center in the overall coordinate system is (3,4,5) km. The translational velocity of the target is zero. A scattering point P rotates around the target center with angular velocity $\boldsymbol{\omega} = (\pi,2\pi,\pi)^T$ rad/s and radius $r = 6.6332$ m, the rotation period is $T = 2\pi/\|\boldsymbol{\omega}\| = 0.8165$ s. Fig. 4.8A shows the micro-Doppler curve on the range–slow time

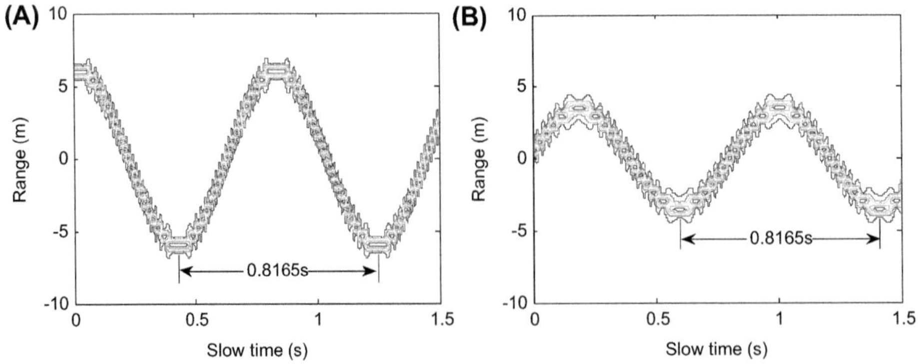

Figure 4.8 The comparison of micro-Doppler curve obtained by monostatic radar and bistatic radar. (A) Monostatic radar; (B) bistatic radar.

plane obtained by the monostatic radar; it uses $f_k = -2\mu\Delta R(t_m)/c$ to complete the range scaling. Fig. 4.7B is the micro-Doppler curve on the range—slow time plane obtained by the bistatic radar consisting of O_T and O_{R2}, which uses $f_k = -\mu\Delta R(t_m)/c$ to finish the range scaling. It can be found that the amplitudes of the two micro-Doppler curves are different from each other, and moreover, none of the amplitudes of the two micro-Doppler curves equals the real rotation radius of the target. On the other hand, the periods of the micro-Doppler curves in Fig. 4.8A and B are the same as the period of the target's rotation.

It can be imagined that since the range resolution of the bistatic (multistatic) radar is related with the position of the target, the amplitude of the micro-Doppler curve is also influenced by the position of the target. In bistatic radar systems, the observation abilities of different transmitters and receivers are not the same. When integrating Doppler feature of multi-based radar, the difference of different radar's observation abilities should be considered.

REFERENCES

[1] Woodbridge K, Banahan CP. Dynamic detection capability of a mobile bi-static weapons locating radar. In: IEEE Radar Conference; 2003. p. 179—83.

[2] Tsao T, Slamani M, Varshney P, et al. Ambiguity function for a bi-static radar. IEEE Transactions on Aerospace and Electronic Systems 1997;33(6):1041—51.

[3] Griffiths HD. From a different perspective-principles, practice and potential of bi-static radar. In: Proc. of the International Conference on Radar Systems; 2003. p. 1—7.

[4] Robert Ogrodnik F. Bi-static laptop radar. In: IEEE National Radar Conference; 1996. p. 369—74.

[5] Skolnik ML. Radar Handbook. Publishing House of Electronics Industry; 2003. p. 932—54.

CHAPTER FIVE

Micro-Doppler Feature Analysis and Extraction

The methods of micro-Doppler feature analysis and extraction are introduced in this chapter, and we will mainly focus on time-frequency analysis method, image processing method, orthogonal matching pursuit (OMP) decomposition method, empirical-mode decomposition (EMD) method, and high-order moment function analysis method. Of all these methods, time-frequency analysis method is the most commonly used. Image processing method is to introduce the algorithms in image processing into micro-Doppler feature extraction, which can be applied to estimate the parameters of a target's micro-Doppler curves combined with time-frequency analysis method. OMP decomposition method is a matching deposition method, which is especially put forward for analysis of sinusoidal frequency-modulation micro-Doppler signal. OMP takes different processing modes in narrowband radar and wideband radar. EMD method is a typical nonlinear analysis method and there is some literature covering the applications of EMD in extracting the micromotion frequencies of targets with rotation [1,2]. In this chapter, applications in micromotion parameters extraction of target with precession are introduced. High-order moment function analysis method aims at micromotion feature fast extraction of target with rotation or vibration. It also takes different processing modes in narrowband radar and wideband radar. For the previously mentioned methods, they are suitable for different applications, and each one of them has its own strong points in micro-Doppler features extraction. These methods will be discussed in detail as follows.

5.1 TIME-FREQUENCY ANALYSIS METHOD

5.1.1 Introduction of Time-Frequency Analysis Method

In the analysis of random signals, if the mathematical expectation of a signal is irrelevant to time, but the autocorrelation function is merely related to the time interval, then the signal is called the wide-sense stationary signal. Otherwise, it is called the nonstationary signal. Micromotion is a nonuniform or nonrigid motion in essence. Micromotion signals possess the characteristic of nonlinearity and nonstationarity. Thus, the core problem of analyzing and handling the micromotion features of a target is to handle the time-frequency signal. For time-varying frequency of nonstationary signal, Fourier transform, which is a traditional signal analysis method, lacks position capabilities in time

Micro-Doppler Characteristics of Radar Targets
ISBN 978-0-12-809861-5, http://dx.doi.org/10.1016/B978-0-12-809861-5.00005-9

or frequency. Time-frequency analysis method aims at nonstationary or time-varying signals. Hence, time-frequency analysis provides an efficient way to study imaging and feature extraction of complex moving targets. The time-frequency features derived from the time-frequency analysis are the bases of micromotion target recognition.

There are mainly two types of time-frequency analysis methods for nonstationary signals. One is the linear time-frequency analysis method, and the other is the quadratic time-frequency method.

5.1.1.1 Linear Time-Frequency Analysis Method

The first type of time-frequency representing method for nonstationary signal is the kernel decomposition method. It is also called the linear time-frequency representation. It decomposes the signal into basal compositions (known as kernels). Typical linear time-frequency representations include the short-time Fourier transform (STFT), Gabor expansion, wavelet transform (WT), and so on.

STFT extracts signal by a very narrow window function and derives its Fourier transform. Denoting $g(t)$ is a window function with a short-time width and it moves along the time axis. Thus, the short-time Fourier transform of signal $x(t)$ is defined as

$$\mathrm{STFT}_x(t,\omega) = \int_{-\infty}^{+\infty} x(\tau)g^*(\tau - t)\mathrm{e}^{-\mathrm{j}\omega\tau}\mathrm{d}\tau \tag{5.1}$$

The smaller the width of the window function is, the better the resolution of time is and the worse the frequency resolution is; the wider the width of the window function is, the better the resolution of frequency is and the worse the time resolution is. If the window function is infinite, such as $g(t) = 1$, the short-time Fourier transform degenerates to the traditional Fourier transform.

Half a century ago, Gabor proposed a method to represent time function simultaneously by time and frequency, which was later called Gabor expansion. And the integral formula of Gabor expansion coefficient is called Gabor transform. Constructing joint time-frequency function of nonstationary signal by time shift and frequency modulation of signal and then partitioning and sampling on the time-frequency plane, Gabor expansion transforms the time-frequency plane (t,f) into a plane of other two discrete sampling grid parameters m and n. The nonstationary signals are presented in 2D plane (m,n). The Gabor expansion of signal $s(t)$ is defined as

$$s(t) = \sum_{m=-\infty}^{\infty}\sum_{n=-\infty}^{\infty} a_{mn}g_{mn}(t) \tag{5.2}$$

where $\gamma_{mn}(t)$ is the dual function of Gabor basic function $g_{mn}(t)$. And

$$g_{mn}(t) = g(t - mT)\exp(\mathrm{j}2\pi(nF)t), \quad m,n = 0,\pm1,\pm2,\cdots \tag{5.3}$$

where T is the time length of grid and F is the frequency length of grid. Then, Gabor expansion coefficients are presented by Gabor transform as follows:

$$a_{mn} = \int_{-\infty}^{\infty} s(t)\gamma_{mn}^{*}(t)\mathrm{d}t \tag{5.4}$$

where $\gamma_{mn}(t)$ is the dual function of $g_{mn}(t)$.

According to the theory of Gabor transform, to recover $s(t)$ through a_{mn}, $g_{mn}(t)$ must be complete and biorthogonal with $\gamma_{mn}(t)$. That is,

$$\int \gamma(t)g^{*}(t - na)\exp(-\mathrm{j}2\pi(nF)t)\mathrm{d}t = \delta_m\delta_n \tag{5.5}$$

Stationary Gabor transform requires critical sampling or oversampling. Since Gaussian window provides the best focusing capability in time-frequency plane, Gaussian window is usually used for $g_{mn}(t)$. In practice, when critically sampling, window function $g_{mn}(t)$ cannot constitute a frame to guarantee its compactly supported performance of time and frequency domain. Meanwhile, it cannot reflect the energy distribution of signal in time-frequency plane. We generally improve redundancy of the expanding coefficient by oversampling the Gabor transform. This makes $g_{mn}(t)$ constitute a complete frame, and at the same time, it smooths the biorthogonal window $g_{mn}(t)$.

Gabor transform has received little attention since it was introduced. Firstly, Gabor expansion coefficients are difficult to calculate. In addition, Gabor transform is the evolution and the reconstruction of STFT, and its time-frequency resolution is closely related to the selection of window function. Until to the 1980s, Gabor transform regained attention. In fact, since Gabor transform adopts Gaussian window, whose Fourier transform is also a Gaussian function the energy in both time and frequency domain can be well focused. Besides, the time-bandwidth product of Gaussian signal meets the lower bound of uncertainty principle, and thus it can get the best time resolution and frequency resolution. Furthermore, the coherent localization feature of Gabor expansion makes it suitable to express transient signal and Gabor expansion. It is successfully applied in the expression of signal and image, speech analysis, target recognition, transient signal detection, and so on.

However, the main drawback of the linear time-frequency analysis technique is called the "window effects." The fixed analysis and comprehensive window functions used in linear time-frequency analysis technique lead to the fact that the time and frequency resolution is limited by window width and cannot be optimized simultaneously. Short—time window function can enhance the time resolution but decrease the frequency resolution. On the contrary, long—time window function can enhance the frequency resolution but decrease the time resolution. This is due to the uncertainty principle, which excludes the possibility of high resolution in time domain and frequency domain simultaneously. Thus, it is hard to get high time and frequency resolution at the same time using STFT or Gabor expansion.

Wavelet transform is a new branch of applied mathematics developed in the late 1980s. In recent years, it has been introduced into engineering vibration signal analysis and other fields. Wavelet transform has the multiresolution characteristic, which can observe signals step by step from coarse to fine. Provided an appropriate basic wavelet is chosen, wavelet transform can express the local characteristic of signal both in time domain and in frequency domain.

Wavelet transform is a time-scale analysis method. It adopts analysis and comprehensive window function with constant Q and can receive a trade-off between time and frequency resolution: low frequency has short bandwidth and long time width; high frequency has long bandwidth and short time width. But the wavelet will not be a proper choice, when it requires the high frequency resolution with high frequency or the high time resolution with low frequency.

5.1.1.2 Quadratic Time-Frequency Method

For representation or analysis of linear frequency modulation (LFM) or other nonstationary signals by time-frequency representation, the quadratic time-frequency method is more audiovisual and rational. The Wigner distribution was first proposed by Wigner in 1932 and applied in quantum mechanics. Until 1948, Ville firstly applied Wigner distribution in signal analysis.

The Wigner-Ville distribution (WVD) is its time-frequency energy density and is defined as

$$W_x(t,\omega) = \int_{-\infty}^{\infty} x\left(t+\frac{\tau}{2}\right)x^*\left(t-\frac{\tau}{2}\right)\mathrm{e}^{-\mathrm{j}\omega\tau}\mathrm{d}\tau \tag{5.6}$$

It is clear that $W_x(t,\omega)$ is a two-dimensional function of t and ω. $W_x(t,\omega)$ possesses many good features like the time edge feature and the frequency edge feature, which can solve the existing problem of short-time Fourier transform to some extent. Besides, it has a specific physical significance that it can be regarded as the signal energy distribution in time domain and frequency domain. It is mainly used as a signal time-frequency analysis method. In 1970, Mark summarized a major weakness of Wigner-Ville distribution—the existence of cross-term interference. Denoting $x(t) = x_1(t) + x_2(t)$, we can get

$$\begin{aligned} W_x(t,\omega) &= \int_{-\infty}^{\infty}\left[x_1\left(t+\frac{\tau}{2}\right)+x_2\left(t+\frac{\tau}{2}\right)\right]\left[x_1^*\left(t-\frac{\tau}{2}\right)+x_2^*\left(t-\frac{\tau}{2}\right)\right]\mathrm{e}^{-\mathrm{j}\omega\tau}\mathrm{d}\tau \\ &= W_{x1}(t,\omega)+W_{x2}(t,\omega)+2\mathrm{Re}\left[W_{x1,x2}(t,\omega)\right] \end{aligned} \tag{5.7}$$

Different from the linear time-frequency representation, the quadratic time-frequency distribution does not meet linear superposition principle. All the quadratic time-frequency distribution should be subject to the so-called "secondary superposition principle": for the quadratic time-frequency distribution of p composition signal $z(t) = \sum_{k=1}^{p} c_k z_k(t)$, each signal composition has a self-composition and each couple of signal composition has a corresponding intercomposition, that is p signal terms and C_p^2 cross-term combined of two. The more the signal compositions are, the more serious the cross-terms are. The cross-terms are oscillating and of no contribution to energy of the signal. The energy is completely included in the self-compositions. However, they bring in ambiguity in time-frequency spectrum. It is a serious drawback of WVD. The cross-terms of WVD are quite serious in general. For example, even though two signal compositions are far from each other in time-frequency plane and not overlapping in the two supporting areas, their cross-terms of WVD distribution still exist. That the cross-terms of any two signal compositions are as small as possible is usually considered as a property any time-frequency distribution wishes to owe. Thus, how to eliminate cross-terms is what we need to consider in designing and using time-frequency distribution.

After applying a window function $h(\tau)$ to WVD of signal, its Fourier transform of τ is called pseudo Wigner-Ville distribution (PWVD)

$$\begin{aligned} \mathrm{PW}_x(t,\omega) &= \int_{-\infty}^{\infty} h(\tau)x\left(t+\frac{\tau}{2}\right)x^*\left(t-\frac{\tau}{2}\right)\mathrm{e}^{-\mathrm{j}\omega\tau}\mathrm{d}\tau \\ &= H(\omega)\overset{\omega}{*}W_x(t,\omega) = \int_{-\infty}^{\infty} H(\theta)W_x(t,\omega-\theta)\mathrm{d}\theta \end{aligned} \tag{5.8}$$

where $H(\omega)$ is the Fourier transform of window function $h(t)$. Applying a window is equivalent to the WVD of no adding window convoluted with spectrum of window function in frequency. It makes the WVD of $x(t)$ smooth and reduces the cross-term interference to some extent. However, it sacrifices the frequency resolution of WVD at the same time.

The other method to reduce cross-terms is to smooth WVD distribution by a smooth function and get the so-called smoothed Wigner-Ville distribution (SWVD)

$$\mathrm{SW}_x(t,\omega) = \int_{-\infty}^{\infty}\int_{-\infty}^{\infty} W_g(t-s,\omega-\theta)W_x(s,\theta)\mathrm{d}s\mathrm{d}\theta = W_g(t,\omega)\overset{t,\omega}{*}W_x(t,\omega) \tag{5.9}$$

where $W_g(t,\omega)$ is a smoothing filter, and "$\overset{t,\omega}{*}$" is the two-dimensional convolution of time and frequency. While smoothing filter $W_g(t,\omega)$ is Wigner-Ville distribution of function $g(t)$, the smoothing WVD is equivalent to spectrogram derived from short-time Fourier of signal $x(t)$ plus window function $g(t)$.

In fact, the time-domain smoothing is usually based on the frequency-domain smoothing of PWVD distribution. The smoothed pseudo Wigner-Ville distribution (SPWVD) can be obtained as

$$\mathrm{SPW}_x(t,\omega,g,h) = \int_{-\infty}^{\infty}\int_{-\infty}^{\infty} g(u)h(\tau)x\left(t-u+\frac{\tau}{2}\right)x^*\left(t-u-\frac{\tau}{2}\right)\mathrm{e}^{-\mathrm{j}\omega\tau}\mathrm{d}\tau\mathrm{d}u \quad (5.10)$$

where $g(t)$ and $h(\tau)$ are both real even number and $g(0) = h(0) = 1$.

Smoothing time-frequency spectrum can lessen cross-terms. However, it reduces time resolution and frequency resolution, and meanwhile, the time-frequency distribution is dispersed. The value $\mathrm{SW}_x(t,\omega)$ of any time-frequency point (t,ω) derived from Eq. (5.9) is the sum of every $W_g(t-s,\omega-\theta)W_x(s,\theta)$. And $\mathrm{SW}_x(t,\omega)$ can be regarded as a weighted Wigner-Ville distribution in the points that are close to (t,ω). That is to say, $\mathrm{SW}_x(t,\omega)$ is the average of signal energy in the neighborhood centered at (t,ω), and it takes the supporting area of kernel function as its supporting area. The average solving destroys the concentration of self-depositions and reduces the resolution. Thus, even though there may not be any energy in some (t,ω) points, nonzero values may arise on the $\mathrm{SW}_x(t,\omega)$ of the point after smoothing of kernel function, if there are nonzero values in the neighborhood of the points. These values distribute symmetrically around the geometric gravity center (t,ω) in the area. Suppose that the averages are not distributed on these points but on the geometric gravity center of the area, which are more representative for the local energy distribution.

Auger and Flandrin once pointed out that the distribution performance will be further improved after proper reassignments (corrections) to SPWVD [3]. Applying the method of time-frequency spectrum reassignment can increase aggregation of the spectrum. Moving the energy on point (t,ω) to the energy center in its neighborhood will impel the distribution of signal in some time-frequency area. It will point with more energy and provide better aggregating property. Rearranging the time-frequency distribution of linear frequency modulation signal can improve the concentration of spectrum. And the key point of reassignment is to assign the distribution of any point (t,ω) to another point $(\hat{t},\hat{\omega})$, which is the geometric gravity center of the energy distribution in the neighborhood of point (t,ω). The rearranged SPWVD (RSOWVD) can be represented as

$$\mathrm{RSPW}_x(t,\omega) = \int_{-\infty}^{+\infty}\int_{-\infty}^{+\infty} \mathrm{SPW}_x(t',\omega')\delta\left(t-\hat{t}(x;t',\omega')\right)\delta\left(\omega'-\hat{\omega}(x;t',\omega')\right)\mathrm{d}t'\mathrm{d}\omega' \quad (5.11)$$

where

$$\widehat{t}(x;t,\omega) = \frac{\iint \tau L(\tau,\theta) SPW_x(t-\tau,\omega-\theta)\mathrm{d}\tau\mathrm{d}\theta}{\iint L(\tau,\theta)\mathrm{SPW}_x(t-\tau,\omega-\theta)\mathrm{d}\tau\mathrm{d}\theta}$$

$$\widehat{\omega}(x;t,\omega) = \frac{\iint \theta L(\tau,\theta)\mathrm{SPW}_x(t-\tau,\omega-\theta)\mathrm{d}\tau\mathrm{d}\theta}{\iint L(\tau,\theta)\mathrm{SPW}_x(t-\tau,\omega-\theta)\mathrm{d}\tau\mathrm{d}\theta}$$

RSPWVD cannot only eliminate the interference of cross-terms but also can improve the time-frequency resolution. While the time-frequency distributions of the multicompositions signal are close to each other, aliasing appears after smoothing the signal spectrum, which cannot guarantee the right concentration of each signal composition.

Aiming at cross-terms of WVD and different needs in practice, serial types of other distribution forms arise, such as Rihaczek distribution, Margenau—Hill distribution, Page distribution, Choi—Williams distribution, Born—Jordan distribution, and Zhao—Atlas—Marks distribution. Cohen found that all those time-frequency distributions are just the transformations of WVD, which can be represented by a uniform time-frequency distribution now customarily called the Cohen time-frequency distribution. In 1966, Cohen gave the uniform equation of all kinds of time-frequency distribution as

$$C_x(t,\omega) \stackrel{\text{def}}{=} \int_{-\infty}^{+\infty} R_x(t,\tau)\mathrm{e}^{-\mathrm{j}\omega\tau}\mathrm{d}\tau = \int_{-\infty}^{+\infty}\int_{-\infty}^{+\infty} A_x(\tau,\theta)\phi(\tau,\theta)\mathrm{e}^{-\mathrm{j}(\theta t+\omega\tau)}\mathrm{d}\tau\mathrm{d}\theta \tag{5.12}$$

where $A_x(\tau,\theta)$ is the ambiguity function of $x(t)$. $\phi(\tau,\theta)$ is the kernel function and another type of the definition is

$$C_x(t,\omega) \stackrel{\text{def}}{=} \int_{-\infty}^{+\infty}\int_{-\infty}^{+\infty} x\left(u+\frac{\tau}{2}\right)x^*\left(u-\frac{\tau}{2}\right)\psi(t-u,\tau)\mathrm{e}^{-\mathrm{j}\omega\tau}\mathrm{d}u\mathrm{d}\tau \tag{5.13}$$

where $\psi(t,\tau)$ is kernel function $\phi(\tau,\theta)$'s inverse Fourier transform of θ. Eqs. (5.12) and (5.13), which are equivalent to each other, are frequently used in the Cohen distribution. These improved distributions conduct two-dimensional smoothing to WVD. They can basically reduce cross-terms interference. However, the cross-terms decreasing and the self-terms maintaining are a couple of contradictions. The cross-terms decreasing will negatively level the self-terms and thus, it reduces the time-frequency concentration. We usually assume that self-terms do not overlap with cross-terms, when discussing the problems and the methods of cross-terms reducing. In fact, there are not any methods of reducing cross-terms applied to the case that the self-terms overlap with the cross-terms.

Besides, the self-terms are located at the origin of ambiguity plane in general, while the cross-terms are often far from the origin. So the kernel function is supposed to be the two-dimensional low pass filter function.

5.1.2 Micro-Doppler Feature Analysis Based on Time-Frequency Analysis

5.1.2.1 Narrowband Radar

In the example given in Chapter 2, we have adopted the time-frequency analysis methods to analyze the micro-Doppler signals. In this section, we will compare the advantages and disadvantages of different time-frequency analysis methods (STFT, Gabor transform, WVD, PWVD, SPWVD, and RSPWVD) in analyzing the micro-Doppler signal.

Simulation parameters are set as follow: Assume that carrier frequency of the transmitted signal is $f_c = 10$ GHz. The origin O of the target-local coordinate system is located at (3,4,5) km of the radar coordinate system. The initial Euler angle of the target-local coordinate system and the reference coordinate system is $(0,\pi/4,\pi/5)$ rad. Assume that there is no translation. The target is rotating with angular velocity $\omega = (\pi,2\pi,\pi)^{\mathrm{T}}$ rad/s and the period of rotation is $T = 2\pi/\|\omega\| = 0.8165$ s.

First, assume that there is only one rotation scatterer in the target, locating at (1,0.5,0.5) in the target-local coordinate system. Fig. 5.1 shows the theoretical curve of the returned signal's micro-Doppler frequency change of the scatterer point. Fig. 5.2A–F show the result of STFT, Gabor transform, WVD, PWVD, SPWVD, and RSPWVD, respectively. And by comparison, STFT and Gabor transform obtain better results. WVD is serious in cross-terms, so it is difficult to recognize the sinusoidal curve reflecting the true micro-Doppler frequency changes of target from Fig. 5.2C. PWVD

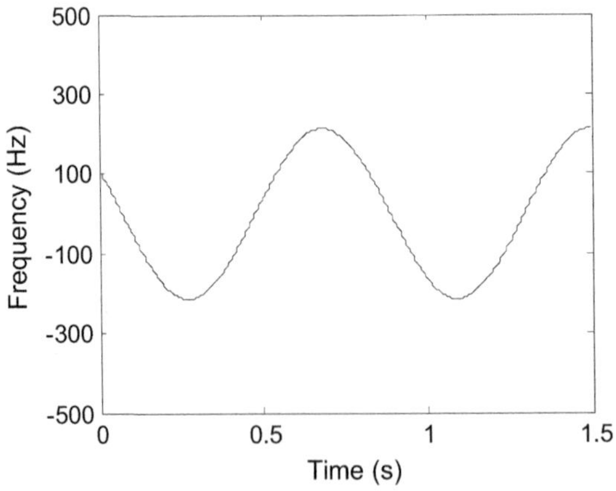

Figure 5.1 Theoretical curve of micro-Doppler frequency when there is single scatterer point.

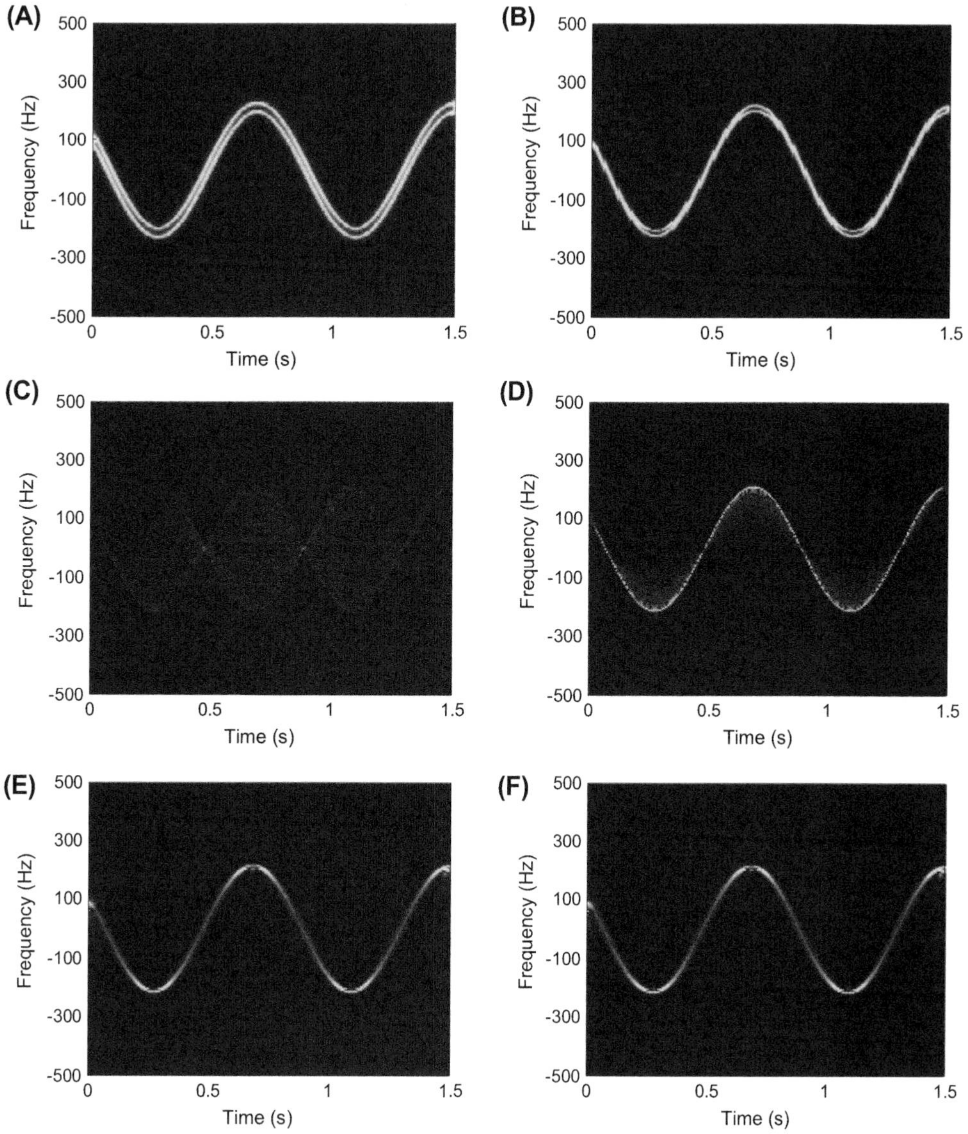

Figure 5.2 Result of time-frequency analysis when there is a single scatterer point. (A) STFT; (B) Gabor transform; (C) WVD; (D) PWVD; (E) SPWVD; (F) RSPWVD.

performs better than WVD. However, the cross-terms are not eliminated completely. SPWVD and RSPWVD perform better results as well.

When the number of the rotation scatterer points increases to two (the newly added scatterer locates at $(-1,-0.5,-0.5)$ m of the target-local coordinate system). The theoretical curve of micro-Doppler frequency changing is shown in Fig. 5.3.

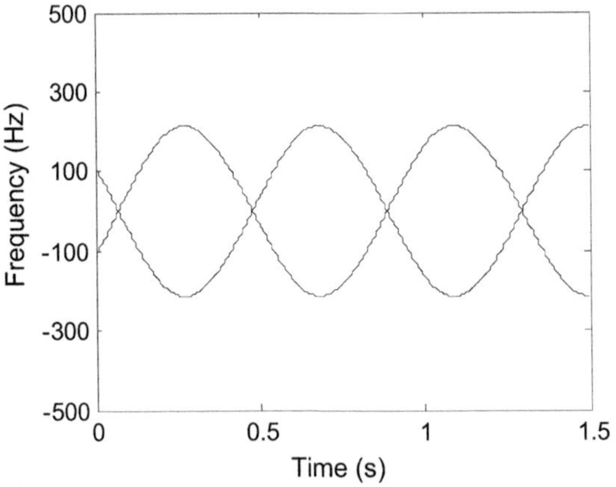

Figure 5.3 Theoretical curve of micro-Doppler frequency when there are two scatterer points.

Fig. 5.4A—F shows the results of all kinds of time-frequency transforms. It shows that STFT and Gabor transform get the best analysis result of the transformation. Results of time-frequency analysis of WVD and PWVD are serious in cross-terms. SPWVD and RSPWVD perform better results as well.

When the number of the rotation scatterer points increases to six, the theoretical curve of micro-Doppler frequency changing is shown in Fig. 5.5. Fig. 5.6A—F shows the results of all kinds of time-frequency transforms. It shows that STFT and Gabor transform get the best analysis result; SPWVD and RSPWVD take the second place; WVD and PWVD cannot get the correct result of time-frequency analysis.

Based on the previous simulation results, the conclusions are given as follows: STFT and Gabor transform get better analysis results than the quadratic time-frequency method, when analyzing micro-Doppler signal in narrowband radar. WVD and PWVD cannot analyze the micro-Doppler effect efficiently influenced by the cross-terms, when there are many micromotion scatterer points of the target. SPWVD and RSPWVD are second to STFT and Gabor transform in analysis performance.

5.1.2.2 Wideband Radar

In wideband radar, if migration through the resolution cells does not occur when micro-motion points are receiving coherent processing in azimuth direction, the micro-motion features can be obtained by extracting and analyzing the signal in the range cells of this micro-motion point, which is shown in Fig. 5.7. It is the same with the analysis method of micro-Doppler signals in narrowband radar. The corresponding analysis is given in Section 3.2.4.

However, micromotions of the target often lead to migration through resolution cells of caterer points for the reason of high-range resolution of wideband radar, and the energy of returned signal is distributed in multiple range cells. Take the simulation in

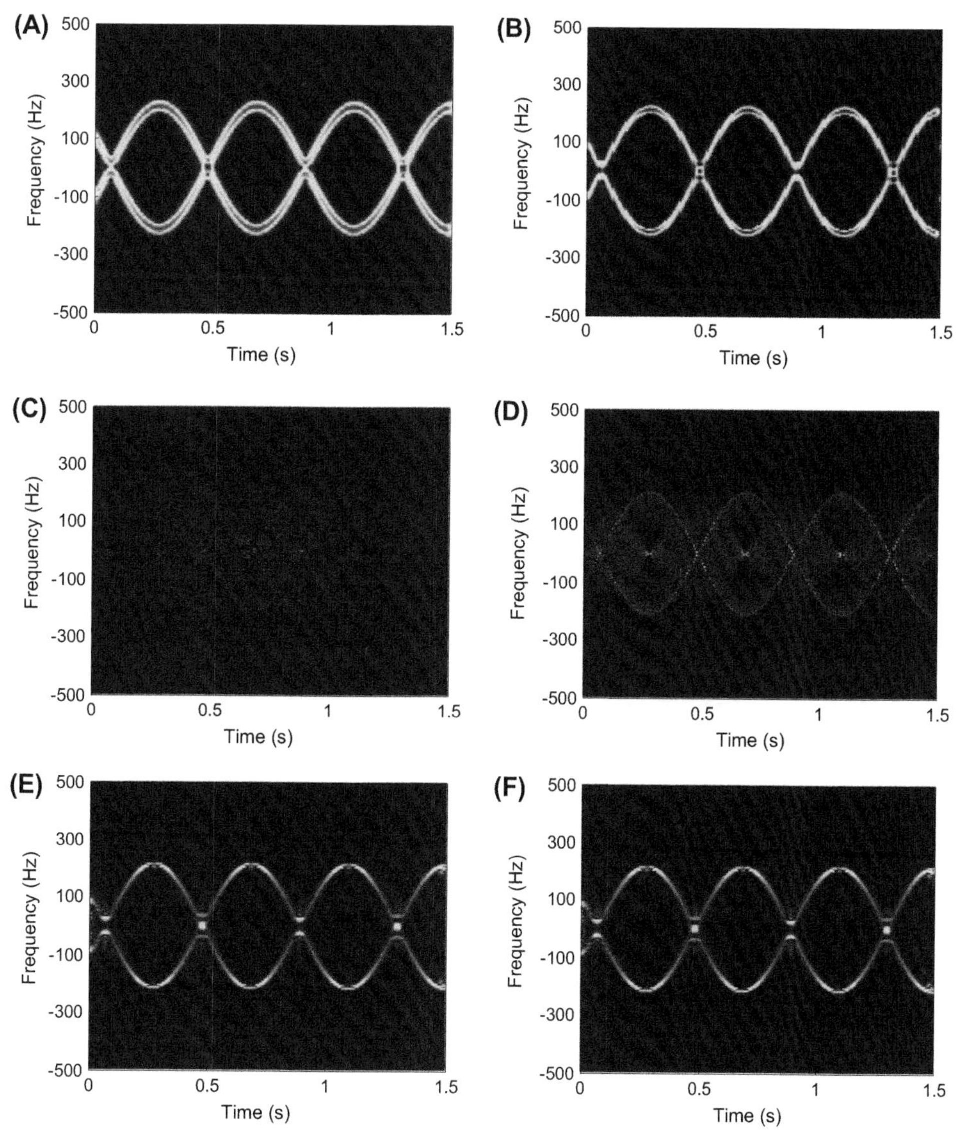

Figure 5.4 Result of time-frequency analysis when there are two scatterer points. (A) STFT; (B) Gabor transform; (C) WVD; (D) PWVD; (E) SPWVD; (F) RSPWVD.

Fig. 3.6B for example, now it is hard to observe a complete micro-Doppler feature curve if merely extracting returned signal of one range cell to conduct time-frequency analysis (as is shown in Fig. 5.8); this is because that even though the phase of signal in one range cell contains micromotion feature of micromotion points, time-frequency method cannot analyze signal efficiently because the amplitude of signal is modulated by sinc function, especially when micro-Doppler signals of multiple scatterer points overlap.

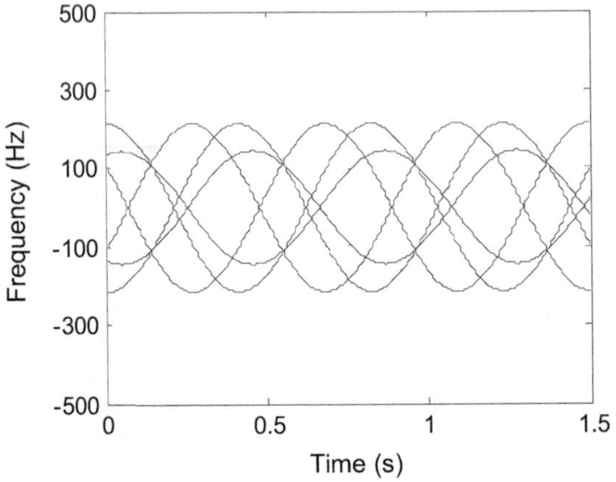

Figure 5.5 Theoretical curve of micro-Doppler frequency when there are six scatterer points.

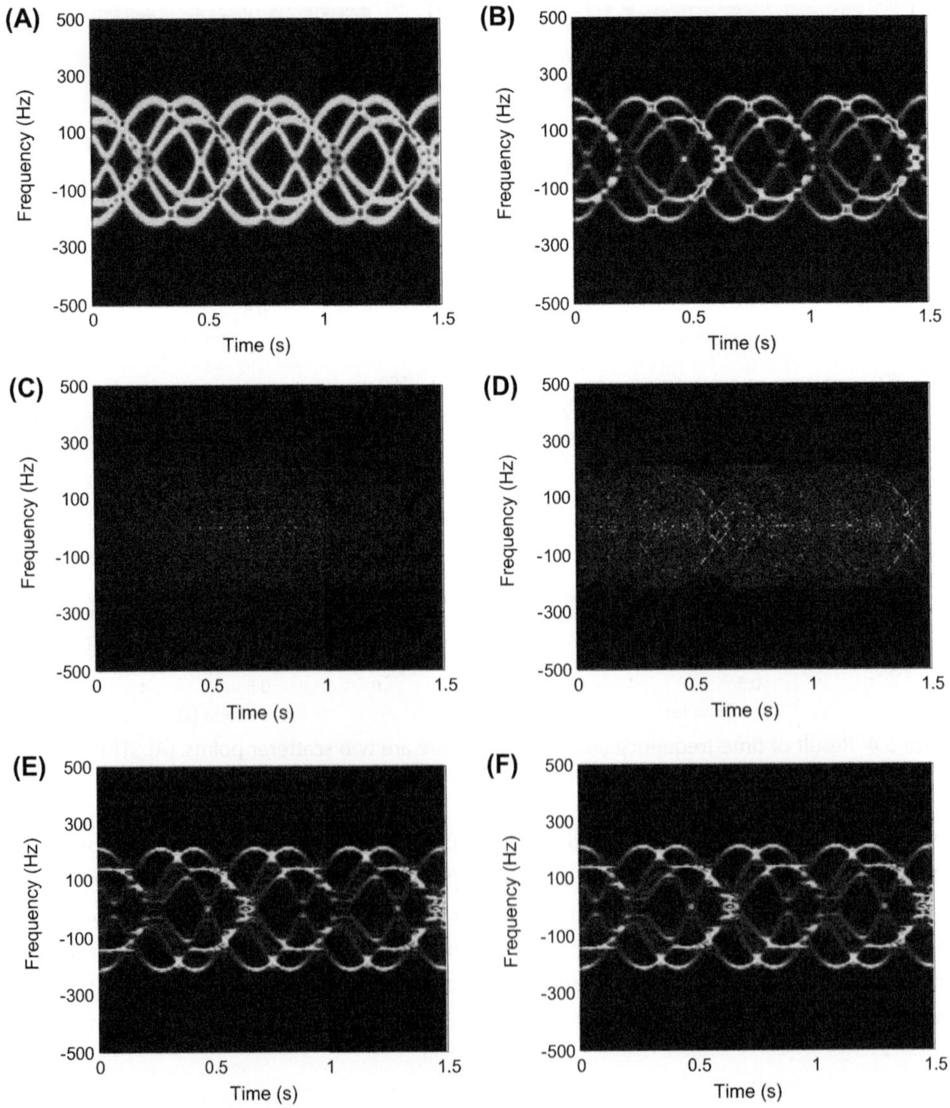

Figure 5.6 Result of time-frequency analysis when there are six scatterer points. (A) STFT; (B) Gabor transform; (C) WVD; (D) PWVD; (E) SPWVD; (F) RSPWVD.

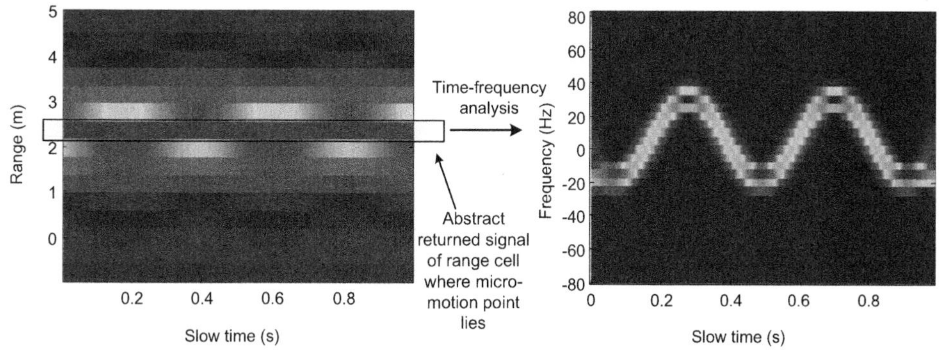

Figure 5.7 Results of time-frequency analysis when migration through resolution cell does not occur during the period that micromotion points are receiving coherent processing in azimuth direction.

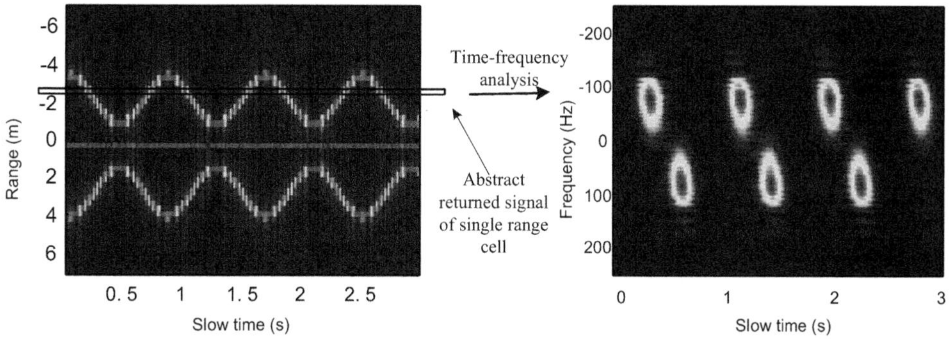

Figure 5.8 Results of time-frequency analysis when migration through resolution cell occurs during the period that micromotion points are receiving coherent processing in azimuth direction.

Besides, when micromotion velocity is relatively high and the micro-Doppler spectral width is relatively wide, the pulse repetition frequency of wideband radar cannot be too high in general. It will lead to micro-Doppler signal subsampled. For example, when the carrier frequency of the transmitted signal is 10 GHz, the rotation frequency of micromotion point is 4 Hz and the rotation radius in radial direction is 6 m, so the micro-Doppler spectral width will reach 20 KHz. And to guarantee Nyquist sampling to micro-Doppler spectrum, the pulse repetition frequency of radar is required highly. When the pulse repetition frequency of radar is less than twice the width of micro-Doppler spectrum, the micro-Doppler signal will overlap in the frequency domain, which brings more difficulties to time-frequency analysis methods. In the simulation shown in Fig. 5.8, the result of the theoretical time-frequency analysis is shown in Fig. 5.9. It is clear that the result of time-frequency analysis in Fig. 5.8 is far from the theoretical result. Thus, new methods of micro-Doppler feature analysis and extraction need to be explored in this case.

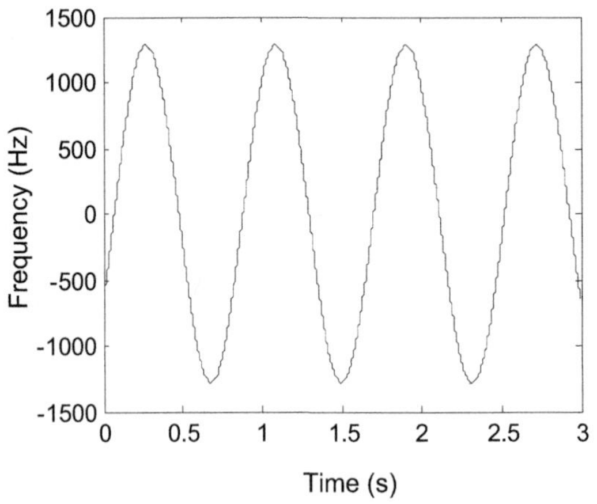

Figure 5.9 Theoretical result of time-frequency analysis of micromotion points.

5.2 IMAGE PROCESSING METHOD

According to the analysis of Chapter 2 to Chapter 4, the target micro-Doppler effects are reflected as the micro-Doppler feature curves in the time-frequency plane by using the narrowband radar; in wideband radar, if radial fretting amplitude of target is greater than a distance resolution cell and the micro-Doppler effects are reflected as micro-Doppler feature curves on range—slow time plane; if radial fretting amplitude of target is smaller than a range resolution cell, we can use time-frequency analysis methods to analyze the echo signal of the range resolution cell and observe the micro-Doppler feature curves on time-frequency plane. Therefore, we can extract the micro-Doppler feature curve parameters of the target from time-frequency plane or range—slow time image. Based on image processing methods, micro-Doppler feature extraction method is proposed in this section. The former problem that generates in the processing of micro-Doppler feature extraction by using wideband radar has been solved through extracting micro-Doppler feature curves from range—slow time image.

5.2.1 Mathematical Morphology Image Processing

Mathematical morphology is an important branch of image signal processing, and it provides a useful tool for solving many image processing problems. The language of mathematical morphology is set theory. For example, the set of all black pixels in a binary image is a complete morphological description of the image. In binary images, the sets in question are members of the 2D integer space Z^2, where each element of a set is a tuple (2D vector) whose coordinates are the (x,y) coordinates of a black (or white, depending on convention) pixel in the image. Gray-scale digital images can be represented as sets whose components are in Z^3. In this case, two components of each element of the set refer to the coordinates of a pixel, and the third corresponds to its discrete gray-level value. Sets in higher dimensional spaces can contain other image attributes, such as color.

Firstly, we give an introduction of some basic concepts to be used [4].

Let A be a set in Z^2. If $a = (x,y)$ is an element of A, then we write

$$a \in A \tag{5.14}$$

Similarly, if a is not an element of A, we write

$$a \notin A \tag{5.15}$$

The set with no elements is called the null or empty set and is denoted by the symbol $\varnothing$.

A set is specified by the contents of two braces: $\{\cdot\}$. For example, when we write an expression of the form $C = \{w|w = -d, d \in D\}$, we mean that set C is the set of elements w, and w is formed by multiplying each of two coordinates of all the elements of set D by -1.

If every element of set A is also an element of another set B, then A is a subset of B, and is denoted as

$$A \subseteq B \tag{5.16}$$

The union of two sets A and B, denoted by

$$A \cup B, \tag{5.17}$$

is the set of all elements belonging to either A, B, or both. Similarly, the intersection of two sets A and B, denoted by

$$A \cap B, \tag{5.18}$$

is the set of all elements belonging to both A and B.

The complement of a set A is the set of elements not contained in A:

$$A^c = \{w|w \notin A\} \tag{5.19}$$

The difference of set A and set B is defined as

$$A - B = \{w|w \in A, w \notin B\} = A \cap B^c \tag{5.20}$$

The reflection of set A, denoted $\widehat{A}$, is defined as

$$\widehat{A} = \{w|w = -a, a \in A\} \tag{5.21}$$

The translation of set A by point $z = (z_1, z_2)$, denoted $(A)_z$, is defined as

$$(A)_z = \{w|w = a + z, a \in A\} \tag{5.22}$$

Then, we introduce two operations of mathematical morphology: dilation and erosion.

Let A and B be sets of Z^2, the dilation of A by B is defined as:

$$A \oplus B = \left\{ z \middle| \left(\widehat{B}\right)_z \cap A \neq \varnothing \right\} \tag{5.23}$$

Set B is commonly called the structuring element in dilation. This equation is based on obtaining the reflection of B about its origin and shifting this reflection by z; meanwhile $\widehat{B}$ and A overlap by at least one element.

The erosion of A by B is defined as

$$\mathrm{A} \ominus \mathrm{B} = \left\{ z \middle| (B)_z \subseteq A \right\} \tag{5.24}$$

This equation indicates that the erosion of A by B is the set of all points z such that B, translated by z, is contained in A.

Based on dilation and erosion, we discuss two other important morphological operations: opening and closing. Opening operation generally smooths the contour of an object, breaks narrow isthmuses, and eliminates thin protrusions. Closing also tends to smooth sections of contours but, as opposed to opening, it generally fuses narrow breaks and long thin gulfs, eliminates small holes, and fills gaps in the contour.

The opening of set A by structuring element B, denoted $A \circ B$, is defined as:

$$A \circ B = (\mathrm{A} \ominus \mathrm{B}) \oplus B \tag{5.25}$$

Thus, the opening A by B is the erosion of A by B, followed by a dilation of the result by B.

The closing of set A by structuring element B, denoted $A \cdot B$, is defined as:

$$A \cdot B = (A \oplus B) \ominus B \tag{5.26}$$

that is to say that the closing of A by B is simply the erosion of A by B, followed by the dilation of the result by B.

The skeleton of A can be expressed in terms of erosions and openings. It can be shown as

$$S(A) = \bigcup_{k=0}^{K} S_k(A) \tag{5.27}$$

where $S_k(A)$ is the subskeleton,

$$S_k(A) = (A \ominus kB) - (A \ominus kB) \circ B \tag{5.28}$$

where B is a structuring element, and $(A \ominus kB)$ indicates k successive erosions of A, and K is the last iterative step before A erodes to an empty set, that is

$$K = \max\{k | (A \ominus kB)\} \neq \varnothing \tag{5.29}$$

The previous are some basic concepts and methods of mathematical morphology. In the case of micro-Doppler feature extraction, in order to accurately extract micro-Doppler curve parameters on time-frequency plane or range—slow time image, we can use the methods of morphology to analyze time-frequency plane or range—slow time image, which can improve the analysis precision.

For example, the precision of time-frequency analysis is improved by using skeleton extraction algorithm to extract the skeleton of micro-Doppler curve on time-frequency plane; the side lobe of high range resolution profile (HRRP) is suppressed as well as by using this method to extract the skeleton of micro-Doppler curve on range—slow time plane. These are useful to improve the parameters estimation precision of micro-Doppler feature extraction algorithm. The skeleton extraction processing of sinusoidal frequency modulation micro-Doppler signal on time-frequency plane is shown in Fig. 5.10. Usually, the curves of time-frequency are not smooth, and we should smooth for time-frequency plane firstly, then set the appropriate threshold to transform it into binary image and extract skeleton of curve. Therefore, we smooth the image to eliminate

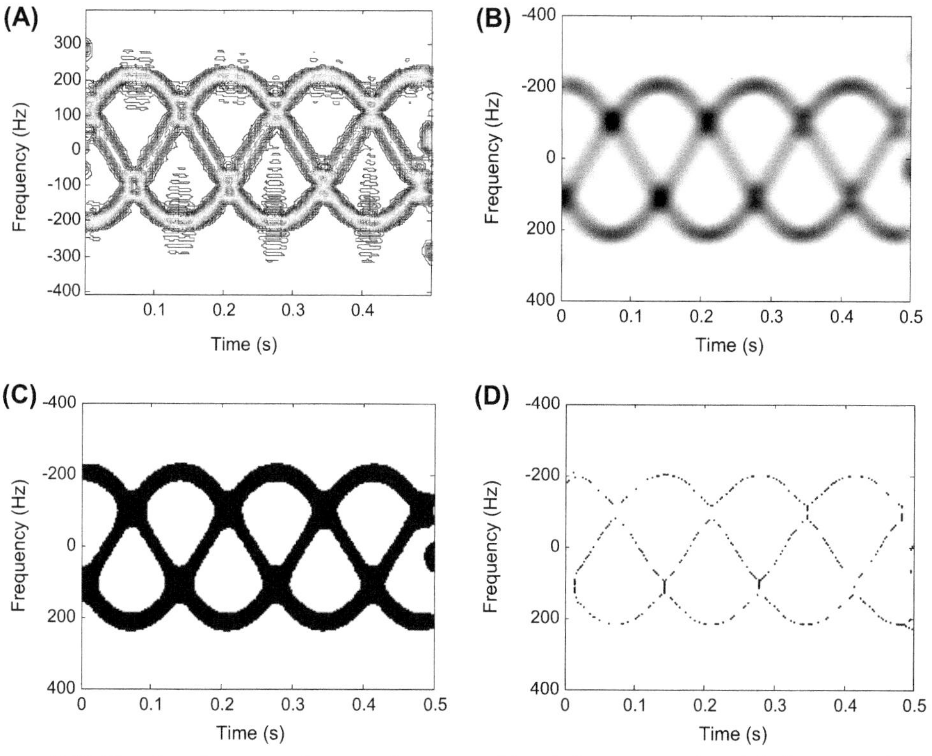

Figure 5.10 The time-frequency image of target and its skeleton extraction. (A) The original time-frequency image; (B) the image after smooth processing; (C) a binary image; (D) the image after skeleton extraction.

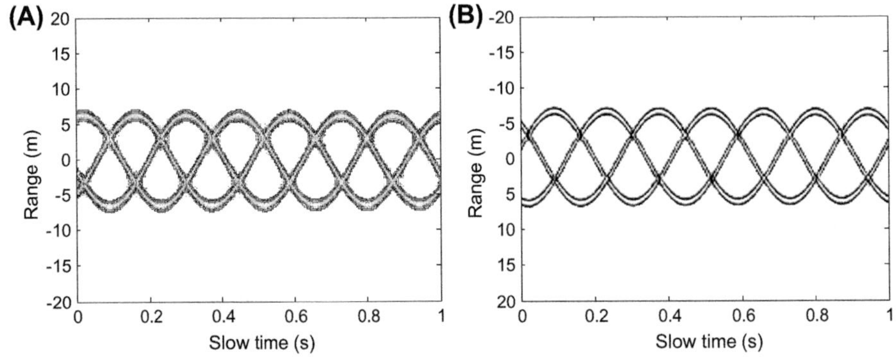

Figure 5.11 The dilation and erosion processing of range–slow time image. (A) The original range–slow time image; (B) the image after dilation and erosion.

influence of noisy points nearby in Fig. 5.10A by using a Gaussian spatial mask, which includes 25 × 25 pixels. The result is shown in Fig. 5.10B, and the noisy points have been eliminated completely. Fig. 5.10B is further converted into a binary image as shown in Fig. 5.10C; black represents 1 and white represents 0. Based on this, the extracted skeleton of curves is shown in Fig. 5.10D and the precision of time-frequency analysis has been improved significantly.

A similar result will result by using dilation and erosion methods. The micro-Doppler image edge information will be extracted by utilizing dilated micro-Doppler image to minus erosive micro-Doppler image. The original "rough" sine curve will be transformed into two "thin" sine curves, but the frequency, frequency offset, and initial phase will not be changed. The baseline position is changed slightly and its right position can be gotten through weighted average of two sine curves. Fig. 5.11 shows the processing of dilation and erosion. The original range–slow time image is shown in Fig. 5.11A and Fig. 5.11B is the image after dilation and erosion. There are six curves in Fig. 5.11B, but the frequency and the amplitude of each sine curve is the same as Fig. 5.11A. It indicates that the image after dilation and erosion will not influence the discriminant about micro-Doppler feature, but it will help to extract micro-Doppler feature parameters when the curves become thin.

5.2.2 Hough Transform [5]

Based on the preprocessing of mathematical morphology, we can obtain the micro-parameters by extracting micro-Doppler curve parameters on time-frequency plane or range–slow time image. Hough transform (HT) is a valid method for the detection of image edge, and a patent for it was proposed and applied for by Paul Hough in 1962 [6]. As HT has evolved for decades, it has been applied in every field of image processing, including graphics recognition, document image tilt correction, target recognition in

SAR/ISAR image, and aviation image interpretation. This method converts the detection of edge in image space to the detection of peak in the parameter space and finishes the detection task by simple accumulative statistics. It uses the parameters form that most boundary points meet to describe the image boundaries and has commendable fault tolerance and robustness.

The basic idea of HT is the duality of point-line. As shown in Fig. 5.12, Fig. 5.12A is the image space *XY* and Fig. 5.12B is the parameters space *PQ*. The linear equation that contains the point (x_i,y_i) in Fig. 5.12A can be written as:

$$y_i = px_i + q \tag{5.30}$$

It also can be written as:

$$q = -px_i + y_i \tag{5.31}$$

This represents a straight line in *PQ*. And the linear equation that contains the point (x_j,y_j) can be written as: $y_j = px_j + q$, and also can be written as: $q = -px_j + y_j$. It represents another straight line in *PQ*. If the two lines intersect at the point (p',q') in *PQ*, the point (p',q') is corresponding to the line that contains (x_i,y_i) and (x_j,y_j) in *XY* because it satisfies the equation $y_i = p'x_i + q'$ and the equation $y_j = p'x_j + q'$. Similarly, each point in the line determined by (x_i,y_i) and (x_j,y_j) in *XY* corresponds to a line in *PQ*, and these lines intersect at the point (p',q'). That is to say, the points that are in the same line in *XY* correspond to the lines that intersect at the same point in *PQ*. On the contrary, the lines that intersect at the same point in *PQ* correspond to the points that are in the same line in *XY*. This is the duality of point-line. According to the relationship, HT converts the detection of line in *XY* to the detection of peak in *PQ*. For example, if there are two lines in *XY*, there will be two peaks in *PQ* after HT and the parameters of these two lines are determined by the positions of the two peaks.

In practical application, Eq. (5.30) cannot represent the line whose equation is: $x = a$ owing to its infinite slope. In order to detect any form of line through HT, we use polar coordinate equation of line:

$$\rho = x\cos\theta + y\sin\theta, \quad \theta \in [0, \pi] \tag{5.32}$$

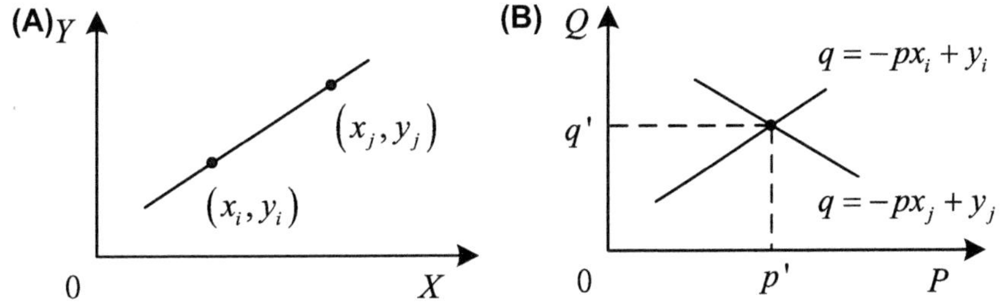

Figure 5.12 The duality of point-line between image space and parameters space.

So, the point in the image space is corresponding to the sine curve in the parameters space (ρ,θ), and the duality of point-line transforms to the duality of point-sine curve. Through the detection of peaks in (ρ,θ), we will obtain the parameters of the lines in the image space.

Based on the previous theories, the steps of HT detection line can be summarized as follows:

Step 1: Set up a discrete accumulator array $A(\rho,\theta)$, $\rho \in [\rho_{\min}, \rho_{\max}]$, $\theta \in [\theta_{\min}, \theta_{\max}]$, where $[\rho_{\min},\rho_{\max}]$ and $[\theta_{\min},\theta_{\max}]$ are the expected value range of ρ and θ, respectively. Zeroize the array $A(\rho,\theta)$;

Step 2: Apply the HT to all the points whose module values are greater than a certain threshold in the image space. It means that the corresponding line will be calculated in (ρ,θ) and the corresponding accumulator increase 1:

$$A(\rho,\theta) = A(\rho,\theta) + 1;$$

Step 3: Search a local maximum of $A(\rho,\theta)$, the position of this maximum represents the parameters of the line in the image space.

HT cannot only extract the parameters of line in image but also extract curves of other shapes, both analytic and nonanalytic. Suppose the parameters $(a_1,a_2,\ldots,a_N)$ are used to represent the curve, and each parameter accumulator simply counts the number of points (x_i,y_i) fitted into the curve equation

$$f(a_1, a_2, \ldots, a_N, x_i, y_i) = 0 \tag{5.33}$$

The accumulator, which contains the local maximum in Hough space, may correspond to the existence of a curve. The micro-Doppler feature curve parameters can be extracted by HT, and the target feature parameters can be obtained according to the relationship of curve parameters and target parameters. For example, in narrowband radar and wideband LFM signal radar, the micro-Doppler effects caused by rotating or vibratory parts behave as the cosine curve on range—slow time plane, so we can use HT to extract micro-Doppler feature. Take rotary motion, for example; we can use the equation of four parameters to describe the cosine curve on range—slow time plane:

$$f = r\cdot\cos(\Omega t_m + \theta_0) + d \tag{5.34}$$

where r is the amplitude of cosine curve, Ω is the angular frequency, $\Omega = 2\pi/T_r$, T_r is the cycle, θ_0 is the initial phase, d is the baseline, and it shows the position of cosine curve in the frequency (range) axis. Owing to extracting the curve feature on range—slow time plane directly, this method is not limited by the condition that the pulse repetition frequency (PRF) is satisfied with Nyquist sampling or not. In the condition that PRF is lower than the required frequency by Nyquist sampling theory, this method can also extract the micro-Doppler feature of target.

However, for the radar with stepped-frequency chirp signal, the micro-Doppler curve is not simple cosine curves on range—slow time plane due to the motion and winding of micro-Doppler domain. We should construct HT equation according to the micro-Doppler curve parameters equation if we use HT to extract micro-Doppler feature, which will be explained in detail in the next section.

5.2.3 Application of Image Processing Method in Micro-Doppler Feature Extraction in SFCS Radar [7]

5.2.3.1 Algorithm

Based on the analysis in Chapter 3, for the radar with stepped-frequency chirp signal, the parameters $(R_Q,\rho,\Omega,\theta_0)$ determine the micro-Doppler curve's equation on the range—slow time plane. We construct the HT equation as follows:

Firstly, Eq. (3.63) is rewritten in the equivalent form

$$\widehat{r}(t_m)+\frac{c}{4\Delta f}+M\cdot\frac{c}{2\Delta f}=R_Q+\rho\sqrt{1+\Omega^2 f_0^2 T_r^2/\Delta f^2}\cos(\Omega t_m+\theta_0+\phi)+\frac{c}{4\Delta f},\quad M=0,\pm1,\pm2,\cdots \tag{5.35}$$

That is

$$R_Q=\widehat{r}(t_m)+M\cdot\frac{c}{2\Delta f}-\rho\sqrt{1+\Omega^2 f_0^2 T_r^2/\Delta f^2}\cos(\Omega t_m+\theta_0+\phi) \tag{5.36}$$

According to the assumption that the length of the target is shorter than the blurry range resolution, it yields

$$-\frac{c}{4\Delta f}<R_Q<\frac{c}{4\Delta f} \tag{5.37}$$

Therefore, Eq. (5.36) can be rewritten as

$$R_Q=\mathrm{mod}\left(\widehat{r}(t_m)-\rho\sqrt{1+\Omega^2 f_0^2 T_r^2/\Delta f^2}\cos(\Omega t_m+\theta_0+\phi)+\frac{c}{4\Delta f},\frac{c}{2\Delta f}\right)-\frac{c}{4\Delta f} \tag{5.38}$$

Let $A=\rho\sqrt{1+\Omega^2 f_0^2 T_r^2/\Delta f^2}$, $\Omega_0=\Omega/(2\pi)$, $\zeta=\theta_0+\phi$, and we have

$$R_Q=\mathrm{mod}\left(\widehat{r}(t_m)-A\cos(2\pi\Omega_0 t_m+\zeta)+\frac{c}{4\Delta f},\frac{c}{2\Delta f}\right)-\frac{c}{4\Delta f} \tag{5.39}$$

Thus, the detection of curves on the range—slow time plane $(t_m,r(t_m))$ is transformed to the peak value detection in the parameters (R_Q,A,Ω,ζ) space by utilizing the HT according to Eq. (5.39), which is the HT transformation equation. The algorithm is as follows:

Step1: Set up a discrete accumulator array $A_c(R_Q,A,\Omega_0,\zeta)$, where $R_Q \in [R_{Q\min}, R_{Q\max}]$, $A \in [A_{\min}, A_{\max}]$, $\Omega_0 \in [\Omega_{0\min}, \Omega_{0\max}]$, $\zeta \in [\zeta_{\min}, \zeta_{\max}]$ are within the prospective value sales. Set all the elements of array $A_c(R_Q,A,\Omega_0,\zeta)$ zero;

Step2: Apply the HT to all the points on the range–slow time plane whose module values are greater than a certain threshold according to Eq. (5.39), ie, calculate the curves $\left(\widehat{R}_Q, \widehat{A}, \widehat{\Omega}, \widehat{\zeta}\right)$ in the HT space corresponding to the point $\left(\widehat{r}\left(\widehat{t}_m\right), \widehat{t}_m\right)$ on the range–slow time plane, and then apply

$$A_c\left(R'_Q, A', \Omega'_0, \zeta'\right) = A_c\left(R'_Q, A', \Omega'_0, \zeta'\right) + 1; \tag{5.40}$$

Step3: Search the local maxima in the HT space and their coordinates $A_c(R_Q,A,\Omega,\zeta)$ are the parameters of the curves on the range–slow time plane.

The rotating or vibrating rate of the micromotion scatterer $\omega_0 = 2\pi\Omega$ can be obtained from the parameters detected by HT. The rotating or vibrating center can be identified from the parameter R_Q. The radius ρ can also be estimated using the following equation: $\rho = A \Big/ \sqrt{1 + \omega_0^2 f_0^2 T_r^2 / \Delta f^2}$.

Generally, the algorithm presented herein can extract the micro-Doppler signatures from most of the targets with rotating or vibrating parts, such as the horizontal rotors on a helicopter and the rotating antenna on a ship. However, two special situations must be taken into consideration.

1. When $A < c/(4N\Delta f)$

In this situation, the amplitude of the curve is less than half of the range resolution. This situation also widely exists in radar target, for example, a truck with vibrating surface induced by a running engine, or a person with chest fluctuation induced by breathing. When $A < c/(4N\Delta f)$, the curve will appear as a straight line on the range–slow time plane and the micro-Doppler signatures cannot be extracted directly by HT.

Reconsider Eq. (3.52), when letting $\omega = -4\pi\mu R_{\Delta 0}/c$ (the reason is the same as inference from Eqs. (3.48) to (3.49)), the HRRP can be synthesized by applying FFT to the right side of Eq. (3.52) with respect to i. Ignoring the influence of the coupling term between v and i, the result of FFT is obtained approximately

$$S(k) = C' \cdot \mathrm{sinc}\left(k + \frac{4\pi\Delta f}{c} R_{\Delta 0} + \frac{4\pi}{c} f_0 T_r v\right) \cdot \exp\left(-\mathrm{j}\frac{4\pi}{c} f_0 R_{\Delta 0}\right) \tag{5.41}$$

where C' is a constant. According to Eqs. (3.59) and (3.60), ie, we use $R(t_m)$ and $v(t_m)$ to replace $R_{\Delta 0}$ and v in Eq. (5.41), and we have

$$S(k, t_m) = C' \cdot \text{sinc}\left(k + \frac{4\pi\Delta f}{c}R(t_m) + \frac{4\pi}{c}f_0 T_r v(t_m)\right) \cdot \exp\left(-\text{j}\frac{4\pi}{c}f_0 R(t_m)\right) \quad (5.42)$$

Because $A < c/(4N\Delta f)$, $|S(k, t_m)|$ appears as a straight line in the range−slow time plane, hence we have the peak location at $k = -4\pi\Delta f R(t_m)/c - 4\pi f_0 T_r v(t_m)/c \approx C''$, where C'' is a constant. Let $k = C''$, and take derivative to phase term of the right side of Eq. (5.42) in terms of slow time t_m, then the variance of the instantaneous frequency of $S(k, t_m)|_{k=C''}$ with respect to t_m is obtained

$$l(t_m) = \frac{1}{2\pi}\psi'(t_m) = \frac{2}{c}f_0\rho\Omega \sin(\Omega t_m + \theta_0) \quad (5.43)$$

It can be found that the instantaneous frequency $l(t_m)$ of $S(k, t_m)|_{k=C''}$ is a sinusoidal function of t_m. Therefore, the micro-Doppler signatures can be extracted by analyzing $S(k, t_m)|_{k=C''}$ in the joint time-frequency domain firstly and then utilizing HT to detect the curve on the time-frequency plane. Similarly, $l(t_m)$ will also be wrapped if $|l(t_m)| > \text{BRF}/2$, where BRF is the burst repetition frequency. Therefore, the curve on the time-frequency plane is depicted as

$$\hat{l}(t_m) = \text{mod}\left(\frac{2}{c}f_0\rho\Omega \sin(\Omega t_m + \theta_0) + \frac{\text{BRF}}{2}, \text{BRF}\right) - \frac{\text{BRF}}{2} \quad (5.44)$$

Then the HT equation can be deduced accordingly

$$\text{mod}\left(\hat{l}(t_m) - A' \sin(2\pi\Omega_0 t_m + \zeta), \text{BRF}\right) = 0 \quad (5.45)$$

where $A' = 2f_0\rho\Omega/c$, $\Omega_0 = \Omega/(2\pi)$, $\zeta = \theta_0$. Thus, the micro-Doppler signature extraction is equivalent to the peak values detection in the parameters (A',Ω_0,ζ) space by HT. Using the detected curve parameters, the motional frequency and radial amplitude of micromotion point can be obtained.

2. When rotating or vibrating rate is extraordinarily high

When BRF is given in a radar system, the higher the rotating or vibrating rate is, the less the slow-time domain samples of the curve in a cycle are. If the number of the samples in a cycle is too low, the detection capability of HT will be depressed and the parameters of the curve cannot be extracted accurately. Simulation results have demonstrated that the number of samples in a cycle should be larger than 10−15 to ensure precise detection ability of HT. For example, in a radar system with BRF = 200 Hz, the parameters of the curve are difficult to be detected directly from the range−slow time plane when the rotating or vibrating rate is higher than 20 Hz. Furthermore, when the value of v increases, the value of the third term in Eq. (3.52) increases accordingly and expands the peaks in HRRP more seriously. Therefore, some supplementary processing is also necessary when extracting the micro-Doppler signatures of micromotion parts with high rotating or vibration rates, such as the tail rotors on a helicopter or missile.

According to the analysis in Section 3.3, the HRRP of the micromotion scatterer is synthesized by taking FFT to Eq. (3.51) with respect to i. When the value of v is quite large, the coupling term between v and i will affect the HRRP more significantly and the Fourier transform is not suitable to describe the micro-Doppler signatures. From Eq. (3.52), the linear relationship between k and i can be observed when assuming v to be a constant approximately in a burst time. Considering k is the instantaneous frequency of $S_d(\omega_k,i)$ in Eq. (3.51) at $\omega_k = -4\pi\mu R_\Delta/c$, when $\omega_k = -4\pi\mu R_\Delta/c$, is an approximate LFM signal. Taking time-frequency analysis to $S_d(\omega_k,i)$, it will appear as a segment of straight line on the time-frequency plane.

After taking short-time Fourier transform (STFT) to $S_d(\omega_k, i)|_{\omega_k=-4\pi\mu R_\Delta/c}$ of each burst, the corresponding time-frequency planes can be obtained, and a new frequency—slow time plane can be constructed by stitching up these planes in order. On that new plane, the micro-Doppler signatures of rotating or vibrating scatterers also appear as the curves depicted by Eq. (3.63), but the slow time sampling number in a cycle has increased to N times. Therefore, the HT can be carried out to extract the micro-Doppler signatures accurately.

The previous algorithm is just for the micro-Doppler signature extraction of rotation and vibration. Nevertheless, if other micromotions can be depicted by certain analytical expressions, one can also extract their micro-Doppler signatures by constructing respective HT equations.

5.2.3.2 Simulation

In this section, we verify the theoretical derivation and the proposed micro-Doppler extraction method using simulated data. The parameters of transmitted radar signal are: $f_0 = 10$ GHz, $T_r = 0.78125$ μs, $\Delta f = 4.6875$ MHz, $N = 64$, BRF = 200 Hz, the total bandwidth $B = 300$ MHz, and the high range resolution $\Delta_R = 0.5$ m.

1. Detect curves using HT on range—slow time plane directly under the most common conditions

Assume two scatterers are both rotating around one center with the distance from the reference point $R_Q = -3$ m. The radii are both 2 m, and the rotating rates are 2 Hz and 8 Hz, respectively. The corresponding curves on range—slow time plane have been shown in Fig. 3.16B. Extract the skeleton for Fig. 3.16B, and two parameter pairs $(R_Q, A, \Omega_0, \zeta)$ detected by HT are shown in Table 5.1. It can be found that the rotating rates of the two scatterers are 2 and 8 Hz, and the amplitude magnification coefficients are 2.3209 and 8.4371 through calculating by $\sqrt{1 + \Omega^2 f_0^2 T_r^2/\Delta f^2}$, respectively. $R_Q = -3$ m is also detected accurately. The value of A of $\Omega_0 = 2$ Hz and $\Omega_0 = 8$ Hz are 4.6 and 17 m, respectively. The radii of rotating scatterers are 2.0466 and 2.0149 m

Table 5.1 The Detected Curve Parameters on Range–Slow Time Plane

Number	R_Q	A	Ω_0	ζ
1	−3	4.6	2	2.6
2	−3	17	8	2.9

by the equation: $\rho = A\Big/\sqrt{1+\omega_0^2 f_0^2 T_r^2/\Delta f^2}$, which agree quite well with the set value of 2 m.

2. Detect curves when $A < c/(4N\Delta f)$

Assume two scatterers rotate around the center with a distance from the reference point $R_Q = -3$ m, the radii are both 0.05 m, and the rotating rates are 2 Hz and 4 Hz, respectively. It can be calculated that $A_1 = 0.1160 < c/(4N\Delta f) = 0.25$, $A_2 = 0.2153 < c/(4N\Delta f) = 0.25$. Because the amplitudes are smaller than half range resolution, the curves appear to be approximately a straight line, as shown in Fig. 5.13A. Extracting the row of straight line, and then implementing Gabor transform to the row, the sinusoidal curves are obtained as shown in Fig. 5.13B. Then, the HT can be implemented to obtain the parameters (A',Ω_0,ζ), and the results are shown in Table 5.2. It can be observed that the rotating rates are detected accurately as $\Omega_0 = 2$ Hz and 4 Hz. The average values of parameter A' are about 43.33 and 84.75, respectively. According to Eq. (5.42), the rotating radii are calculated as $\rho = A'c/(2f_0 \cdot 2\pi\Omega_0) = 0.0517$ m and 0.0506 m, which are quite close to the set value.

3. Detect curves when rotating or vibrating rate is extraordinarily high

Assume two scatterers both rotate around one center with the distance from the reference point $R_Q = -3$ m, the radii are both $\rho = 1$ m, and the rotating rates are the same at 30 Hz. The curves on the range–slow time plane are shown in Fig. 5.14A. The

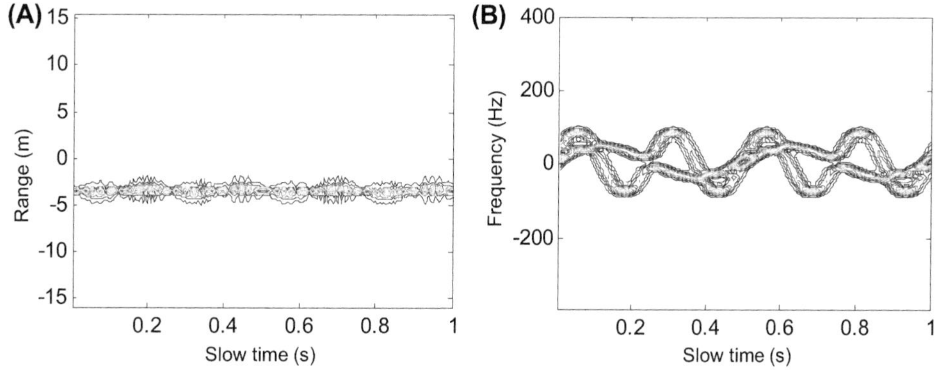

Figure 5.13 The micromotion characteristics of rotation scatterer when $A < c/(4N\Delta f)$. (A) Range–slow time image; (B) the result of time-frequency analysis.

Table 5.2 The Curve Parameters When $A < c/(4N\Delta f)$

Index	**1**	**2**	**3**	**4**	**5**	**6**	**7**
A'	41	44	45	85	84	83	87
Ω_0	2	2	2	4	4	4	4
ζ	0.1	0.0	0.0	0.0	0.1	0.0	0.1

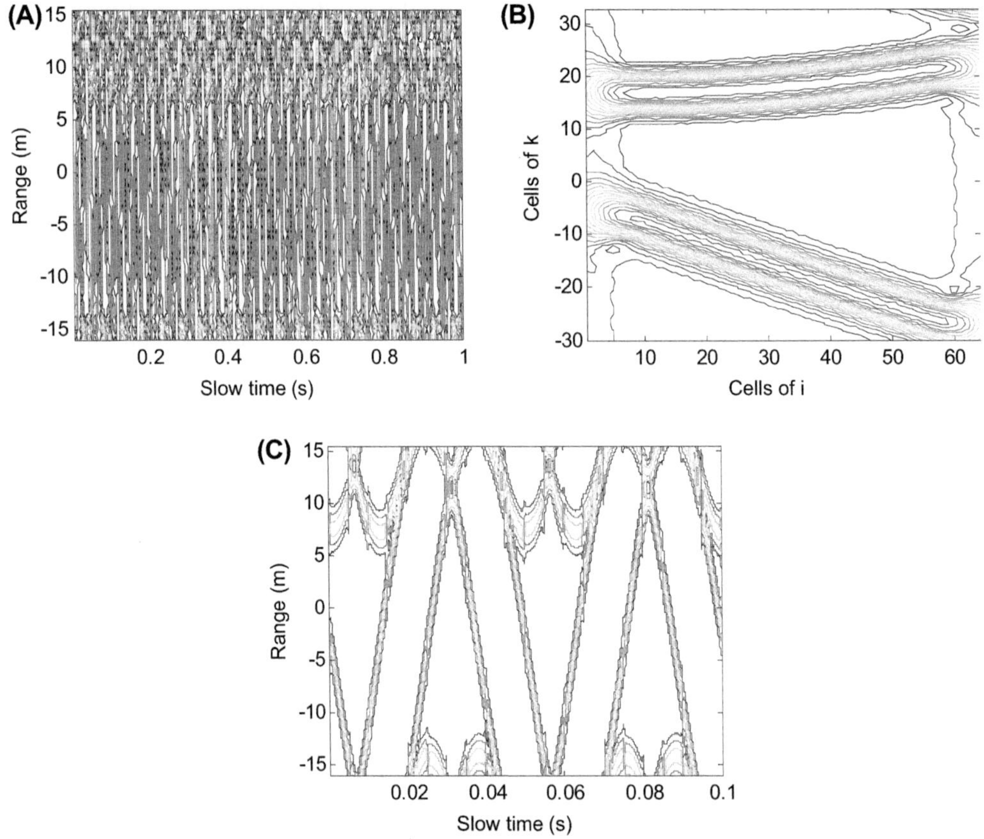

Figure 5.14 Rotation scatterer micro-Doppler characteristics with high rotation frequency. (A) Range–slow time image; (B) time-frequency analysis result of high-resolution image; (C) assembling result of time-frequency image.

curves are very blurry because of the high rotating rates; as a result, the parameters of the curves cannot be detected exactly by the HT from the range–slow time plane. Fig. 5.14B shows the STFT result. Stitching up the planes of all the bursts in order, a new range–slow time plane can be constructed, as shown in Fig. 5.14C. For the clearness of the resultant figure and convenience for comprehension, Fig. 5.14C just presents a part of the new plane from 0 to 0.1 s in the slow time, and the frequency-axis is transformed

Table 5.3 Curve Parameter Extracted in Time-Frequency Plane With High Rotation Frequency

Index	1	2	3	4
R_Q	−3	−3.5	−3	−2.5
A	31.5	32	32	32
Ω_0	30	30	30	30
ζ	3.1	3.1	3.1	3.1

to the range-axis for the consistency. Conduct HT to new image and the detected parameter pairs (R_Q,A,Ω_0,ζ) are shown in Table 5.3. The value of $\Omega_0 = 30$ Hz is detected accurately, and the amplitude magnification coefficient is 31.4318. $A = 32$, and the radius is 1.0181 m.

5.2.3.3 Robustness Discussion

In actual situations, the micromotion parts are always masked by the target's body. As a result, the induced micro-Doppler signals are contaminated by the target body's returns. In the processing of micro-Doppler feature extraction, these body returns must be considered as "noise" because they will bring in mistakes. Meanwhile, for simplicity, we have assumed the target consists of point scatterers with unit scattering coefficients. However, for a rotating target with large radius or high frequency in real-world situations, it is highly possible that the back-scattering coefficients have time-varying complex values [7]. In what follows, we further study the performance of the proposed algorithm in these situations.

1. Signal to noise

In order to extract micro-Doppler feature from the echo of target, we must restrain the echo of target and increase signal to noise (SNR) of micro-Doppler signal. After sufficiently accurate translation compensation, the body scatterers can be considered as micromotion scatterers with approximately zero rotating frequency, therefore, the corresponding curves on the range—slow time plane or the time-frequency plane appear to be straight lines. According to the difference between the micro-Doppler curves and the body's straight lines, a primary and simple noise suppression method is designed as follows: assuming the data matrix on the range—slow time plane or the frequency—slow time plane is $\mathbf{M}$ with $X \times Y$ points, the straight line removing processing can be expressed as:

$$\mathbf{M}'(x,y) = |\mathbf{M}(x,y)| - \frac{1}{Y}\left(\sum_{y=1}^{Y}|\mathbf{M}(x,y)|\right)\cdot \mathbf{M}_Y \tag{5.46}$$

where $\mathbf{M}_Y$ is a $1 \times Y$ vector with all elements equal to one. Then, the HT can be implemented to $\boldsymbol{M}'(x,y)$ for micro-Doppler feature extraction.

As shown in Fig. 5.15A, the simulation conditions are the same with those in 1 of Section 5.2.3.2, and the body's returns are added with SNR = −20 dB. It can be found that the micro-Doppler curves are relatively weak and the HT may have errors when detecting the parameter pairs of curves. The result of the noise suppression processing is shown in Fig. 5.16B. Applying the HT to this figure, the parameter pairs can be detected accurately. However, the micro-Doppler extraction algorithm presented herewith cannot work effectively when the noise is overwhelming. As shown in Fig. 5.16, when the SNR decreases to −30 dB, the noise suppression does not work effectively so that the curve with frequency 8 Hz cannot be detected by HT anymore.

When $A < c/(4N\Delta f)$ or the rotating rate is extraordinarily high, the Gabor transform or the STFT needs to be implemented before the HT. In these cases, the body's returns

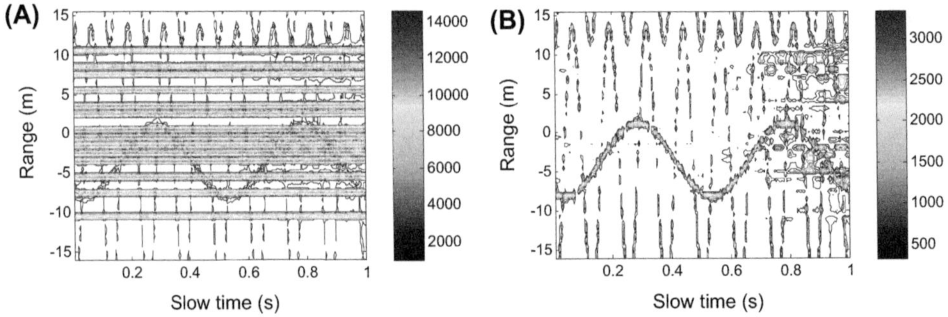

Figure 5.15 The noise suppression processing when SNR = −20 dB. (A) The original range–slow time image; (B) the result after the noise suppression processing.

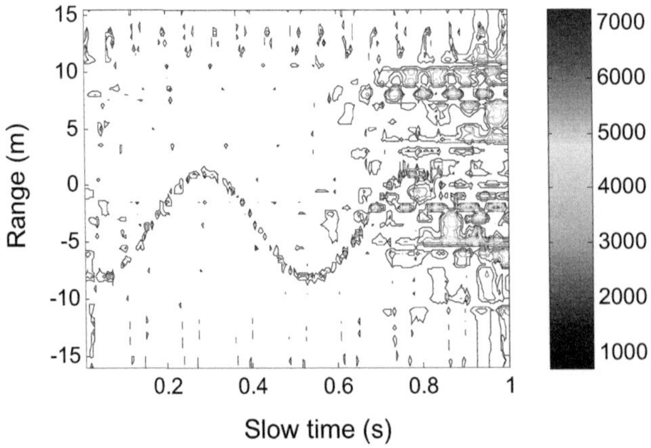

Figure 5.16 The noise suppression processing when SNR = −30 dB.

will bring in adverse effects because of the limitation of the time-frequency resolution. Therefore, the antinoise ability of the algorithm will be degraded to some extent. Especially for STFT, the multicomponent signal of the body's returns will seriously contaminate the micro-Doppler curves on the time-frequency plane.

Fig. 5.17 shows the robustness of the proposed algorithm when $A < c/(4N\Delta f)$ and SNR $= -16$ dB. The results of Gabor transform and the noise suppression processing are shown in Fig. 5.17A and B, respectively, and the HT can detect the parameters of the curves accurately.

Fig. 5.18 shows the robustness of the algorithm when rotating rate is extraordinarily high and SNR $= -10$ dB. Fig. 5.18A shows the result on the range—slow time plane, Fig. 5.18B shows the STFT result, and Fig. 5.18C shows the result after the noise suppression processing. It can be learned that the efficiency of algorithm is affected because micro-Doppler feature curve is polluted.

2. Time-varying complex scattering coefficients

It is highly possible that the scattering coefficients are time-varying complex values if the rotating scatterers have large radii or high frequencies. When the rotating scatterers have large radii, we detect the curves from the range—slow time plane directly; therefore, the time-varying complex scattering coefficients just lead to the varying moduli of the curves. When the rotating rates of scatterers are extraordinarily high, we detect the curves on the frequency (range)—slow time plane after taking STFT to each burst and stitching up the time-frequency planes in order, and in each burst time, the scattering coefficients can be considered as constants approximately. In such case, the time-varying complex scattering coefficients also just lead to the varying moduli of the curves on the range—slow time plane. Due to the robustness of the HT to the discontinuities of the pattern, the adverse effects of the curves' varying moduli are very limited.

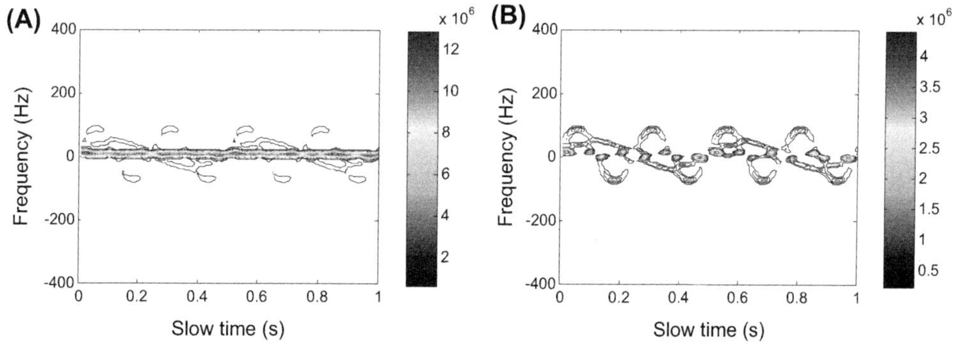

Figure 5.17 The result of the noise suppression processing when $A < c/(4N\Delta f)$ and SNR $= -16$ dB. (A) The result of Gabor transform; (B) the result of the noise suppression processing.

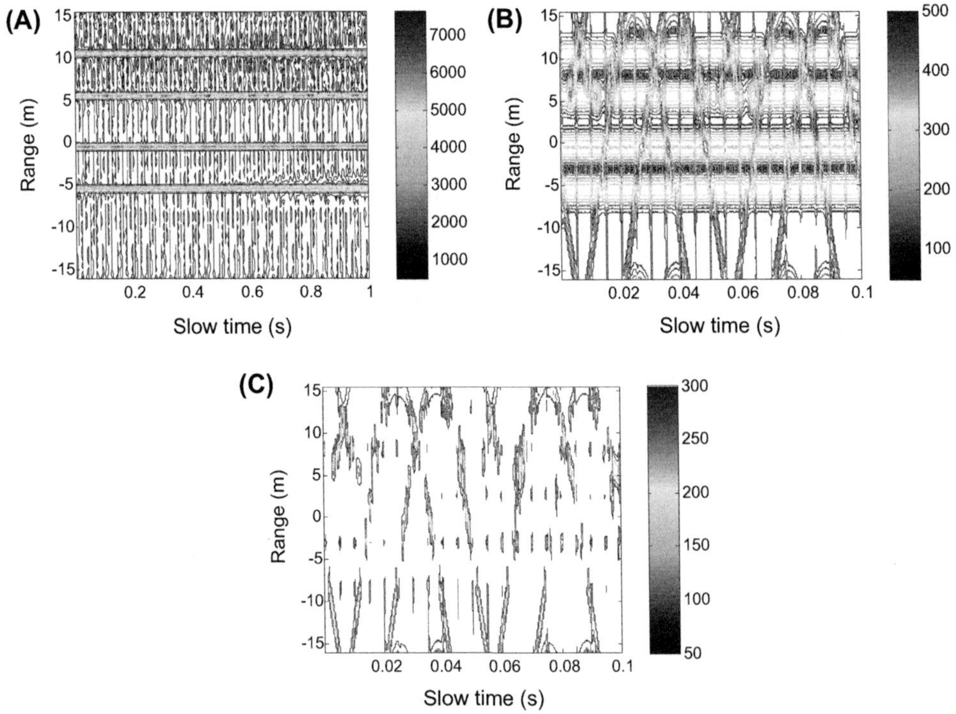

Figure 5.18 The result of the noise suppression processing when rotating frequencies of scatterers are high and SNR = −10 dB. (A) The original range–slow time image; (B) the result after jointing time-frequency images; (C) the result of noise suppression processing.

Fig. 5.19 shows the curves on the range–slow time plane. The moduli and phases of scattering coefficients randomly vary within [0,1] and [0,2π], respectively. Fig. 5.20 shows the curves on the range–slow time plane. The moduli and phases of scattering coefficients also randomly vary within [0,1] and [0,2π], respectively. It can be learned that time-varying complex scattering coefficients mainly cause the amplitude fluctuation of micro-Doppler feature curves. In this case, the parameter pairs of the curves can also be accurately detected by HT.

5.3 ORTHOGONAL MATCHING PURSUIT DECOMPOSITION METHOD

The signal decomposition and reconstruction are very important ways for signal analysis. There are many methods for signal decomposition and reconstruction, such as the Fourier transform, Wavelet transform, discrete cosine transform (DCT), matching

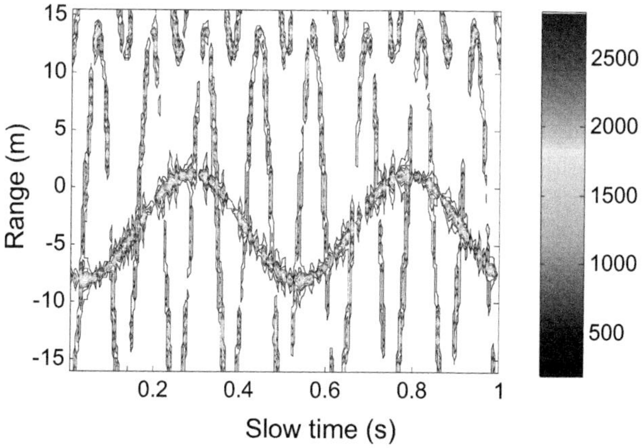

Figure 5.19 The impact for range–slow time image when scattering coefficients are time-varying complex.

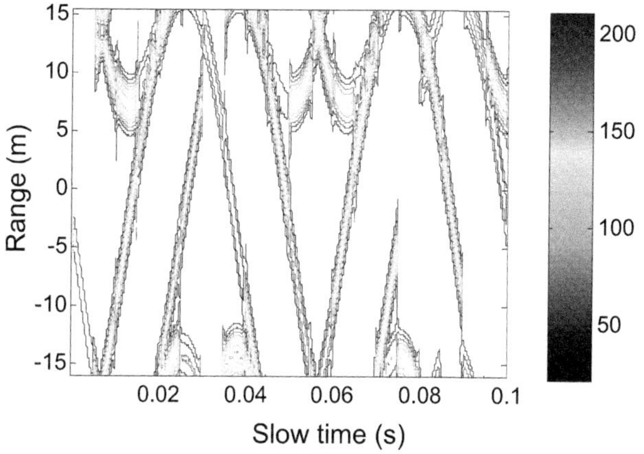

Figure 5.20 The impact for jointing time-frequency images when scattering coefficients are time-varying complex.

pursuit (MP) decomposition, etc. Compared with the other methods, MP decomposition, which is proposed by Mallat and Zhang in Ref. [8], is a much more flexible method for decomposing a signal into a linear expansion of waveforms that are selected from a redundant dictionary of functions. As a refinement of MP, the orthogonal matching pursuit (OMP) algorithm is proposed by Y.C. Pati et al. [9]. By maintaining full backward orthogonality of the residual at every step, OMP converges faster than MP when utilizing nonorthogonal dictionaries. Due to its advantages in adaptive signal representing, OMP has been widely applied in signal processing, especially in sparse signal processing [10].

We intend to utilize OMP to extract the micro-Doppler features in this section. According to the principle of OMP algorithm, the atoms in the dictionary should be designed corresponding to the intrinsic characteristics of the signal to be decomposed. In the following, the applications of OMP algorithm for micro-Doppler feature extraction in narrowband radar and wideband radar are expatiated, respectively.

5.3.1 Application in Narrowband Radar

From Eq. (2.24), it can be found that the micro-Doppler signal in narrowband radar echoed by a rotating scatterer possesses a sinusoidal frequency modulated (SFM) form. Therefore, the SFM atoms are suitable for constructing the dictionary $\mathbf{D} = \{\boldsymbol{d}_i\}$ for OMP decomposition, where the i-th SFM atom can be expressed as

$$\boldsymbol{d}_i = \exp\left(-\mathrm{j}2\pi \widehat{f}_i t - \mathrm{j}2\pi \widehat{r}_i \sin\left(\widehat{\Omega}_i t + \widehat{\theta}_i \right) \right) \tag{5.47}$$

where $\widehat{f}_i$ denotes the carrier frequency with expected range of $\left[\widehat{f}_{\min}, \widehat{f}_{\max}\right]$, $\widehat{r}_i$ denotes the amplitude with expected range of $\left[\widehat{r}_{\min}, \widehat{r}_{\max}\right]$, $\widehat{\Omega}_i$ denotes the rotating frequency with expected range of $\left[\widehat{\Omega}_{\min}, \widehat{\Omega}_{\max}\right]$, and $\widehat{\theta}_i$ denotes the initial phase with expected range of $\left[\widehat{\theta}_{\min}, \widehat{\theta}_{\max}\right]$. Then normalize the power of the atom:

$$\boldsymbol{d}_i \leftarrow \boldsymbol{d}_i / \|\boldsymbol{d}_i\| \tag{5.48}$$

Let the vector $\boldsymbol{S}$ denote the target echo signal after digital sampling, then the implementation of micro-Doppler feature extraction based on OMP algorithm in narrowband radar is as follows:

Step (1) Initialize the residual $\boldsymbol{R}_{S0} = \boldsymbol{S}$, the index set $\varsigma_0 = [\,]$, the matched atoms set $\mathbf{D}_0 = \varnothing$, the matched atoms recorder matrix $\boldsymbol{\Pi}_0 = [\,]$, the power threshold $\delta > 0$, the iteration counter $g = 1$, and the maximum iteration number G;

Step (2) Calculate $\{\langle \boldsymbol{R}_{S g-1}, \boldsymbol{d}_i \rangle; \boldsymbol{d}_i \in \mathbf{D} \setminus \mathbf{D}_{g-1}\}$;

Step (3) Find $\boldsymbol{d}_{i'} \in \mathbf{D} \setminus \mathbf{D}_{g-1}$ such that

$$|\langle \boldsymbol{R}_{S g-1}, \boldsymbol{d}_{i'} \rangle| \geq \alpha \sup_i |\langle \boldsymbol{R}_{S g-1}, \boldsymbol{d}_i \rangle| \tag{5.49}$$

where $\boldsymbol{d}_i \in \mathbf{D} \setminus \mathbf{D}_{g-1}$ and $0 < \alpha \leq 1$;

Step (4) If $|\langle \boldsymbol{R}_{S g-1}, \boldsymbol{d}_{i'} \rangle| < \delta$ then stop;

Step (5) Record the index of $\boldsymbol{d}_{i'}$ in the dictionary $\mathbf{D}$: $\varsigma_g = [\varsigma_{g-1} \quad i']$; then set $\mathbf{D}_g = \mathbf{D}_{g-1} \cup \{\boldsymbol{d}_{i'}\}$ and $\boldsymbol{\Pi}_g = [\boldsymbol{\Pi}_{g-1}, \boldsymbol{d}_{i'}]$;

Step (6) Solve $\widehat{\boldsymbol{x}} = \arg\min_{\boldsymbol{x}} \|\boldsymbol{S} - \boldsymbol{\Pi}_g \boldsymbol{x}\|_2$. From least-square method, it yields $\widehat{\boldsymbol{x}} = \left(\boldsymbol{\Pi}_g^{\mathrm{H}} \boldsymbol{\Pi}_g \right)^{-1} \boldsymbol{\Pi}_g^{\mathrm{H}} \boldsymbol{S}$;

Step (7) Update the residual $\boldsymbol{R}_{S g} = \boldsymbol{S} - \boldsymbol{\Pi}_g \widehat{\boldsymbol{x}}$; set $g \leftarrow g + 1$. If $g \leq G$, repeat Steps (2)–(7); if $g > G$, then stop.

After the iteration, the micro-Doppler signal $\boldsymbol{S}$ is decomposed as follows

$$\boldsymbol{S} = \boldsymbol{\Pi}_g \widehat{\boldsymbol{x}} + \boldsymbol{R}_{Sg} \tag{5.50}$$

where the parameters $\left(\widehat{f}_i, \widehat{r}_i, \widehat{\Omega}_i, \widehat{\theta}_i\right)$ of the atoms in $\mathbf{D}_g$ indicate the micro-Doppler feature parameters of the target.

5.3.2 Application in Wideband Radar

5.3.2.1 Algorithm

In wideband radar, it is known that, if the amplitude of micromotion is close to or even less than the range resolution, and if the micromotional scatterers do not migrate through range cells during the imaging time, then the micro-Doppler features can be obtained by analyzing the echoes from the range cells where micromotional scatterers are located [7]. In this case, the micro-Doppler feature extraction algorithms in Section 5.3.1 can be directly applied to wideband radar systems. When the amplitude of micromotion is much larger than the range resolution, the echo of the micromotional scatterer is distributed in several range cells. As a result, when utilizing OMP decomposition, a new dictionary is required.

According to the analyses in Chapter 3, the micro-Doppler signal in wideband radar echoed by a rotating scatterer can be described by Eqs. (3.27) and (3.29). From Eq. (3.29), it can be found that $R_\Delta(t_m)$ consists of two terms: $\left\|\overrightarrow{GO''}\right\|$ and $r\cos(\Omega t_m + \theta)\sin\varepsilon$. When only using the module information of Eq. (3.27) to extract micro-Doppler features (such as the imaging processing algorithms in Section 5.2), $\left\|\overrightarrow{GO''}\right\|$ can be considered as a constant approximately if the rotation center scatterer O'' does not migrate through range cells during the imaging time. As a result, the echo signal of a micromotional scatterer appears as a sinusoidal curve on the range—slow time domain. However, when using OMP decomposition for micro-Doppler feature extraction, both the module and phase information of Eq. (3.27) are important. Therefore, $\left\|\overrightarrow{GO''}\right\|$ cannot be considered as a constant in this case.

As shown in Fig. 5.21, $OXYZ$ is the global coordinate system, $oxyz$ is the local coordinate system of the target with translations with respect to the global coordinates. The radar is located at the origin O of the global coordinate system. Assume the target translates from $oxyz$ to $o'x'y'z'$ with velocity $\boldsymbol{v}$. Simultaneously, a scatterer P on the target rotates around the center C with angular velocity $\boldsymbol{\omega} = (\omega_x, \omega_y, \omega_z)^{\mathrm{T}}$, where the superscript "T" means the transpose of a vector or a matrix. Let $\Omega = \|\boldsymbol{\omega}\|$. The coordinates of point o in the global coordinate system $OXYZ$ at the initial time is (X_o, Y_o, Z_o), the coordinates of point C in the local coordinate system $oxyz$ at the initial time are

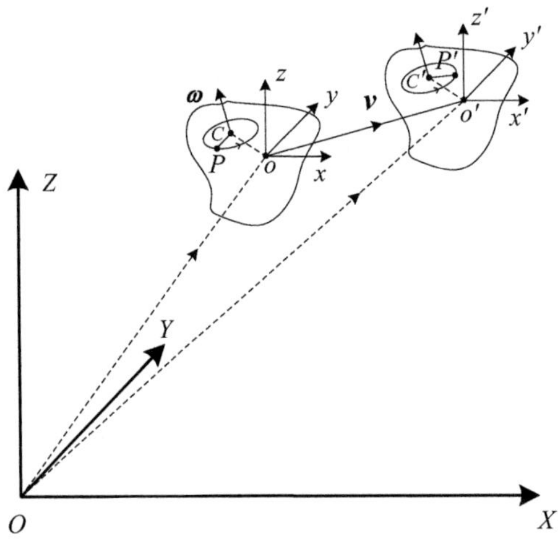

Figure 5.21 Geometry of a radar and a target with rotating parts.

(x_C, y_C, z_C), and the target velocity $\boldsymbol{v}$ is expressed as $(v_X, v_Y, v_Z)^T$ in the $OXYZ$ coordinate system. According to the geometry, one has

$$R_O(t_m) = \left\| \overrightarrow{Oo} + \boldsymbol{v} t_m \right\| = \sqrt{(X_o + v_X t_m)^2 + (Y_o + v_Y t_m)^2 + (Z_o + v_Z t_m)^2} \quad (5.51)$$

Expanding Eq. (5.51) in Taylor's series and neglecting the higher order terms, $R_O(t_m)$ can be represented as

$$R_O(t_m) = R_O(0) + \frac{X_o v_X + Y_o v_Y + Z_o v_Z}{\sqrt{X_o^2 + Y_o^2 + Z_o^2}} t_m \quad (5.52)$$

Let $R_C(t_m)$ be the distance between the rotation center C' and radar at slow time t_m, we have

$$\begin{aligned} R_C(t_m) &= \left\| \overrightarrow{Oo} + \overrightarrow{oC} + \boldsymbol{v} t_m \right\| \\ &= \sqrt{(X_o + x_C + v_X t_m)^2 + (Y_o + y_C + v_Y t_m)^2 + (Z_o + z_C + v_Z t_m)^2} \\ &\approx R_C(0) + \frac{(X_o + x_C) v_X + (Y_o + y_C) v_Y + (Z_o + z_C) v_Z}{\sqrt{(X_o + x_C)^2 + (Y_o + y_C)^2 + (Z_o + z_C)^2}} t_m \end{aligned} \quad (5.53)$$

When the scatterer P is rotating around the center C, the geometry of the target and the radar at slow time t_m in Fig. 5.21 can be redrawn as shown in Fig. 5.22. The radar is located at O in the far field, and the lengths of segments $\overline{OC'}$ and $\overline{OP'}$ are $R_C(t_m)$ and $R_P(t_m)$, respectively. $\boldsymbol{n}$ is the unit vector of the LOS and $\boldsymbol{n}'$ is the projection of $\boldsymbol{n}$ onto the

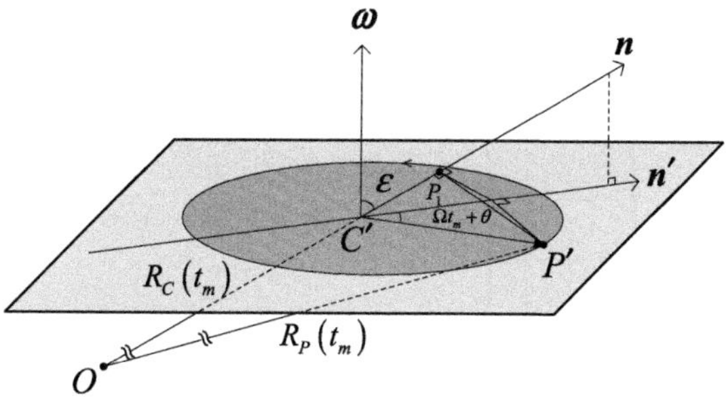

Figure 5.22 Geometry of the rotating target and radar.

plane spanned by the trajectory of scatterer P. Because $\left\|\overrightarrow{C'P'}\right\| \le \left\|\overrightarrow{OP'}\right\|$ and $\left\|\overrightarrow{C'P'}\right\| \le \left\|\overrightarrow{OC'}\right\|$, it yields

$$\begin{aligned} R_P(t_m) &= \left\|\overrightarrow{OC'} + \overrightarrow{C'P'}\right\| \approx \left\|\overrightarrow{OC'}\right\| + \left\|\overrightarrow{C'P_1}\right\| \\ &= R_C(t_m) + \left\|\overrightarrow{C'P'}\right\|\cos(\Omega t_m + \theta)\sin \boldsymbol{\varepsilon} \end{aligned} \tag{5.54}$$

where P_1 is the projection of P' onto the vector $\boldsymbol{n}$, $\boldsymbol{\varepsilon}$ is the angle between $\boldsymbol{n}$ and $\boldsymbol{\omega}$, and θ is the initial phase term. Therefore, it can be calculated that

$$\begin{aligned} \Delta R(t_m) &= R_P(t_m) - R_O(t_m) \\ &= R_C(0) + \frac{(X_o + x_C)v_X + (Y_o + y_C)v_Y + (Z_o + z_C)v_Z}{\sqrt{(X_o + x_C)^2 + (Y_o + y_C)^2 + (Z_o + z_C)^2}} t_m \\ &\quad + \left\|\overrightarrow{C'P'}\right\|\cos(\Omega t_m + \theta)\sin \boldsymbol{\varepsilon} - R_O(0) - \frac{X_o v_X + Y_o v_Y + Z_o v_Z}{\sqrt{X_o^2 + Y_o^2 + Z_o^2}} t_m \\ &\approx (R_C(0) - R_O(0)) + \frac{x_C v_X + y_C v_Y + z_C v_Z}{\sqrt{X_o^2 + Y_o^2 + Z_o^2}} t_m + \left\|\overrightarrow{C'P'}\right\|\cos(\Omega t_m + \theta)\sin \boldsymbol{\varepsilon} \end{aligned} \tag{5.55}$$

It can be found from Eq. (5.55) that $\Delta R(t_m)$ consists of three terms: the first term $(R_C(0) - R_O(0))$ is a constant; the second term is a first-order term with respect to t_m, which is induced by the movement of the target body, and in fact, this term is crucial for focusing the azimuthal image of C in ISAR imaging; the last term is a sinusoidal function with respect to t_m, which is induced by the micromotion of P.

Substitute Eq. (5.55) to Eq. (3.27) and assume scatterer P with a unit scattering coefficient, then it yields

$$S_d(f_k, t_m) = T_p \mathrm{sinc}\left(T_p \left(f_k + \frac{2\mu}{c} \left(\begin{array}{l} R_C(0) - R_O(0) + \dfrac{x_C v_X + y_C v_Y + z_C v_Z}{\sqrt{X_o^2 + Y_o^2 + Z_o^2}} t_m \\ + \left\| \overrightarrow{C'P'} \right\| \cos(\Omega t_m + \theta) \sin \varepsilon \end{array} \right) \right) \right)$$

$$\cdot \exp\left(-\mathrm{j}\frac{4\pi}{\lambda} \left(\begin{array}{l} R_C(0) - R_O(0) + \dfrac{x_C v_X + y_C v_Y + z_C v_Z}{\sqrt{X_o^2 + Y_o^2 + Z_o^2}} t_m \\ + \left\| \overrightarrow{C'P'} \right\| \cos(\Omega t_m + \theta) \sin \varepsilon \end{array} \right) \right) \tag{5.56}$$

Eq. (5.56) depicts the range profiles variation with respect to t_m. If the micromotion of P is limited within a range cell, $|S_d(f, t_m)|$ can be considered as independent of t_m. Then, for a given value of f, taking the derivative to the phase term on the right side of Eq. (5.56) in terms of t_m, the instantaneous frequency is obtained as

$$f_d(t_m) = -\frac{2}{\lambda} \cdot \frac{x_C v_X + y_C v_Y + z_C v_Z}{\sqrt{X_o^2 + Y_o^2 + Z_o^2}} + \frac{2\Omega \left\| \overrightarrow{C'P'} \right\| \sin \varepsilon}{\lambda} \sin(\Omega t_m + \theta) \tag{5.57}$$

where the first term is the target body's Doppler frequency and the second term is the micro-Doppler frequency induced by the micromotion. Let

$$\hat{d} = R_C(0) - R_O(0),\ \hat{r} = \left\| \overrightarrow{C'P'} \right\| \sin \varepsilon,\ \hat{\Omega} = \Omega,\ \hat{\theta} = \theta,\ \hat{f} = \frac{x_C v_X + y_C v_Y + z_C v_Z}{\sqrt{X_o^2 + Y_o^2 + Z_o^2}} \tag{5.58}$$

Then Eq. (5.56) is rewritten as

$$S_d(f_k, t_m) = T_p \mathrm{sinc}\left(T_p \left(f_k + \frac{2\mu}{c} \left(\hat{d} + \hat{f} t_m + \hat{r} \cos\left(\hat{\Omega} t_m + \hat{\theta} \right) \right) \right) \right) \cdot \exp\left(-\mathrm{j}\frac{4\pi}{\lambda} \left(\hat{d} + \hat{f} t_m + \hat{r} \cos\left(\hat{\Omega} t_m + \hat{\theta} \right) \right) \right) \tag{5.59}$$

When $\hat{r} = 0$, Eq. (5.59) also denotes the echo of target body scatterers.

In digital signal processing, $S_d(f_k,t_m)$ is sampled to be a matrix. Assuming there are N_A and N_B samples in f_k domain and t_m domain, respectively, $S_d(f_k,t_m)$ is expressed as a $N_A \times N_B$ matrix:

$$\mathbf{S} = \left(s_{a,b}\right) = \left(S_d(f_k, t_m)\right) \tag{5.60}$$

When the target echoes consist of micromotional parts echo and body echo, it is better to eliminate the body echo first. From Eq. (5.59), it can be found that the body echo in t_m domain is a single-frequency signal with carrier frequency $-2\widehat{f}/\lambda$ Therefore, a dictionary $\mathbf{A}_b$ can be constructed according to Eq. (5.59), where the l-th atom in the dictionary can be expressed as

$$\mathbf{A}_b^{(l)} = \left[A_b^{(l)}(f_k, t_m)\right] = \text{sinc}\left(T_p\left(f_k + \frac{2\mu}{c}\left(\widehat{d}_l + \widehat{f}_l t_m\right)\right)\right)\exp\left(-\mathrm{j}\frac{4\pi}{\lambda}\left(\widehat{d}_l + \widehat{f}_l t_m\right)\right) \tag{5.61}$$

where the superscript (l) denotes the index of atoms in the dictionary. And then normalize the power of the atom: $\mathbf{A}_b^{(l)} \leftarrow \mathbf{A}_b^{(l)} \Big/ \left\|\mathbf{A}_b^{(l)}\right\|_F$, where $\|\cdot\|_F$ denotes the Frobenius norm of matrix. Let $\widehat{d}_l \in \left[\widehat{d}_{\min}, \widehat{d}_{\max}\right]$ and $\widehat{f}_l \in \left[\widehat{f}_{\min}, \widehat{f}_{\max}\right]$, and the sampling intervals are $\Delta\widehat{d}$ and $\Delta\widehat{f}$, respectively. The number of atoms in the dictionary $\mathbf{A}_b$ is

$$K_b = \left(\left\lfloor\frac{\widehat{d}_{\max} - \widehat{d}_{\min}}{\Delta\widehat{d}}\right\rfloor + 1\right)\left(\left\lfloor\frac{\widehat{f}_{\max} - \widehat{f}_{\min}}{\Delta\widehat{f}}\right\rfloor + 1\right) \tag{5.62}$$

Generally, compared with the micro-Doppler frequency, the target body's Doppler frequency is much smaller. Considering a general aerial target, it can be calculated $\widehat{f} = 0.0493$ when setting $(x_C,y_C,z_C) = (5\text{m},5\text{m},5\text{m})$, $(X_o,Y_o,Z_o) = (30\text{km},0\text{km},5\text{km})$ and $\boldsymbol{v} = (0,300,0)^{\mathrm{T}}$ m/s. Even for a space target with high translational speed, the value of $\widehat{f}$ is also small. For example, when setting $(x_C,y_C,z_C) = (5\text{m},5\text{m},5\text{m})$, $(X_o,Y_o,Z_o) = (300\text{km},0\text{km},500\text{km})$ and $\boldsymbol{v} = (0,7800,0)^{\mathrm{T}}$ m/s, it can be calculated that $\widehat{f} = 0.0669$. Therefore, compared with $\widehat{d}$ and $\widehat{r}$, $\widehat{f}$ can be approximate to zero during the imaging time. Eq. (5.59) is then approximate to

$$\begin{aligned} S_d(f_k, t_m) \approx\ & T_p\text{sinc}\left(T_p\left(f_k + \frac{2\mu}{c}\left(\widehat{d} + \widehat{r}\cos\left(\widehat{\Omega}t_m + \widehat{\theta}\right)\right)\right)\right) \\ & \times \exp\left(-\mathrm{j}\frac{4\pi}{\lambda}\left(\widehat{d} + \widehat{r}\cos\left(\widehat{\Omega}t_m + \widehat{\theta}\right)\right)\right) \end{aligned} \tag{5.63}$$

It can be found that the range—slow time image $S_d(f_k,t_m)$ is determined by $\left(\widehat{d}, \widehat{r}, \widehat{\Omega}, \widehat{\theta}\right)$. In order to decompose the micro-Doppler signal by OMP, a dictionary $\mathbf{A}_m$ can be constructed by setting values of $\left(\widehat{d},\widehat{r},\widehat{\Omega},\widehat{\theta}\right)$ in atoms different to each other. Let $\widehat{d}_l \in \left[\widehat{d}_{\min},\widehat{d}_{\max}\right]$, $\widehat{r}_l \in \left[\widehat{r}_{\min},\widehat{r}_{\max}\right]$, $\widehat{\Omega}_l \in \left[\widehat{\Omega}_{\min},\widehat{\Omega}_{\max}\right]$, $\widehat{\theta}_l \in \left[\widehat{\theta}_{\min},\widehat{\theta}_{\max}\right]$, and the sampling intervals of $\left(\widehat{d},\widehat{r},\widehat{\Omega},\widehat{\theta}\right)$ are $\Delta\widehat{d}$, $\Delta\widehat{r}$, $\Delta\widehat{\Omega}$, and $\Delta\widehat{\theta}$, respectively. The number of atoms in the dictionary is

$$K_m = \left(\left\lfloor\frac{\widehat{d}_{\max}-\widehat{d}_{\min}}{\Delta\widehat{d}}\right\rfloor+1\right)\left(\left\lfloor\frac{\widehat{r}_{\max}-\widehat{r}_{\min}}{\Delta\widehat{r}}\right\rfloor+1\right)\left(\left\lfloor\frac{\widehat{\Omega}_{\max}-\widehat{\Omega}_{\min}}{\Delta\widehat{\Omega}}\right\rfloor+1\right) \times \left(\left\lfloor\frac{\widehat{\theta}_{\max}-\widehat{\theta}_{\min}}{\Delta\widehat{\theta}}\right\rfloor+1\right) \tag{5.64}$$

The l-th atom in the dictionary can be expressed as

$$\mathbf{A}_m^{(l)} = \left[A_m^{(l)}(f_k,t_m)\right] = \operatorname{sinc}\left(T_p\left(f_k+\frac{2\mu}{c}\left(\widehat{d}_l+\widehat{r}_l\cos\left(\widehat{\Omega}_l t_m+\widehat{\theta}_l\right)\right)\right)\right) \times \exp\left(-\mathrm{j}\frac{4\pi}{\lambda}\left(\widehat{d}_l+\widehat{r}_l\cos\left(\widehat{\Omega}_l t_m+\widehat{\theta}_l\right)\right)\right) \tag{5.65}$$

When the target echo only contains the echo of micromotional scatterers, OMP decomposition with dictionary $\mathbf{A}_m$ can be directly utilized to extract micro-Doppler features; when the target echo contains both body echo and micromotional parts echo, OMP decomposition with dictionary $\mathbf{A}_b$ can be utilized to eliminate the body echo first, and then OMP decomposition with dictionary $\mathbf{A}_m$ can be used to extract micro-Doppler features.

It can be found from Eqs. (5.61) and (5.65) that each atom in $\mathbf{A}_b$ and $\mathbf{A}_m$ is in fact a complex image. In order to extend the OMP algorithm in vector space to the complex image space to decompose the radar echoes in range–slow time domain and to obtain the micro-Doppler features of the target, define the inner product operation in $C^{N_A\times N_B}$ space as

$$\langle\mathbf{U},\mathbf{V}\rangle = \sum_{a=1}^{N_A}\sum_{b=1}^{N_B} u_{a,b}v_{a,b}^* = tr\left(\mathbf{U}\mathbf{V}^{\mathrm{H}}\right) \tag{5.66}$$

where both $\mathbf{U}$ and $\mathbf{V}$ are $N_A \times N_B$ matrixes, and the superscript "H" denotes the conjugate transpose of a matrix. It is easy to obtain

$$\langle\mathbf{U},\mathbf{U}\rangle = \sum_{a=1}^{N_A}\sum_{b=1}^{N_B} u_{a,b}u_{a,b}^* = \|\mathbf{U}\|_F^2 \tag{5.67}$$

The complex image orthogonal matching pursuit decomposition algorithm for micro-Doppler feature extraction is implemented as follows:

Step (1) Initialize the residual $\mathbf{R_S}_0 = \mathbf{S}$, the index set $\varsigma_0 = [\,]$, the matched atoms recorder matrix $\mathbf{H}_0 = \varnothing$, the power threshold $\delta > 0$, the iteration counter $g = 1$, and the maximum iteration number G;

Step (2) Calculate $\left\{\left\langle \mathbf{R_S}_{g-1}, \mathbf{A}^{(l')}\right\rangle; \mathbf{A}^{(l')} \in \left\{\mathbf{A}^{(l)}\right\} \setminus \mathbf{H}_{g-1}\right\}$, where $\mathbf{A}$ denotes the dictionary;

Step (3) Find $\mathbf{A}^{(g')} \in \left\{\mathbf{A}^{(l)}\right\} \setminus \mathbf{H}_{g-1}$ such that

$$\left|\left\langle \mathbf{R_S}_{g-1}, \mathbf{A}^{(g')}\right\rangle\right| \geq \alpha \sup_{l'} \left|\left\langle \mathbf{R_S}_{g-1}, \mathbf{A}^{(l')}\right\rangle\right| \tag{5.68}$$

where $\mathbf{A}^{(l')} \in \left\{\mathbf{A}^{(l)}\right\} \setminus \mathbf{H}_{g-1}$ and $0 < \alpha \leq 1$;

Step (4) If $\left|\left\langle \mathbf{R_S}_{g-1}, \mathbf{A}^{(g')}\right\rangle\right| < \delta$ then stop;

Step (5) Record the index of $\mathbf{A}^{(g')}$ in the dictionary $\left\{\mathbf{A}^{(l)}\right\}$: $\varsigma_g = [\,\varsigma_{g-1} \quad g'\,]$; then set $\mathbf{h}_g = \mathbf{A}^{(g')}$ and $\mathbf{H}_g = \mathbf{H}_{g-1} \cup \{\mathbf{h}_g\}$;

Step (6) Solve $\boldsymbol{x} = (x_i; i = 1, 2, \cdots, g)$, satisfying

$$\min \left\| \mathbf{S} - \sum_{i=1}^{g} x_i \mathbf{h}_i \right\|_F^2 \tag{5.69}$$

From least-square method (please refer to Appendix 5-A), it yields

$$\boldsymbol{x} = \begin{bmatrix} \langle \mathbf{h}_1, \mathbf{h}_1 \rangle & \langle \mathbf{h}_2, \mathbf{h}_1 \rangle & \cdots & \langle \mathbf{h}_g, \mathbf{h}_1 \rangle \\ \langle \mathbf{h}_1, \mathbf{h}_2 \rangle & \langle \mathbf{h}_2, \mathbf{h}_2 \rangle & \cdots & \langle \mathbf{h}_g, \mathbf{h}_2 \rangle \\ \vdots & \vdots & \ddots & \vdots \\ \langle \mathbf{h}_1, \mathbf{h}_g \rangle & \langle \mathbf{h}_2, \mathbf{h}_g \rangle & \cdots & \langle \mathbf{h}_g, \mathbf{h}_g \rangle \end{bmatrix}^{-1} \begin{bmatrix} \langle \mathbf{S}, \mathbf{h}_1 \rangle \\ \langle \mathbf{S}, \mathbf{h}_2 \rangle \\ \vdots \\ \langle \mathbf{S}, \mathbf{h}_g \rangle \end{bmatrix} \tag{5.70}$$

Step (7) Updates the residual $\mathbf{R_S}_g = \mathbf{S} - \sum_{i=1}^{g} x_i \mathbf{h}_i$, where $\mathbf{R_S}_g$ is orthogonal to every atom in $\mathbf{H}_g$ (please see Appendix 5-B for the proof);

Step (8) Sets $g \leftarrow g + 1$. If $g \leq G$, repeat Steps (2)−(7); if $g > G$, then stop.

It is noted that when $N_A = 1$, the previous algorithm is reverted back to the traditional OMP algorithm in vector space.

When the target echo only contains the echo of micromotional scatterers, let $\mathbf{A} = \mathbf{A}_m$ in this algorithm, then after the iteration, the micro-Doppler signal $\mathbf{S}$ is decomposed as follows:

$$\mathbf{S} = \sum_{i=1}^{G} x_i \mathbf{h}_i + \mathbf{R_S}_g \tag{5.71}$$

where the parameters $\left(\widehat{d}, \widehat{r}, \widehat{\Omega}, \widehat{\theta}\right)$ of the atoms $\mathbf{h}_i$ $(i = 1, 2, \ldots, G)$ indicate the micro-Doppler feature parameters of the target.

It can be found that each atom contains an amplitude term and a phase term. Both the amplitude term and the phase term contain the micro-Doppler features of the micromotional scatterers. When PRF is lower than the Nyquist rate of micro-Doppler signal, although the phase terms of both the micro-Doppler signal and atoms are wrapped, the amplitude terms with sinc function are not affected by the undersampling in slow time domain. Therefore, theoretically the previous algorithm can work well in this case. However, when the amplitude of micro-Doppler curve is less than the range resolution of radar, the sinc function is approximate to a constant with respect to t_m, then complex image OMP decomposition algorithm is reverted back to the traditional OMP algorithm in vector space. In this case, the algorithm cannot deal with the micro-Doppler feature extraction when PRF is lower than the Nyquist rate. Fortunately, when the amplitude of micro-Doppler curve is less than the range resolution of radar, the spectral width of micro-Doppler signal is usually small. For example, when the carrier frequency and bandwidth of transmitted signal are 10 GHz and 500 MHz, respectively, the rotating frequency and radius of micromotional scatterer are 4 Hz and 0.1 m, respectively, it can be calculated that the spectral width of micro-Doppler signal is just about 335 Hz. In this case, the PRF of general radars is able to satisfy the Nyquist sampling principle.

5.3.2.2 Simulation

1. Effectiveness validation

In the simulation, the radar is located at the origin O of the global coordinate system. The carrier frequency of transmitted signal is 10 GHz. The bandwidth of the transmitted signal is 500 MHz and $T_p = 4$ μs. The range resolution is 0.3 m and PRF = 1000 Hz. The initial location of the target center o is (30km,0km,5km) in the global coordinate system. The translational velocity of the target is $\boldsymbol{v} = (0{,}300{,}0)^{\mathrm{T}}$ m/s. The coordinates of the rotation center C in the local coordinate system are (5m,5m,5m). Three scatterers rotate around C and all of their angular velocities and radii are $\boldsymbol{\omega} = (0{,}0{,}2\pi)^{\mathrm{T}}$ rad/s and $\left\|\overrightarrow{C'P'}\right\| = 2.8284$ m, respectively, as shown in Fig. 5.23A. It can be calculated that $\Omega = 2\pi$ rad/s and $\left\|\overrightarrow{C'P'}\right\| \sin\varepsilon = 2.7899$ m, then the bandwidth of micro-Doppler signal is 2337.3 Hz, which is much larger than PRF. Fig. 5.23B shows the range–slow time image of the target.

The complex image OMP algorithm is utilized to decompose the range–slow time image, and the results are shown in Table 5.4. The value of $\widehat{d}$ is between 5.7 and 6.0, and the average value is 5.8; the value of $\widehat{r}$ is between 2.6 and 3.0, and the average

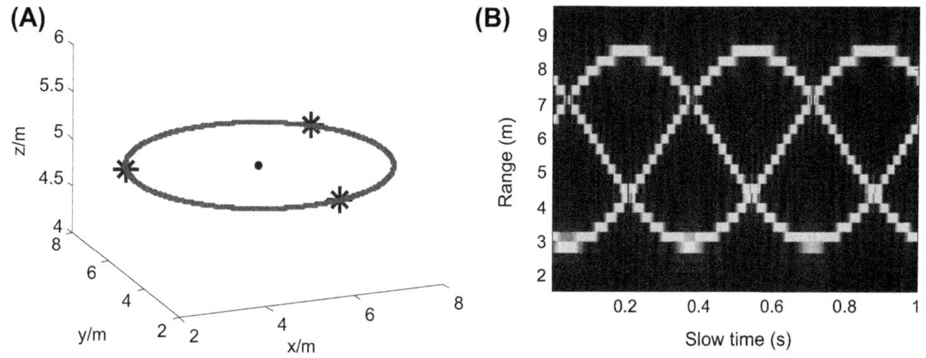

Figure 5.23 Target model and its range–slow time image. (A) The geometry of the target with three rotating scatterers (each "*" denotes a rotating scatterer); (B) the range–slow time image of the target.

Table 5.4 Micro-Doppler Features Extracted by Complex Image OMP Decomposition

Iteration Number	**1**	**2**	**3**	**4**	**5**	**6**	**7**	**8**	**9**
$\widehat{d}$	5.7	5.7	5.7	5.7	6.0	6.0	5.7	6.0	5.7
$\widehat{r}$	3.0	2.7	2.7	2.8	2.7	3.0	2.7	2.6	2.8
$(2\pi)/\widehat{\Omega}$	1.0	1.0	1.0	1.0	1.0	1.0	1.0	1.0	1.0
$\widehat{\theta}$	5.0	0.8	5.0	0.8	0.8	0.8	2.9	2.9	5.0
Coefficient ($\boldsymbol{x}$)	806.8	660.3	675.1	747.2	586.4	611.4	510.9	578.5	500.5

OMP, orthogonal matching pursuit.

value is 2.7778; the value of $(2\pi)/\widehat{\Omega}$ is 1.0. It can be found that the micro-Doppler features corresponding to the three rotational scatterers are extracted successfully. The micro-Doppler atoms in Table 5.4 can be used to reconstruct a new range–slow time image shown as Fig. 5.24, which is very close to the original range–slow time image of Fig. 5.23B. It demonstrates that the algorithm can extract the micro-Doppler features effectively.

2. Robustness analysis

In this section, we analyze the robustness of the proposed algorithm by three simulations corresponding to three degradation situations. The first situation is that the micro-Doppler signal is contaminated by Gaussian white noise. In this situation, the whole target's echoes are modulated by micromotions, eg, the echoes of a spinning space target. The second situation is that the micro-Doppler signal is contaminated by the body echoes of the target. In the processing of micro-Doppler feature extraction, these body returns must be considered as "noise." The last situation is that the scattering coefficients of micromotional scatterers are complex and time varying. In this situation, the

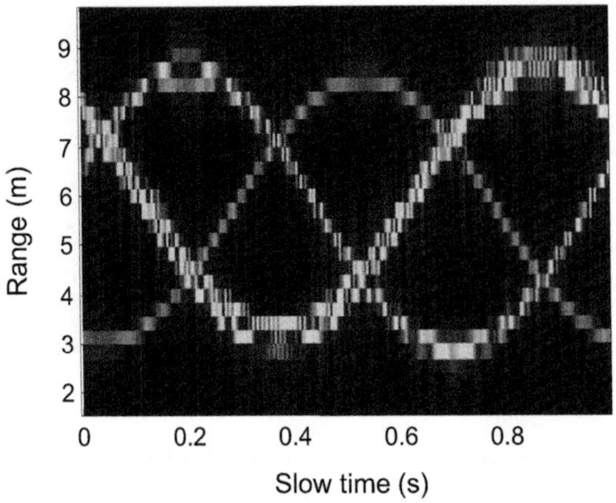

Figure 5.24 The reconstructed range–slow time image with the micro-Doppler atoms in Table 5.4.

additional modulation induced by time-varying scattering coefficients will bring negative influences to the micro-Doppler feature extraction.

a. The first situation

Assume that the parameters of radar are the same as those in the former simulation. The initial location of the target center o is (300km, 0km, 500km) in the global coordinate system. The translational velocity of the target is $\boldsymbol{v} = (0,7800,0)^{\mathrm{T}}$ m/s. The coordinates of the rotation center C in the local coordinate system are (5m,5m,5m). Three scatterers rotate around C and all of their angular velocities and radii are $\omega = (0,0,4\pi)^{\mathrm{T}}$ rad/s and $\left\|\overrightarrow{C'P'}\right\| = 5.6569$ m, respectively. It can be calculated that $\Omega = 4\pi$ rad/s and $\left\|\overrightarrow{C'P'}\right\| \sin\varepsilon = 2.9109$ m, then the bandwidth of micro-Doppler signal is 4877.3 Hz, which is much larger than PRF. The Gaussian white noise is then added to the echoes.

Fig. 5.25A shows the range–slow time image of the target when the SNR is −10 dB. The OMP algorithm is then utilized to decompose the range–slow time image, and the results are shown in Table 5.5. From the table, it can be found that the extracted micro-Doppler feature parameters of the first six iterations are close to the set values, while the micro-Doppler feature parameters of the latter iterations are not very precise due to the interference of noise. Taking the micro-Doppler feature parameters of the first six iterations, the value of $\widehat{d}$ is between 6.8 and 7.1, and the average value is 6.9; the value of $\widehat{r}$ is between 2.7 and 3.0, and the average value is 2.8667; the value of $(2\pi)/\widehat{\Omega}$ is 0.5. Therefore, the micro-Doppler features corresponding to the three rotational scatterers are extracted reasonably successfully. These six micro-Doppler atoms can reconstruct a

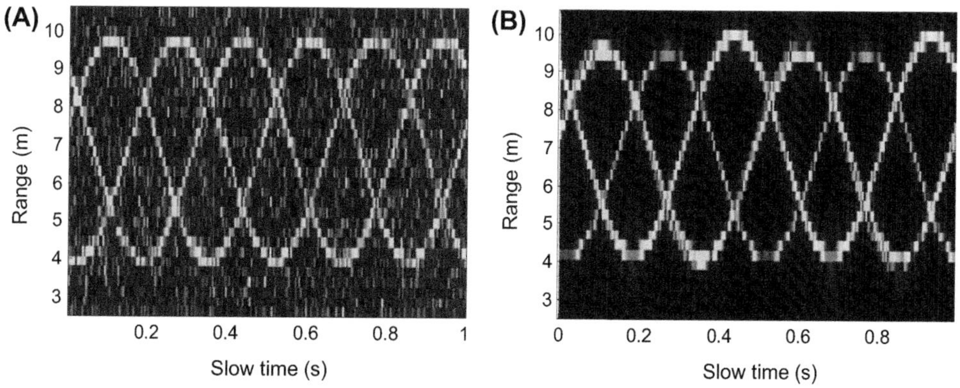

Figure 5.25 Micro-Doppler feature extraction when Gaussian white noise is added to the target echoes with SNR = −10 dB. (A) The range–slow time image of the target; (B) the reconstructed range–slow time image with the micro-Doppler atoms of the first six iterations in Table 5.5.

Table 5.5 Micro-Doppler Features Extracted by Complex Image OMP Decomposition (SNR = −10 dB, in the First Situation)

Iteration Number	1	2	3	4	5	6	7	8	9
$\hat{d}$	6.8	7.1	6.8	6.8	7.1	6.8	6.6	7.5	6.1
$\hat{r}$	2.9	2.9	2.7	2.7	3.0	3.0	3.1	2.5	2.6
$(2\pi)/\hat{\Omega}$	0.5	0.5	0.5	0.5	0.5	0.5	0.6	0.8	0.2
$\hat{\theta}$	5.0	0.8	2.9	4.9	0.8	2.9	4.9	5.8	1.9
Coefficient ($\boldsymbol{x}$)	672.3	633.1	613.6	552.8	548.6	535.9	486.6	458.5	407.9

OMP, orthogonal matching pursuit.

new range–slow time image shown as Fig. 5.25B, which is very close to the original range–slow time image of Fig. 5.25A.

Fig. 5.26A shows the range–slow time image of the target when the SNR is −15 dB. The micro-Doppler curves are severely submerged by noise. The results of OMP decomposition are shown in Table 5.6. From the table, it can be found that the extracted micro-Doppler feature parameters of the 1, 2, 3, 5, 6 iterations are close to the set values, while the micro-Doppler feature parameters of the other iterations are not very precise due to the interference of noise. The micro-Doppler atoms in Table 5.6 are used to reconstruct a new range–slow time image shown as Fig. 5.26B, where some false micro-Doppler curves are generated by noise. It should be noted that the Hough transform method in Section 5.2.2 cannot obtain the micro-Doppler features from the image of Fig. 5.26A.

When SNR is lower than −15 dB, the errors become too serious and the micro-Doppler features can no longer be correctly extracted.

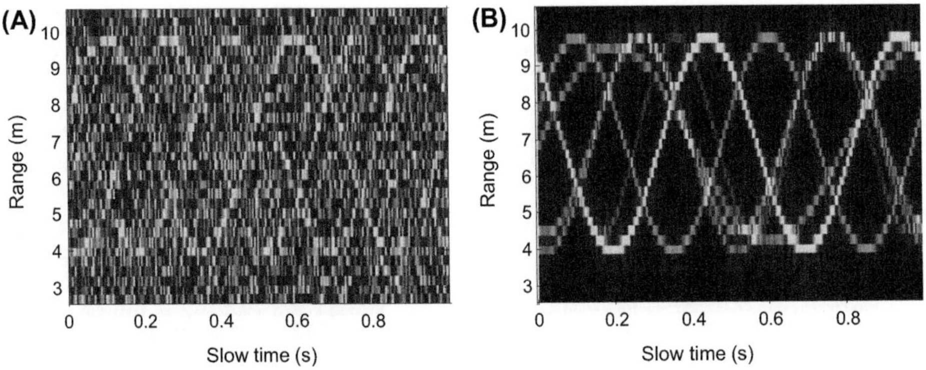

Figure 5.26 Micro-Doppler feature extraction when Gaussian white noise is added to the target echoes with SNR = −15 dB. (A) The range–slow time image of the target; (B) the reconstructed range–slow time image with the micro-Doppler atoms in Table 5.6.

Table 5.6 Micro-Doppler Features Extracted by Complex Image OMP Decomposition (SNR = −15 dB, in the First Situation)

Iteration Number	1	2	3	4	5	6	7	8	9
$\hat{d}$	6.8	6.8	7.1	6.8	7.1	6.8	7.2	7.5	7.1
$\hat{r}$	2.9	2.9	2.7	2.7	3.1	2.7	2.7	2.8	3.3
$(2\pi)/\hat{\Omega}$	0.5	0.5	0.5	0.8	0.5	0.5	0.4	0.3	0.3
$\hat{\theta}$	5.0	2.9	2.9	4.7	2.8	5.0	0.6	3.6	4.2
Coefficient ($\boldsymbol{x}$)	712.9	527.5	526.1	461.1	456.0	455.0	451.9	406.9	404.9

OMP, orthogonal matching pursuit.

b. The second situation

In this situation, the micromotion parts are masked by the target's body. As a result, the induced micro-Doppler signals are contaminated by the target body's returns. The parameters of radar are the same as those in the former simulation. The initial location of the target center o is (30km,0km,5km) in the global coordinate system. The translational velocity of the target is $\boldsymbol{v} = (0,300,0)^{\mathrm{T}}$ m/s. The coordinates of the rotation center C in the local coordinate system are (5m,5m,5m). Three scatterers rotate around C and all of their angular velocities and radii are $\boldsymbol{\omega} = (0,0,4\pi)^{\mathrm{T}}$ rad/s and $\left\|\overrightarrow{C'P'}\right\| = 2.8284$ m, respectively. It can be calculated that $\Omega = 4\pi$ rad/s and $\left\|\overrightarrow{C'P'}\right\|\sin\varepsilon = 2.7899$ m, then the bandwidth of micro-Doppler signal is 4674.5 Hz, which is much larger than PRF. Several body scatterers randomly distribute around the rotation center. When considering the body echoes as "noise," the SNR of micro-Doppler signal is −20 dB.

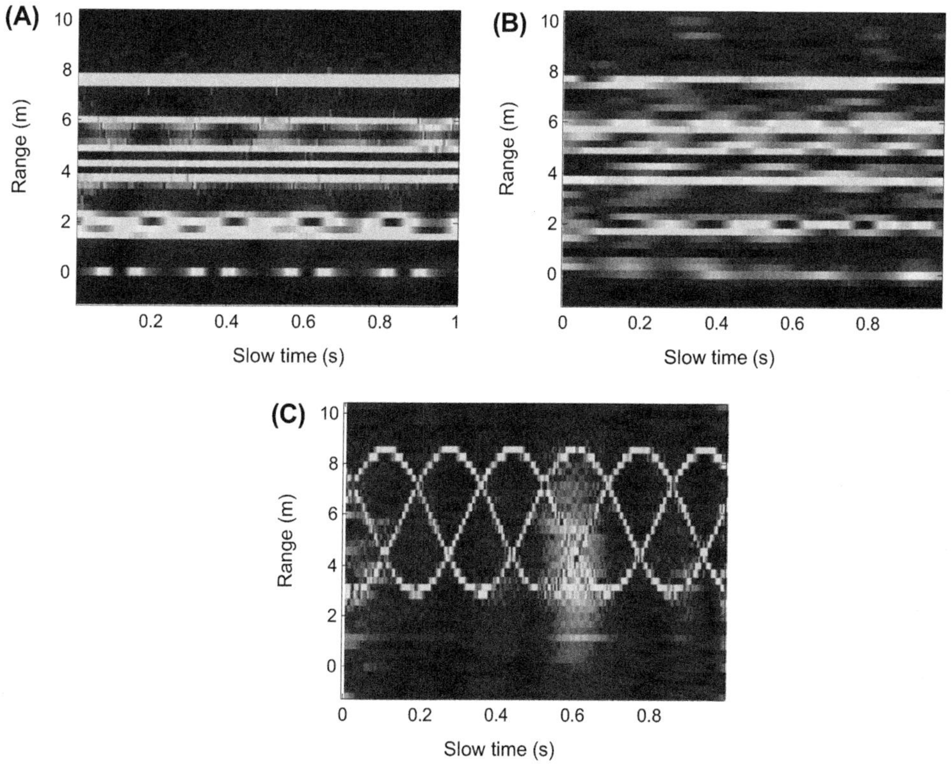

Figure 5.27 Micro-Doppler feature extraction in the second situation with SNR = −20 dB. (A) The range–slow time image of the target; (B) the reconstructed range–slow time image of the body echoes; (C) the residual signal after eliminating the body echoes from the target echoes.

Fig. 5.27A shows the range–slow time image of the target. Because the power of the micro-Doppler signal is much lower than that of the body echoes, the micro-Doppler curves in the figure are much obscured.

Via constructing the dictionary according to Eq. (5.61), the body echoes are extracted by utilizing the complex image OMP algorithm. Fig. 5.27B shows the reconstructed range–slow time image of the extracted body echoes. After eliminating the body echoes from the target echoes, the residual signal is shown in Fig. 5.27C. Then the residual signal is decomposed with the dictionary constructed according to Eq. (5.65), and the results are shown in Table 5.7. Obviously, the micro-Doppler features are extracted correctly.

When SNR is low to −25 dB, the range–slow time image of the target is given in Fig. 5.28A, and the separated body signal and micro-Doppler signal are shown in Fig. 5.28B and C, respectively. The OMP decomposition results of micro-Doppler signal are shown in Table 5.8. From the table, it can be found that the extracted micro-Doppler

Table 5.7 Micro-Doppler Features Extracted by Complex Image OMP Decomposition After Eliminating the Body Echoes (SNR = −20 dB, in the Second Situation)

Iteration Number	1	2	3	4	5	6	7	8	9
$\widehat{d}$	5.7	5.7	5.7	6.0	6.0	5.7	6.0	5.7	5.7
$\widehat{r}$	2.8	3.0	2.7	3.0	2.9	3.0	2.7	2.8	3.0
$(2\pi)/\widehat{\Omega}$	0.5	0.5	0.5	0.5	0.5	0.5	0.5	0.5	0.5
$\widehat{\theta}$	2.9	2.9	0.8	0.8	5.0	5.0	5.1	0.9	3.0
Coefficient ($\boldsymbol{x}$)	454.5	365.8	330.0	319.8	310.6	302.9	294.9	292.4	288.2

OMP, orthogonal matching pursuit.

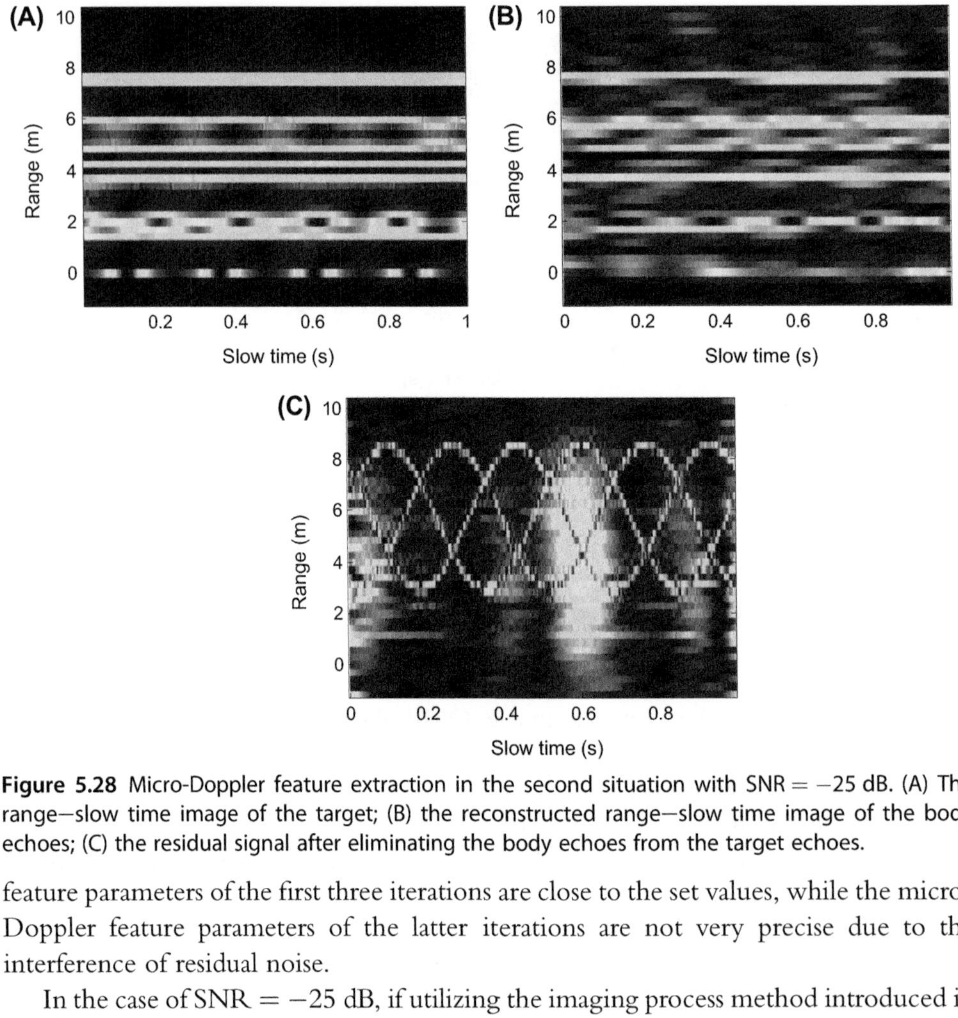

Figure 5.28 Micro-Doppler feature extraction in the second situation with SNR = −25 dB. (A) The range–slow time image of the target; (B) the reconstructed range–slow time image of the body echoes; (C) the residual signal after eliminating the body echoes from the target echoes.

feature parameters of the first three iterations are close to the set values, while the micro-Doppler feature parameters of the latter iterations are not very precise due to the interference of residual noise.

In the case of SNR = −25 dB, if utilizing the imaging process method introduced in Section 5.2, the body echo suppression process in Eq. (5.46) is used to eliminate the

Table 5.8 Micro-Doppler Features Extracted by Complex Image OMP Decomposition After Eliminating the Body Echoes (SNR = −25 dB, in the Second Situation)

Iteration Number	1	2	3	4	5	6	7	8	9
$\widehat{d}$	5.7	5.4	5.7	6.0	5.4	5.1	8.1	8.1	8.1
$\widehat{r}$	3.0	2.8	2.7	3.0	2.7	2.2	1.7	1.3	1.9
$(2\pi)/\widehat{\Omega}$	0.5	0.5	0.5	0.4	0.4	0.8	0.2	0.3	1.0
$\widehat{\theta}$	4.9	2.9	0.8	6.0	5.3	3.8	4.2	0.7	3.2
Coefficient ($\boldsymbol{x}$)	598.0	589.5	550.3	503.9	429.8	395.8	393.9	386.5	380.3

OMP, orthogonal matching pursuit.

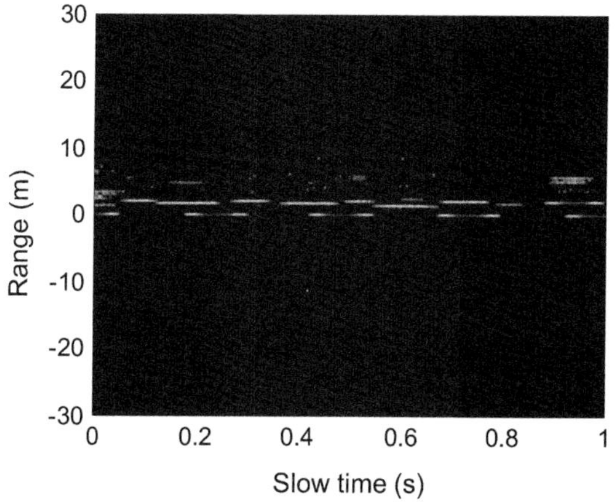

Figure 5.29 Body echo suppression process in Eq. (5.46) when SNR = −25 dB.

negative influence of body echo, which results in Fig. 5.29. It can be found that, the body echo cannot be suppressed effectively due to its high power. As a result, the latter Hough transform is unable to obtain the correct micro-Doppler features. It indicates that the complex image OMP algorithm has better antinoise performance.

c. The third situation

The parameters of radar and target are the same as those in the former simulation. The target echo only contains the micro-Doppler signals. Assume the module and phase of the scattering coefficients vary within [0,1] and [0,2π], respectively. The range−slow time image of the target is shown in Fig. 5.30A. The micro-Doppler features extracted by complex image OMP decomposition are given in Table 5.9. It can be found that the extracted micro-Doppler feature parameters of the first six iterations are close to the set values, while the micro-Doppler feature parameters of the latter iterations are not very precise. These six micro-Doppler atoms can reconstruct a new range−slow time image shown as Fig. 5.30B. Obviously, the micro-Doppler features of one of the three scatterers

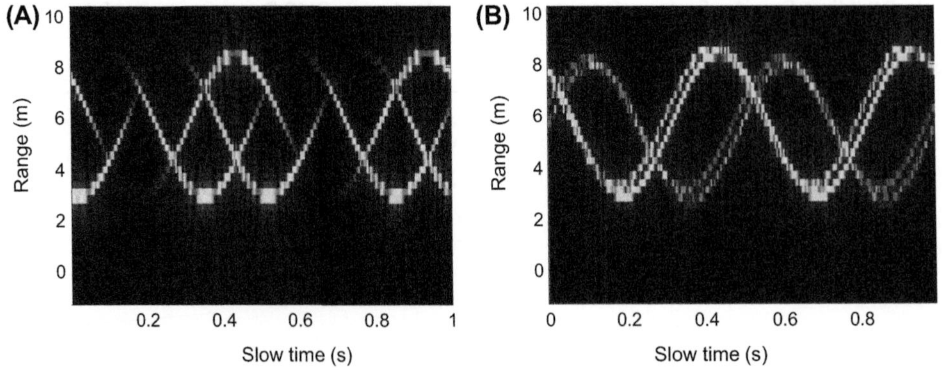

Figure 5.30 Micro-Doppler feature extraction when scattering coefficients are time varying. (A) The range–slow time image of the target; (B) the reconstructed range–slow time image with the micro-Doppler atoms in Table 5.9.

Table 5.9 Micro-Doppler Features Extracted by Complex Image OMP Decomposition (in the Third Situation)

Iteration Number	1	2	3	4	5	6	7	8	9
$\widehat{d}$	6.0	5.7	5.7	5.4	5.4	6.0	5.7	5.7	6.0
$\widehat{r}$	2.8	2.5	2.8	2.7	2.5	2.7	3.0	2.5	2.8
$(2\pi)/\widehat{\Omega}$	0.5	0.5	0.5	0.5	0.5	0.5	0.4	0.4	0.4
$\widehat{\theta}$	0.9	5.0	0.7	4.8	4.8	0.9	3.5	2.6	4.3
Coefficient ($\boldsymbol{x}$)	162.2	158.4	157.4	151.1	150.8	144.1	141.7	140.2	136.8

OMP, orthogonal matching pursuit.

failed to be obtained. It indicates that the time varying of scattering coefficients negatively influences the micro-Doppler feature extraction.

From the previous simulations, the complex image OMP algorithm possesses very robust performance in micro-Doppler feature extraction.

5.4 EMPIRICAL-MODE DECOMPOSITION METHOD

EMD was proposed by Huang etc. of NASA in 1998 [11]. The characteristic of EMD is to decompose the signal into a serial of intrinsic mode functions (IMFs) by "shifting." The signal that has the following two properties is defined as IMF by Huang: (1) the number of the maximum value and the minimum value equals the number of zero crossing, or the difference of the two numbers is one at most; (2) the envelopes of the maximum value and the minimum value are symmetric of t -axis. For IMF, its phase function is obtained by Hilbert transform, and then the instantaneous frequency is obtained by the derivative of the phase. However, the signals do not have the properties of IMF in general. So the IMF components can be shifted by EMD solution method. Firstly, the average of the upper envelope and the lower envelope m is obtained, according to the maximum value and the minimum value of $a(t)$:

$$m = \frac{1}{2}(v_1(t) + v_2(t)) \tag{5.72}$$

where h is the difference between $a(t)$ and m. Regard h as new $a(t)$ and repeat the above operation until h meets the requirements of IMF when we set $\mathrm{IMF}_1 = h$. And then regard $r(t)$ as a new $a(t)$. Repeat the above operations, and IMF_2, IMF_3, ... are obtained successively. Stop the decomposition until $r(t)$ is monotonous mainly or $|r(t)|$ is quite small. Thus, the original signal is decomposed into many IMF components and a residual component $r(t)$, that is

$$a(t) = \sum_{l=1}^{L} \mathrm{IMF}_l + r(t) \tag{5.73}$$

After the decomposition of the signal $a(t)$ by EMD, the first IMF component IMF_1 is the component of the highest frequency in $a(t)$. With the orders of IMF components increasing, the frequency of IMF is decreasing. The residual $r(t)$ is the component of the lowest frequency, which is the monotonous term of the signal. The property of EMD makes the IMF components have the distinctive physical meaning, which is beneficial to the analysis of the components of each frequency.

As stated previously, in wideband radar, when the pulse repetition frequency of radar is twice larger than the spectrum of micro-Doppler signal, and the micromotion point does not migrate through resolution cell, the micro-Doppler feature of the target can be obtained by extracting and processing the echo in the range unit where the micromotion point locates with time-frequency analysis or by EMD. However, in wideband radar, the more common problem we face is the micro-Doppler feature extraction when the pulse repetition frequency of radar is smaller than twice the spectrum of micro-Doppler signal, and the micromotion point migrates through resolution cell. Though in the previous sections the micro-Doppler feature extraction solution method is based on Hough transform and complex image OMP decomposition, these solutions are put forward on rotation targets. For more complicated motions, such as precession, the calculated amount of the algorithms will be quite large. Because the expression of micro-Doppler signal induced by precession is rather complicated and contains many parameters (shown as Eqs. (2.51) and (3.39)), when constructing the Hough transform equation or the OMP dictionary, many more parameters will cause the geometric progression addition of the calculated amount of Hough transform. It will cause great increase of the quantity of the atoms in OMP dictionary, which will also cause great increase of the quantity of OMP decomposition. Thus, a more proper method solution should be researched further to accomplish of the feature extraction of the precession target. A solution method based on the combining of micro-Doppler curves separation and EMD is given.

From Eq. (3.39) we can see that, in range—slow time image, the micro-Doppler curves induced by precession mainly contain four angular frequency components, that is $\Omega_c + \Omega_s$, Ω_c, Ω_s, and $|\Omega_c - \Omega_s|$. Thus, if the micro-Doppler curves are decomposed by EMD in ideal conditions, IMF_1 is the component whose angular frequency is $\Omega_c + \Omega_s$. IMF_2 is the component whose angular frequency is the larger one in Ω_c and Ω_s. IMF_3 is the component whose angular frequency is the smaller one in Ω_c and Ω_s. IMF_4 is the component whose angular frequency is $|\Omega_c - \Omega_s|$. However, what the micro-Doppler curves reflect is the amplitude information of the range—slow time matrix $S_d(f_k,t_m)$. If we want to decompose the micro-Doppler curves by EMD, we should first separate the micro-Doppler curves from the range—slow time image. When a target contains many scattering points, we should also separate the micro-Doppler curves from each other.

Because inevitably sidebands exist in the range image, this will bring the negative effects to the separation research of the micro-Doppler curves. So the sidebands should be suppressed before curve separation. Sidebands can be suppressed by the mathematic morphology method stated in Section 5.2.1. For example, the micro-Doppler curves turning "thick" caused by the sidebands will be suppressed by extracting the micro-Doppler curve skeleton in $|S_d(f_k, t_m)|$.

When the sideband of the micro-Doppler curves of range image is suppressed and the skeleton of the curves is extracted, the micro-Doppler curves separation can be processed further. It is easy to prove that the derivative function of the micro-Doppler curves determined by $\Delta R(t_m)$ is a continuous function on t_m. That is to say, the micro-Doppler curve relative to each scattering point is smooth. Thus, the separation of the micro-Doppler curves can be accomplished by the smoothness of each curve. $S_k(|S_d(f_k, t_m)|)$, the skeleton of $|S_d(f_k, t_m)|$, is expressed as a $M_s \times N_s$ matrix. M_s is the range sampling amount, and N_s is the azimuth sampling amount. Assume some curves contain N' samples, and the serial number of the No. n' sampling point is $(x_{n'}, y_{n'})$, whose amplitude is $S_{n'}$. Because of the smoothness of the micro-Doppler curves, the difference of the derivatives of the adjacent two points is quite small. So the micro-Doppler curve will be searched and separated by solving the following optimization model:

$$\min f(x_{n'}, y_{n'}) = \left\{ \sum_{n'=M'+2}^{N'} \left| \frac{x_{n'} - x_{n'-1}}{y_{n'} - y_{n'-1}} - \frac{x_{n'-1} - x_{n'-1-M'}}{y_{n'-1} - y_{n'-1-M'}} \right| \right\} \tag{5.74}$$

$$\text{s.t.} \quad 1 \le x_{n'} \le M_s, 1 \le y_{n'} \le N_s, y_{n'} - y_{n'-1} > 0, S_{n'} > \zeta, x_{n'} \in \mathbf{N}, y_{n'} \in \mathbf{N}$$

where $\zeta > 0$, and M' is a small natural number preset. When the sampling interval of slow time is very small, the micro-Doppler curves on several adjacent slow time sampling units may present as a straight line segment. It makes $(x_{n'-1} - x_{n'-2})/(y_{n'-1} - y_{n'-2}) = 0$. To solve this problem, we set that $M' > 1$, by calculating the value of $(x_{n'-1} - x_{n'-1-M'})/(y_{n'-1} - y_{n'-1-M'})$ to substitute $(x_{n'-1} - x_{n'-2})/(y_{n'-1} - y_{n'-2})$. The limiting condition $y_{n'} - y_{n'-1} > 0$ ensures that the search direction in slow-time direction is unidirectional. $S_{n'} > \zeta$ limits that only when the energy is larger than the threshold, ζ can be searched. When $f(x_{n'}, y_{n'})$ in the optimization model takes the minimum value, the micro-Doppler curve separated will be most smooth. To reduce the computational complexity, the following simplifying model can be used when solving the optimization specifically:

$$\min f(x_{n'}, y_{n'}) = \left| \frac{x_{n'} - x_{n'-1}}{y_{n'} - y_{n'-1}} - \frac{x_{n'-1} - x_{n'-1-M'}}{y_{n'-1} - y_{n'-1-M'}} \right|,$$
$$\text{s.t.} \quad M' + 2 \le n' \le N', 1 \le x_{n'} \le M_s, 1 \le y_{n'} \le N_s, \tag{5.75}$$
$$y_{n'} - y_{n'-1} > 0, S_{n'} > \zeta, x_{n'} \in \mathbf{N}, y_{n'} \in \mathbf{N}$$

Specially, the steps of separating each micro-Doppler curve from $S_k(|S_d(f_k, t_m)|)$ are stated as follows:

Step (1) Initialization: $i = 1$, $d = 0$, and $\mathbf{c}^{N_s \times 1}$ is a zero vector.

Step (2) Solve the local maxima in the column i, which is larger than the threshold ζ in $S_k(|S_d(f_k, t_m)|)$. The sequence $\mathbf{r}_m$ consists of the serial row numbers in $S_k(|S_d(f_k, t_m)|)$. Assume that there are p elements in $\mathbf{r}_m$, $i_p = 1$, $\mathbf{c}(1) = \mathbf{r}_m(i_p)$, and construct the zero matrix $\mathbf{C}^{N_s \times p}$.

Step (3) $i = i + 1$. Search the points larger than the threshold ζ in column i in $S_k(|S_d(f_k, t_m)|)$, where the sequence $\mathbf{r}'_m$ consists of the serial row numbers, and the serial order of the element is i'_p. If the number of elements in $\mathbf{r}'_m$ is more than 0, turn to Step (4). If the number of the elements in $\mathbf{r}'_m$ equals to 0, $d = d + 1$, and turn to Step (5).

Step (4) Judgment: when $i = 2$,

$$\mathbf{c}(i) = \arg\min_{i'_p} \left| \mathbf{r}'_m\left(i'_p\right) - \mathbf{c}(i-1) \right| \tag{5.76}$$

When $2 < i < i_0$, let

$$\mathbf{c}(i) = \arg\min_{i'_p} \left| \left[\mathbf{r}'_m\left(i'_p\right) - \mathbf{c}(i-1) \right] - \frac{1}{i-2} \sum_{i_d=3}^{i} [\mathbf{c}(i_d - 1) - \mathbf{c}(i_d - 2)] \right| \tag{5.77}$$

When $i_0 \le i \le N_s$, let

$$\mathbf{c}(i) = \arg\min_{i'_p} \left| \left[\mathbf{r}'_m\left(i'_p\right) - \mathbf{c}(i-1) \right] - \frac{1}{i_0 - 2} \sum_{i_d = i - i_0 + 3}^{i} \left[\mathbf{c}(i_d - 1) - \mathbf{c}(i_d - 2) \right] \right| \quad (5.78)$$

Then turn to Step (6).

Step (5) Judgment: When $d \ge D$, stop the searching of this curve and turn to Step (7). When $d < D$, set $\mathbf{r}'_m = [1, 2, \ldots, M_s]$ and turn to Step (4).

Step (6) Judgment: When $i < N_s$, turn to Step (3). When $i = N_s$, turn to Step (7).

Step (7) Suppose the column i_p of $\mathbf{C}^{N_s \times p}$ is $\mathbf{c}^{N_s \times 1}$. All points whose sequential number is $(i_d, \mathbf{c}(i_d))(\ i_d = 1, 2, \ldots, N_s)$ in $S_k(|S_d(f_k, t_m)|)$ consist of a micro-Doppler curve extracted. Set all the points in $S_k(|S_d(f_k, t_m)|)$ whose sequential number is $(i_d, \mathbf{c}(i_d))(i_d = 1, 2, \ldots, N_s)$. Set $i_p = i_p + 1$, $i = 1$, $d = 0$, and all the elements in $\mathbf{c}^{N_s \times 1}$ equal to zero. When $i_p \le p$, set $\mathbf{c}(1) = \mathbf{r}_m(i_p)$ and then turn to Step (3).

Step (8) When $i_p > p$, stop searching. Each column in matrix $\mathbf{C}^{N_s \times p}$ represents a micro-Doppler curve.

From the previous steps we can see that, when separating the micro-Doppler curves in $S_k\left(|S_d(f_k, t_m)|\right)$, the thought of "Clean" [12] is used. When a micro-Doppler curve is extracted, it will be removed from $S_k\left(|S_d(f_k, t_m)|\right)$, which is described in Step (7), so as to avoid the interaction as possible. The aim to set i_0 in the algorithm is to investigate the smoothness in i_0 interval. Predict the next position of point according to the average of each derivative of each point on curves. The aim to set parameter d is to overcome the discontinuity that may appear on curves after "Clean." The aim to set threshold ζ is to ensure that only the point whose energy is larger than ζ will be searched, so as to avoid the influence of sidebands and clutters. The value of ζ is confirmed according to the practical application. If the image is converted into binary image, the value of threshold ζ is set at zero, that is $\zeta = 0$.

Then the micro-Doppler curves that are separated are decomposed by EMD. Because IMF components have their own specific physical meaning by EMD, some micromotion parameters of the precession target can be obtained from the results of EMD. From Eq. (3.39), the micro-Doppler curves mainly contain the components of four frequencies $\Omega_c + \Omega_s$, Ω_c, Ω_s, and $|\Omega_c - \Omega_s|$. So in an ideal condition, IMF_1 decomposed by EMD is the component whose angular frequency is $\Omega_c + \Omega_s$. IMF_2 is the component whose angular frequency is the larger one of Ω_c and Ω_s. IMF_3 is the component whose angular frequency is the smaller one of Ω_c and Ω_s. IMF_4 decomposed by EMD is the component whose angular frequency is $|\Omega_c - \Omega_s|$. However, quantization errors are brought into practical applications because of the ranging sampling. So the micro-Doppler curve obtained is not ideally smooth. $\mathbf{c}^{N_s \times 1}$ in the micro-Doppler curve separation algorithm has quantization errors depending on the interval of the range profile sampling unit. The quantization errors lead to the local micro-Doppler curves presenting ladder-like, so that the high-frequency component is brought in. So when the micro-Doppler curves are decomposed by EMD, the components obtained first are useless high-frequency components. After that, the useful micro-Doppler signal components are decomposed. Increasing the range sampling rate and smooth processing of the micro-Doppler curves will reduce the negative influence of quantization errors on EMD to some extent. However, it is difficult to remove the high-frequency components completely. Consequently, when the micro-Doppler curves are decomposed by EMD, the component of low-angular frequency, such as $|\Omega_c - \Omega_s|$, cannot be extracted.

Processing the range—slow time image shown in Fig. 3.12 by the above solution method, the results are shown in Fig. 5.31, in which Figure (A) is the skeleton of the range—slow time image in Fig. 3.12, and Figure (B) is the micro-Doppler curves separated, which is shown in different color. It is obvious that three micro-Doppler curves are extracted successfully by the curve separation algorithm. The micro-Doppler curves of the scattering points on the precession target are decomposed by EMD, and the result is shown in Fig. 5.31C, in which $IMF_1 \sim IMF_3$are high-frequency cluster components, and IMF_4 is a single frequency signal whose frequency is 8 Hz, that is, $\Omega_c + \Omega_s$. IMF_5 is a single frequency signal whose frequency is 5 Hz, that is, Ω_c. IMF_6 is a single frequency signal whose frequency is 3 Hz, that is, Ω_s. Others are the residual low-frequency components.

EMD is to decompose the micro-Doppler curves after curve separation. So the accuracy of curve separation is the premise and the guarantee for accurate extraction of the precession feature for the subsequent EMD process. The micro-Doppler curve separation algorithm stated in this section is quite robust in low SNR. There two main reasons: the first is that when obtaining the range image by "dechirp," high processing gain is brought in by pulse compression; the second is that the image processing algorithms are used before the curve separation. Besides, Gaussian smoothing will filter quantities of irregular noise of the range—slow time image. The influence of the noise can be suppressed distinctively when producing the binary image by setting a proper threshold. The result of the micro-Doppler curve separation is shown in Fig. 5.32 when SNR is -15 dB. We can see that the irregular points are suppressed quite well of the image by image processing algorithms, so that the micro-Doppler curves are separated accurately.

The algorithm has robustness to some extent on radar cross-section (RCS) fluctuation phenomenon of the scattering point. The two main reasons are as follows. One is that the threshold of the binary image before skeleton extraction can be reduced properly so as to keep the continuity of the micro-Doppler curves in the binary image, when it is transformed into a binary image, the amplitude fluctuation effect of the micro-Doppler curves will be removed. The other is that in curve separation algorithm, the number of the discontinuous point of the curve exceeding the threshold D will be regarded as the terminal of the curve. So increasing the threshold D properly will reduce the

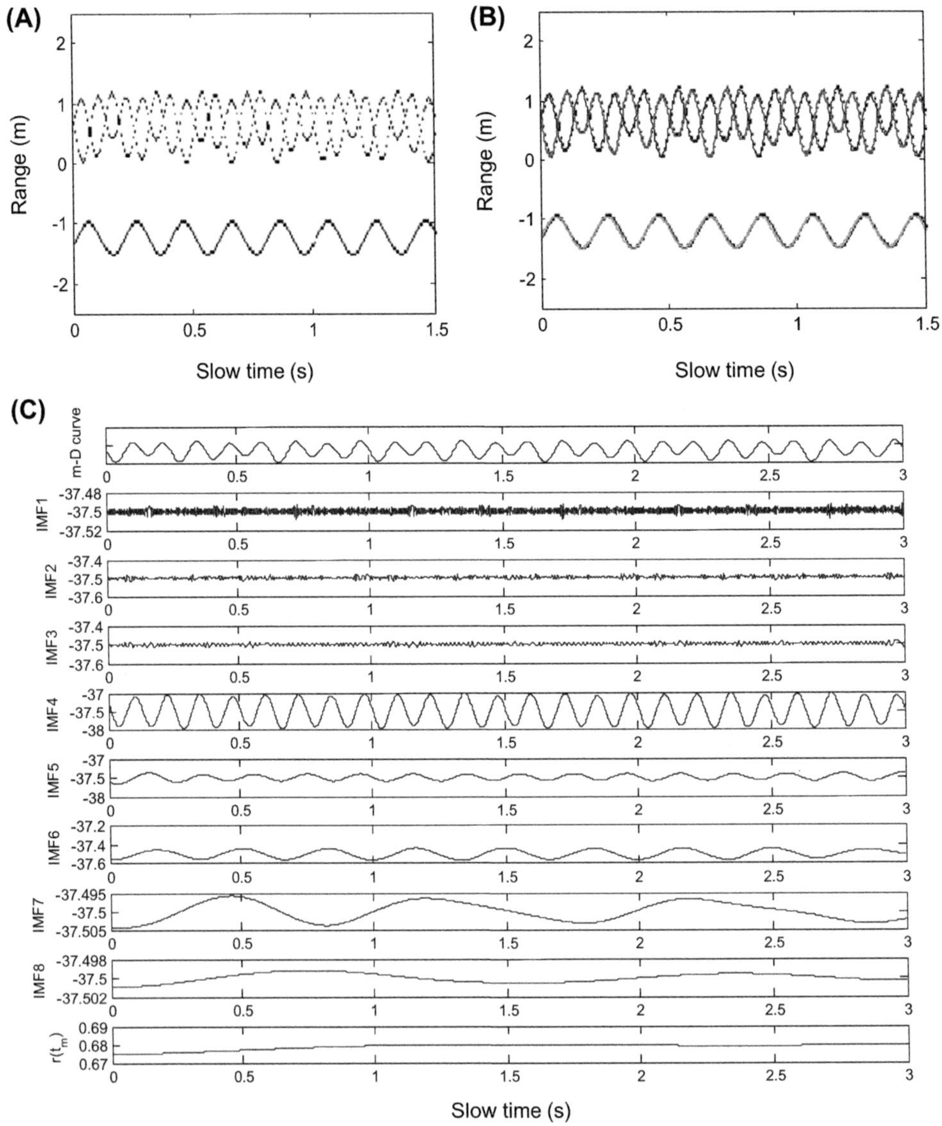

Figure 5.31 EMD processing of the micro-Doppler curves of the precession target in wideband radar. (A) Skeleton extraction result; (B) micro-Doppler curves separation result; (C) decomposition result by EMD.

discontinuity of the micro-Doppler curves caused by RCS fluctuation. The result of micro-Doppler curves separation when the scattering coefficient of the scattering points fluctuates between 0.2 and 1 is shown in Fig. 5.33, and the threshold is reduced to half of its original value when the binary image transforms. The micro-Doppler curves in range—slow time profile is shown in Fig. 5.33A. The separation result of the micro-

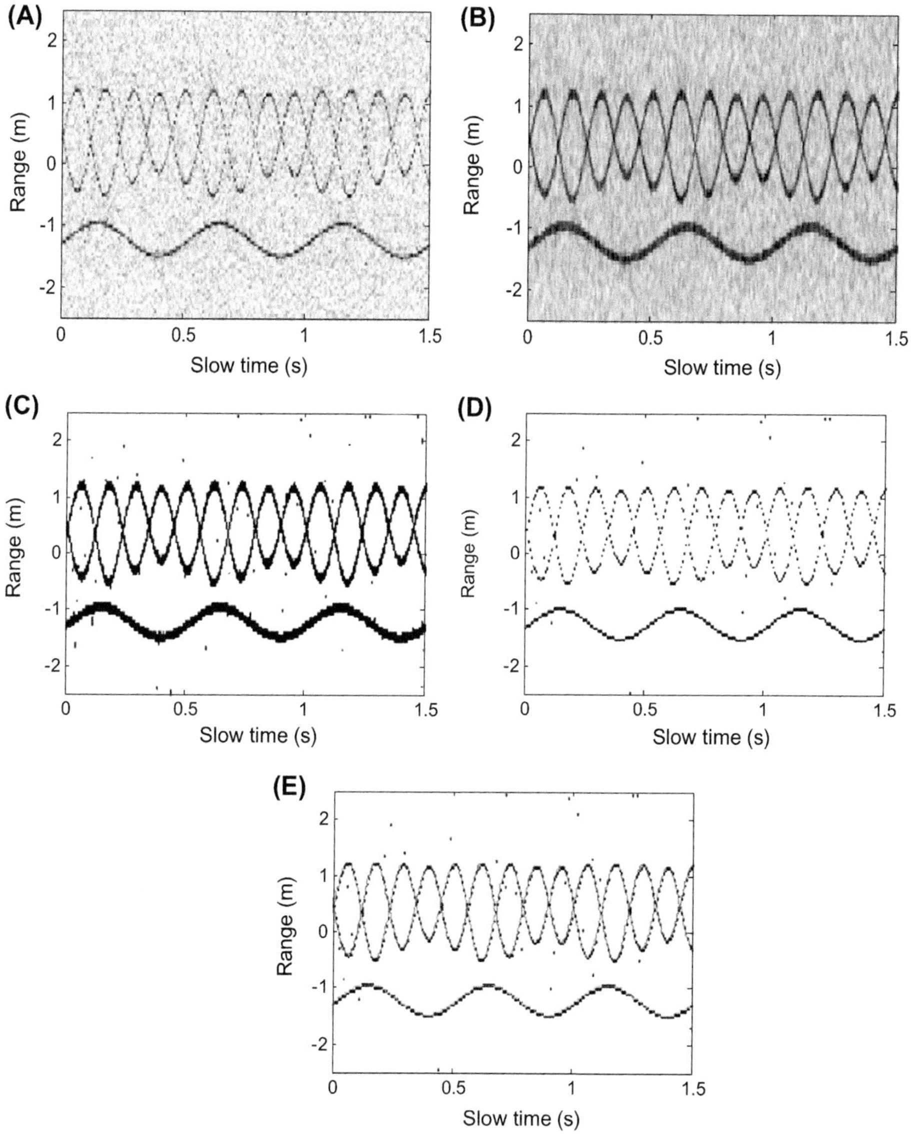

Figure 5.32 The micro-Doppler curves separation processing when SNR = −15 dB. (A) The initial range–slow time image; (B) the result by Gaussian filtering; (C) binary image; (D) skeleton extraction; (E) the result of micro-Doppler curves separation.

Doppler curves is shown in Fig. 5.33B, which perfectly overcomes the RCS fluctuation effect of the scattering points. When the scattering coefficient of the scattering points fluctuates between 0.2 and 1 and the echo SNR is −10 dB, the separation result of the micro-Doppler curves is shown in Fig. 5.34. The micro-Doppler curves are separated successfully as well in this condition.

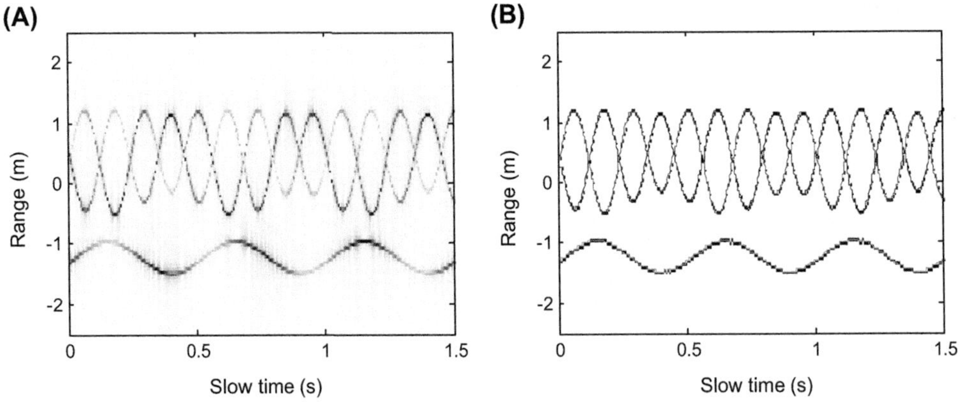

Figure 5.33 The micro-Doppler curve separation process when the scattering point RCS is fluctuating. (A) The initial range–slow time image; (B) micro-Doppler curves separation result.

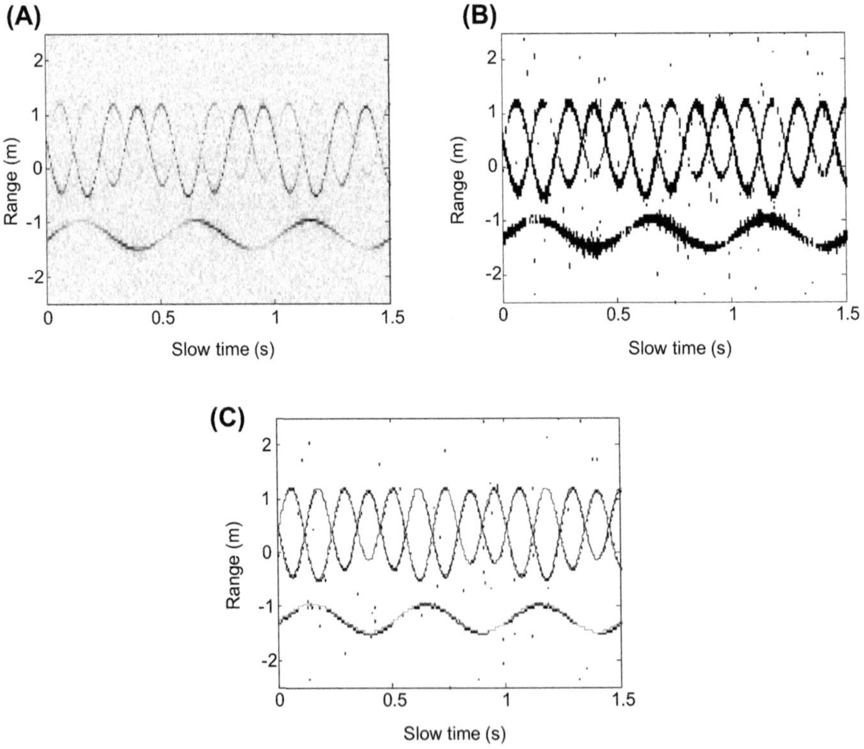

Figure 5.34 Micro-Doppler curves separation process when the scattering point RCS is fluctuating and SNR = −10 dB. (A) The initial range–slow time image; (B) binary process; (C) micro-Doppler curves separation result.

5.5 HIGH-ORDER MOMENT FUNCTION ANALYSIS METHOD

This section introduces the micro-Doppler feature extraction method of rotating or vibrating target based on high-order moment function (HOMF). In this method, the HOMF is calculated by processing the echo signal of the rotating (vibration) scatterer. According to the characteristic that the imaginary part of the HOMF can be approximated to a single frequency signal with the frequency of rotation (vibration), the rotation (vibration) frequency of target can be extracted. And then the estimation of the rotation radius (amplitude) is accomplished by using the modulus accumulation function of the M-th order real part of the Fourier transform results of the search signal. Next the theoretical basis and the concrete implementation method of the HOMF analysis will be described from narrowband radar applications and broadband radar applications.

5.5.1 Application in Narrowband Radar

5.5.1.1 Algorithm

In order to facilitate the analysis, we take the form of rotating micromotion as an example, the micro-Doppler signal of a single rotating scattered point is written as follows:

$$s_{\text{micro-Doppler}}(t) = \sigma \exp(\mathrm{j}2\pi r \sin(\Omega t + \theta)) \tag{5.79}$$

where σ is the scattering coefficient of rotating point, θ is the initial phase. For a given signal delay value τ, the first-order and second-order instantaneous moments of this signal are defined respectively as:

$$P_1(s_{\text{micro-Doppler}}(t), \tau) = s_{\text{micro-Doppler}}(t) \tag{5.80}$$

$$\begin{aligned} P_2(c(t), \tau) &= s_{\text{micro-Doppler}}(t) s^*_{\text{micro-Doppler}}(t) \\ &= \sigma^2 \exp(\mathrm{j}2\pi r(\sin(\Omega t + \theta) - \sin(\Omega(t+\tau) + \theta))) \\ &= \sigma^2 \exp\left(\mathrm{j}4\pi r \sin\left(\frac{\Omega}{2}\tau\right)\cos\left(\Omega t + \theta + \frac{\Omega}{2}\tau\right)\right) \end{aligned} \tag{5.81}$$

Through calculation, the M-th order instantaneous moment of $s_{\text{micro-Doppler}}(t)$ is:

$$\begin{aligned} P_M(s_{\text{micro-Doppler}}(t), \tau) &= P_2(P_{M-1}(s_{\text{micro-Doppler}}(t), \tau), \tau) \\ &= \sigma^{2^{M-1}} \exp\left(\mathrm{j}2^M \pi r\left(\sin\left(\frac{\Omega}{2}\tau\right)\right)^{M-1} \right. \\ &\qquad \left. \times \sin\left(\Omega t + \theta + (M-1)\cdot\left(\frac{\Omega}{2}\tau + \frac{\pi}{2}\right)\right)\right) \end{aligned} \tag{5.82}$$

Define

$$a(M) = 2^M \pi r\left(\sin\left(\frac{\Omega}{2}\tau\right)\right)^{M-1} = 2\pi r\left(2\sin\left(\frac{\Omega}{2}\tau\right)\right)^{M-1} \tag{5.83}$$

Assuming $0 < \tau < \pi/(3\Omega)$ (generally the sampling frequency of radar echo signal is $f_s \gg \Omega/(2\pi)$, this premise is easy to be satisfied), we can get the derivation of $a(M)$:

$$\frac{\mathrm{d}a(M)}{\mathrm{d}M} = 2\pi r\left(2\sin\left(\frac{\Omega}{2}\tau\right)\right)^{M-1}\ln\left(2\sin\left(\frac{\Omega}{2}\tau\right)\right) \tag{5.84}$$

Since $0 < \tau < \pi/(3\Omega)$, we can get

$$\ln\left(2\sin\left(\frac{\Omega}{2}\tau\right)\right) < 0 \tag{5.85a}$$

$$1 > \left(2\sin\left(\frac{\Omega}{2}\tau\right)\right)^{M-1} > 0 \tag{5.85b}$$

$$\lim_{M\to+\infty} a(M) = 2\pi r \lim_{M\to+\infty}\left(2\sin\left(\frac{\Omega}{2}\tau\right)\right)^{M-1} = 0 \tag{5.85c}$$

From Eqs. (5.85a) and (5.85b) we can know

$$\frac{\mathrm{d}a(M)}{\mathrm{d}M} < 0 \tag{5.86}$$

Therefore, combined with Eq. (5.85c) under this premise, there possibly exists an M making $a(M) << 1$. Since $\left|\sin\left(\Omega t + \theta + (M-1)\cdot\left(\frac{\Omega}{2}\tau + \frac{\pi}{2}\right)\right)\right| \le 1$, the imaginary part of $P_M(s_{\text{micro-Doppler}}, \tau)$ can be approximately written as:

$$\begin{aligned}\mathrm{Im}[P_M(c(t),\tau)] \approx \sigma^{2^{M-1}} 2^M \pi r\left(\sin\left(\frac{\Omega}{2}\tau\right)\right)^{M-1} \\ \times \sin\left(\Omega t + \theta + (M-1)\cdot\left(\frac{\Omega}{2}\tau + \frac{\pi}{2}\right)\right)\end{aligned} \tag{5.87}$$

From Eq. (5.87), it is a sinusoidal signal relative to the time t that the imaginary part of the M-th order moment function indicated by Eq. (5.79), and its frequency is $\Omega/(2\pi)$. The modulus function (MF) of Fourier transform results of the imaginary part of the M-th order instantaneous moment function is defined as:

$$\begin{aligned} \mathrm{MF\_}I_M(s_{\text{micro-Doppler}}(t), f, \tau) &= |\mathrm{FFT}[\mathrm{Im}[P_M(s_{\text{micro-Doppler}}(t), \tau)]]| \\ &= \sigma^{2^{M-1}} 2^M \pi^2 r \left(\sin\left(\frac{\Omega}{2}\tau\right)\right)^{M-1} \left(\delta\left(f + \frac{\Omega}{2\pi}\right) \right. \\ &\quad \left. - \delta\left(f - \frac{\Omega}{2\pi}\right)\right) \end{aligned} \tag{5.88}$$

where $|\cdot|$ represents modulus, FFT[$\cdot$] represents Fourier transform, and $\delta(\cdot)$ represents impulse function. By the nature of the impulse function, the frequency of the positive frequency part of the peak point of $\mathrm{MF\_}I_M(s_{\text{micro-Doppler}}(t), f, \tau)$ is $\Omega/(2\pi)$. Therefore, the estimation of the rotation frequency can be accomplished by the peak point detection of the function $\mathrm{MF\_}I_M(s_{\text{micro-Doppler}}(t), f, \tau)$.

However, by the definition of the higher-order moment function Eq. (5.82), when there are multiple rotational scattering points on the target, the high-order moments will exist cross-terms. Because of the interference of cross-terms, the $\mathrm{MF\_}I_M(s_{\text{micro-Doppler}}(t), f, \tau)$ value of echo signal is very complex. Considering the different time delay values τ, signal's auto item produces a peak in the corresponding rotation frequency, while the cross-term is not always the peak value in a fixed nonrotating frequency. Therefore, the function $\mathrm{MF\_}I_M(s_{\text{micro-Doppler}}(t), f, \tau)$ of different time delay τ can be accumulated by the cumulative multiplication, and the product of modulus function (PMF) of the signal Fourier transform results of imaginary part of the M-th order instantaneous moment the signal is obtained:

$$\mathrm{PMF\_}I_M(s_{\text{micro-Doppler}}(t), f, \tau) = \prod_{\tau} \mathrm{MF\_}I_M(s_{\text{micro-Doppler}}(t), f, \tau) \tag{5.89}$$

Eq. (5.89) can make the $\mathrm{PMF\_}I_M(s_{\text{micro-Doppler}}(t), f, \tau)$ value of signal's auto item to be amplified in the corresponding peak values of the rotation frequency since there is always a peak. That is, the function $\mathrm{PMF\_}I_M(s_{\text{micro-Doppler}}(t), f, \tau)$ will still have a higher peak at the corresponding rotating frequency. For the cross-terms, the maximum value is not always obtained at a fixed nonrotating frequency, which makes the increasing velocity of the amplitude accumulation of $\mathrm{PMF\_}I_M(s_{\text{micro-Doppler}}(t), f, \tau)$ at the nonrotation frequency much smaller than that of the rotating frequency. Based on this principle, in the function

$\mathrm{PMF\_}I_M(s_{\text{micro-Doppler}}(t),f,\tau)$ of the amplitude of the results, we will see the obvious peak in the rotating frequency.

In the analysis of Eq. (5.79) it can be found that the spectral broadening of the signal $B = 4\Omega r/\lambda$ after the DC component is removed, that is, the bandwidth of the signal is determined by the wavelength of the signal and the rotating frequency and radius of the scattering point. Since the wavelength is known, the rotation frequency has been estimated, so the estimation of the rotation radius can be accomplished by detecting the signal's bandwidth. The signal bandwidth of Eq. (5.79) can be represented as [13]:

$$B = 2\sqrt{2\int_{-\infty}^{+\infty} f^2|F(f)|^2 \mathrm{d}f}, \tag{5.90}$$

where $F(f)$ is the normalized power spectrum of the signal. Thus, the estimated value of the radius of rotation is:

$$\widehat{r} = \frac{\lambda B}{4\widehat{\Omega}}, \tag{5.91}$$

where $\widehat{\Omega}$ is the estimated rotational frequency.

However, if there are multiple rotational scattering points on the target, the bandwidth of the signal obtained by using Eq. (5.90) is not necessarily the bandwidth of the echo signal corresponding to the n-th rotating scatter point, and the calculated radius of rotation can also be wrong. In order to solve this problem, the method of serial elimination detection combined with high-order moment function can be used. The specific process is as follows:

Step (1) By detecting the peak point of the function $\mathrm{PMF\_}I_M(s_{\text{micro-Doppler}}(t),f,\tau)$ of the signal after the DC component, the collection of the rotation frequency is obtained and written as $\mathbf{F}(\Omega') = \{\Omega'_1, \Omega'_2, \cdots, \Omega'_N\}$;

Step (2) Using the obtained frequency, combined in Eq. (5.91), calculating all possible radius set $r'_i (i = 1, 2, \cdots, N)$ successively. Note that because of the existence of noise, the signal bandwidth obtained by using Eq. (5.91) may have a certain bias, it is necessary to use neighborhood $\mathbf{U}(r'_i)$ of radius r'_i to represent (defining the j-th element in the neighborhood as $r'_{i,j}$) in the collection of statistical radius, and rewrite the collection $\mathbf{\Gamma}$ to $\mathbf{\Gamma} = \{\mathbf{U}(r'_1), \mathbf{U}(r'_2), \cdots, \mathbf{U}(r'_N)\}$.

Step (3) Defining reference signal as

$$s_{i,j}(t, \theta_q) = \exp\left(\mathrm{j}2\pi r'_{i,j} \sin\left(\Omega'_i t + \theta_q\right)\right) \tag{5.92}$$

where $\Omega' \in \mathbf{F}(\Omega')$, $\theta_q \in [0, 2\pi)$ and m is the number of rotational scattering point. A new signal is obtained by conjugate multiplication of signal $s_{i,j}(t,\theta_q)$ with signal $s_{\text{micro-Doppler}}(t)$:

$$\begin{aligned} s'_{i,j}(t,\theta_q) &= s_{\text{micro-Doppler}}(t)s^*_{i,j}(t,\theta_q) \\ &= \sigma \exp\left(\mathrm{j}2\pi\left(r\sin(\Omega t+\theta) - r'_{i,j}\sin\left(\Omega'_i t+\theta_q\right)\right)\right) \end{aligned} \tag{5.93}$$

Step (4) Calculating the function $P_M\left(s'_{i,j}(t,\theta_q),\tau\right)$ of signal $s'_{i,j}(t,\theta_q)$ and Fourier transforming of the real part and take the mod, we can get the function $\text{MF\_}R_M\left(s'_{i,j}(t,\theta_q),f,\tau\right)$. Accumulating the function $\text{MF\_}R_M\left(s'_{i,j}(t,\theta_q),f,\tau\right)$ for different time delay values τ, we can get the function $\text{PMF\_}R_M\left(s'_{i,j}(t,\theta_q),f,\tau\right)$. By the nature of the function $\text{PMF\_}R_M\left(s'_{i,j}(t,\theta_q),f,\tau\right)$ it can be learned that if $r'_{i,j}$, f'_i, and θ_q are the same scattering points of the rotation radius, frequency, and initial phase, respectively, $\text{PMF\_}R_M\left(s'_{i,j}(t,\theta_q),f,\tau\right)$ will not produce a peak at the frequency Ω'_i/π. Otherwise, the peak will continue to exist. Correspondingly, the rotation radius, frequency, and initial phase of the B of the echo signal bandwidth are:

$$\left(\hat{r},\Omega',\hat{\theta}\right) = \underset{r'_{i,j},\Omega'_i,\theta_q}{\arg\min}\,\text{PMF\_}R_M\left(s'_{i,j}(t,\theta_q),\Omega'_i/\pi\right) \tag{5.94}$$

where $\Omega' \in \mathbf{F}(\Omega')$, $r'_{i,j} \in \mathbf{U}\left(r'_i\right)$, $\theta \in [0, 2\pi)$, and arg min $[\cdot]$ indicates the minimal value of the number in the brackets.

By calculation, the M-th order instantaneous moment of signal $s'_{i,j}(t,\theta_q)$ is:

$$\begin{aligned} P_M\left(s'_{i,j}(t,\theta_q),\tau\right) &= P_2\left(P_{M-1}\left(s'_{i,j}(t,\theta_q),\tau\right),\tau\right) \\ &= \sigma^{2^{M-1}} \exp\left(\mathrm{j}2^M\pi r\left(\sin\left(\frac{\Omega}{2}\tau\right)\right)^{M-1}\right. \\ &\quad \left.\times \sin\left(\Omega t+\theta+(M-1)\cdot\left(\frac{\Omega}{2}\tau+\frac{\pi}{2}\right)\right)\right) \\ &\quad \times \exp\left(-\mathrm{j}2^M\pi r'_{i,j}\left(\sin\left(\frac{\Omega'_i}{2}\tau\right)\right)^{M-1}\right. \\ &\quad \left.\times \sin\left(\Omega'_i t+\theta_q+(M-1)\cdot\left(\frac{\Omega'_i}{2}\tau+\frac{\pi}{2}\right)\right)\right) \end{aligned} \tag{5.95}$$

and the real part is:

$$\begin{aligned}\mathrm{Re}\left[P_M\left(s'_{i,j}(t,\theta_q),\tau\right)\right] &= \sigma^{2^{M-1}}\sin\left(2^M\pi r\left(\sin\left(\frac{\Omega}{2}\tau\right)\right)^{M-1}\right.\\ &\quad\times\sin\left(\Omega t+\theta+(M-1)\cdot\left(\frac{\Omega}{2}\tau+\frac{\pi}{2}\right)\right)\\ &\quad-2^M\pi r'_{i,j}\left(\sin\left(\frac{\Omega'_i}{2}\tau\right)\right)^{M-1}\\ &\quad\left.\times\sin\left(\Omega'_i t+\theta_q+(M-1)\cdot\left(\frac{\Omega'_i}{2}\tau+\frac{\pi}{2}\right)\right)\right)\end{aligned} \tag{5.96}$$

For narrowband radar target, there must be a value M, which can make both $2^M\pi r\left(\sin\left(\frac{\Omega}{2}\tau\right)\right)^{M-1}$ and $2^M\pi r'_{i,j}\left(\sin\left(\frac{\Omega'_i}{2}\tau\right)\right)^{M-1}$ approach zero. The process of the proof is approximate to the Eq. (5.83) to the Eq. (5.86). We can do the following approximation according to the nature of the sinusoidal function:

$$\begin{aligned}&\mathrm{Re}\left[P_M\left(s'_{i,j}(t,\theta_q),\tau\right)\right]\\ &\approx 1-\left(\sigma^{2^{M-1}}2^M\pi r\left(\sin\left(\frac{\Omega}{2}\tau\right)\right)^{M-1}\sin\left(\Omega t+\theta+(M-1)\cdot\left(\frac{\Omega}{2}\tau+\frac{\pi}{2}\right)\right)\right.\\ &\quad\left.-\sigma^{2^{M-1}}2^M\pi r'_{i,j}\left(\sin\left(\frac{\Omega'_i}{2}\tau\right)\right)^{M-1}\sin\left(\Omega'_i t+\theta_q+(M-1)\cdot\left(\frac{\Omega'_i}{2}\tau+\frac{\pi}{2}\right)\right)\right)^2\\ &= 1-\left(\sigma^{2^{M-1}}2^M\pi\right)^2\left(\frac{r^2}{2}\left(\sin\left(\frac{\Omega}{2}\tau\right)\right)^{2M-2}\left(1-\cos\left(2\Omega t+2\theta+2(M-1)\cdot\left(\frac{\Omega}{2}\tau+\frac{\pi}{2}\right)\right)\right)\right.\\ &\quad+rr'_{i,j}\left(\sin\left(\frac{\Omega}{2}\tau\right)\sin\left(\frac{\Omega'_i}{2}\tau\right)\right)^{M-1}\cos\left((\Omega+\Omega'_i)t+\theta+\theta_q+(M-1)\cdot\left(\frac{\Omega+\Omega'_i}{2}\tau+\pi\right)\right)\\ &\quad-rr'_{i,j}\left(\sin\left(\frac{\Omega}{2}\tau\right)\sin\left(\frac{\Omega'_i}{2}\tau\right)\right)^{M-1}\cos\left((\Omega-\Omega'_i)t+\theta-\theta_q+(M-1)\cdot\frac{\Omega-\Omega'_i}{2}\tau\right)\\ &\quad\left.+\frac{\left(r'_{i,j}\right)^2}{2}\left(\sin\left(\frac{\Omega'_i}{2}\tau\right)\right)^{2M-2}\left(1-\cos\left(2\Omega'_i t+2\theta_q+2(M-1)\cdot\left(\frac{\Omega'_i}{2}\tau+\frac{\pi}{2}\right)\right)\right)\right)\end{aligned} \tag{5.97}$$

On the analysis of Eq. (5.97), the signal $\mathrm{Re}\left[P_M\left(s'_{i,j}(t,\theta_q),\tau\right)\right]$ is the frequency of Ω/π, Ω'_i/π, $(\Omega+\Omega'_i)/(2\pi)$, and $(\Omega-\Omega'_i)/(2\pi)$ of the sinusoidal signal and a DC signal superposition. Fourier transform process is as follows:

$$\begin{aligned}
&\mathrm{FFT}\left[\mathrm{Re}\left[P_M\left(s'_{i,j}(t,\theta_q),\tau\right)\right]\right]\\
&=2\pi\delta(f)-\frac{1}{2}\pi\left(\sigma^{2^{M-1}}2^M\pi\right)^2\\
&\times\left(r^2\left(\sin\left(\frac{\Omega}{2}\tau\right)\right)^{2M-2}\left(2\delta(f)-\left(\delta\left(f+\frac{\Omega}{\pi}\right)+\delta\left(f-\frac{\Omega}{\pi}\right)\right)\exp\left(\mathrm{j}2\pi f\frac{\theta+(M-1)\cdot\left(\frac{\Omega}{2}\tau+\frac{\pi}{2}\right)}{\Omega}\right)\right)\right.\\
&-\left(r'_{i,j}\right)^2\left(\sin\left(\frac{\Omega_i}{2}\tau\right)\right)^{2M-2}\left(2\delta(f)-\left(\delta\left(f+\frac{\Omega'_i}{\pi}\right)+\delta\left(f-\frac{\Omega'_i}{\pi}\right)\right)\right.\\
&\times\exp\left(\mathrm{j}2\pi f\frac{\theta_q+(M-1)\cdot\left(\frac{\Omega'_i}{2}\tau+\frac{\pi}{2}\right)}{\Omega'_i}\right)\\
&+2rr'_{i,j}\left(\sin\left(\frac{\Omega}{2}\tau\right)\right)^{M-1}\left(\sin\left(\frac{\Omega_i}{2}\tau\right)\right)^{M-1}\left(\delta\left(f+\frac{\Omega+\Omega'_i}{\pi}\right)+\delta\left(f-\frac{\Omega+\Omega'_i}{\pi}\right)\right)\\
&\times\exp\left(\mathrm{j}2\pi f\frac{\theta+\theta_q+(M-1)\cdot\left(\frac{\Omega+\Omega'_i}{2}\tau+\pi\right)}{\Omega+\Omega'_i}\right)-2rr'_{i,j}\left(\sin\left(\frac{\Omega}{2}\tau\right)\right)^{M-1}\left(\sin\left(\frac{\Omega_i}{2}\tau\right)\right)^{M-1}\\
&\left.\times\left(\delta\left(f+\frac{\Omega-\Omega'_i}{\pi}\right)+\delta\left(f-\frac{\Omega-\Omega'_i}{\pi}\right)\right)\exp\left(\mathrm{j}2\pi f\frac{\theta-\theta_q+(M-1)\cdot\frac{\Omega-\Omega'_i}{2}\tau}{\Omega-\Omega'_i}\right)\right)
\end{aligned}\tag{5.98}$$

Defining function $\mathrm{MF\_}R_M\left(s'_{i,j}(t,\theta_q),\Omega'_i/\pi,\tau\right)$ as the value at frequency Ω'_i/π of the mod value $\mathrm{MF\_}R_M\left(s'_{i,j}(t,\theta_q),f,\tau\right)$ of the Fourier transform result of signal $\mathrm{Re}\left[P_M\left(s'_{i,j}(t,\theta_q),\tau\right)\right]$. Then analyzing $\mathrm{MF\_}R_M\left(s'_{i,j}(t,\theta_q),f,\tau\right)$ as follows:

1. if $\Omega\neq\Omega'_i$, then

$$\begin{aligned}
\mathrm{MF_}R_M\left(s'_{i,j}(t,\theta_q),\Omega'_i/\pi,\tau\right)&=\frac{\left(r'_{i,j}\right)^2}{2}\pi\left(\sigma^{2^{M-1}}2^M\pi\left(\sin\left(\frac{\Omega'_i}{2}\tau\right)\right)^{M-1}\right)^2\\
&=\frac{\left(r'_{i,j}\right)^2}{2}\pi\left(\sigma^{2^{M-1}}2^M\pi\left(\sin\left(\frac{\Omega'_i}{2}\tau\right)\right)^{M-1}\right)^2
\end{aligned}\tag{5.99}$$

2. if $\Omega = \Omega_i'$, then

$$\begin{aligned}\mathrm{MF\_}R_M\left(s_{i,j}'(t,\theta_q),\Omega_i'/\pi,\tau\right) &= \frac{\pi}{2}\left(\sigma^{2^{M-1}}2^M\pi\left(\sin\left(\frac{\Omega_i'}{2}\tau\right)\right)^{M-1}\right)^2 \\ &\quad\times\left|r^2+r_{i,j}'^2-2rr_{i,j}'\cos\frac{\theta-\theta_q}{2}\right| \\ &\geq \frac{\pi}{2}\left(\sigma^{2^{M-1}}2^M\pi\left(\sin\left(\frac{\Omega_i'}{2}\tau\right)\right)^{M-1}\right)^2\left(r-r_{i,j}'\right)^2\end{aligned} \tag{5.100}$$

When $\theta = \theta_q$, symbol "=" in the formula "≥" sets up.

From comprehensive analysis of Eqs. (5.99) and (5.100) we can know that when $r_{i,j}' = r$, $\Omega_i' = \Omega$, and $\theta_q = \theta$ the function $\mathrm{MF\_}R_M\left(s_{i,j}'(t,\theta_q),\Omega_i'/\pi,\tau\right)$ will obtain the minimum value. That is when the estimation of the rotation radius and the initial phase of the scattering point at rotating frequency $\Omega_i'/(2\pi)$ is accurate, the function value of $\mathrm{MF\_}R_M\left(s_{i,j}'(t,\theta_q),f\right)$ at Ω_i'/π will be smaller than that of $\mathrm{MF\_}R_M\left(s_{i,j}'(t,\theta_q),\Omega_i'/\pi,\tau\right)$ when estimation is wrong. In order to weaken the influence of cross-terms, we can accumulate function $\mathrm{MF\_}R_M\left(s_{i,j}'(t,\theta_q),f,\tau\right)$ for different time delay τ. By detecting the minimal radius and the initial phase of the $\mathrm{PMF\_}R_M\left(s_{i,j}'(t,\theta_q),\Omega_i'/\pi\right)$, the extraction of the rotation radius and the initial phase can be accomplished. In order to avoid the influence of different time delay value τ on the weights of $\mathrm{MF\_}R_M\left(s_{i,j}'(t,\theta_q),\Omega_i'/\pi,\tau\right)$ in the process of calculating function $\mathrm{PMF\_}R_M\left(s_{i,j}'(t,\theta_q),f\right)$, the $\mathrm{MF\_}R_M\left(s_{i,j}'(t,\theta_q),f,\tau\right)$ can be divided by $\left(\sin\left(\Omega_i'\tau/2\right)\right)^{2M-2}$ then cumulative multiply the results, and we can get $\mathrm{PMF\_}R_M\left(s_{i,j}'(t,\theta_q),f\right)$.

For the case of multiple scattering points, the "clean" algorithm can be used to eliminate the echo signal from the echo signal, and then repeat this step to complete the micromotion feature extraction of all rotating scattered points.

5.5.1.2 Simulation

The simulation parameters are the same as the simulation parameters in Section 5.1.2.1. Suppose there are still two rotational scattering points on the target, and the coordinates of the local coordinate system are (1,0.5,0.5) m and (−1,−0.5,−0.5) m respectively, and the reflection coefficient of the scattered point is 1. The duration of echo signal is $T_p = 3.2688$ s, and the sampling rate is $f_s = 2000$ Hz.

The normalized amplitude of the echo signal function $\mathrm{MF\_}I_3(s_{\text{micro-Doppler}}(t),f,\tau)$ obtained by using Eq. (5.88) is shown in Fig. 5.35. From Fig. 5.35, it can be found that the corresponding frequency at the peak value of the function $\mathrm{MF\_}I_3(s_{\text{micro-Doppler}}(t),f,\tau)$

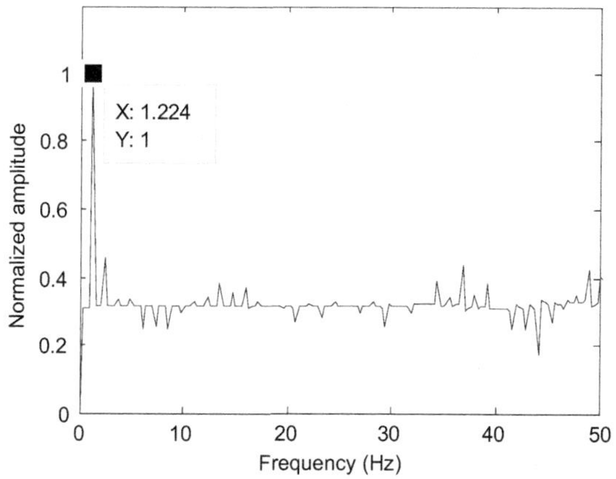

Figure 5.35 Normalized amplitude results of echo signal function MF_I_3.

is 1.224 Hz, which is close to the theoretical value, but there are still many false peak points, and these are very likely to affect the detection of the rotation frequency. By Eq. (5.82) the definition of HOMFs we can find that these false peak point is also very likely to be caused by the interference of cross-terms. In order to reduce the effect of cross-terms, the function PMF_$I_3(s_{\text{micro-Doppler}}(t),f,\tau)$ can be obtained by using different time delay. Fig. 5.36 shows the value of the normalized amplitude of the echo signal function PMF_$I_3(s_{\text{micro-Doppler}}(t),f,\tau)$ (the time delay τ range is $[1/f_s,10/f_s]$, the sampling interval is $1/f_s$). In Fig. 5.36 as the cause of the cumulative multiplication, the peak value of the self-term is effectively amplified and the peak of the cross-term is reduced, so the

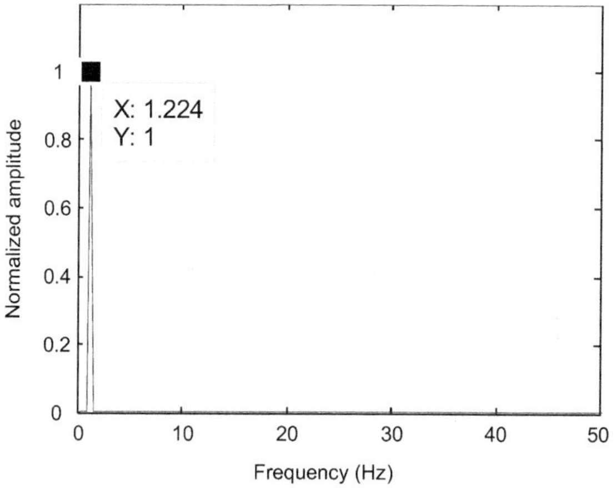

Figure 5.36 Normalized amplitude results of echo signal function PMF_I_3.

rotation frequency can be determined as 1.224 Hz. It is clear that using HOMF analysis method can effectively achieve the extraction of the rotation frequency.

In order to extract the rotation radius and the initial phase of the rotation point further, the bandwidth of the signal $s_{\text{micro-Doppler}}(t)$ calculated by using the Eq. (5.90) is 450 Hz. At the same time, the rotation radius corresponding to the rotation frequency of the 1.224 Hz by using Eq. (5.91) is 0.44 m. Setting the frequency corresponding to the rotation radius of the neighborhood ranges from 0.3 to 0.5 m, and the sampling interval is 0.01 m. At the same time, setting the selection range of the initial phase θ_q of the reference signal in Eq. (5.92) is [0,2π) rad, and the sampling interval is 0.1 rad. Then, after the echo signal is conjugated multiplied by the reference signal, the normalized amplitude value of the signal $1/\text{PMF\_}R_3\left(s'_{i,j}(\tau,\theta_q),f\right)$ in Eq. (5.93) is shown in Fig. 5.37 (reciprocal the function $\text{PMF\_}R_3\left(s'_{i,j}(\tau,\theta_q),f\right)$ and detecting the peak value, which will have the same effect with Eq. (5.94)). By detecting the peak position, we will find that the corresponding rotation radius and initial phase are (1 rad, 0.42 m) respectively. Then using the "clean" algorithm to eliminate the echo signal of the rotating scattered point in the signal $s_{\text{micro-Doppler}}(t)$, and then calculating the $\text{PMF\_}I_3$ function of the eliminated signal, the frequency is still 1.224 Hz by detecting the peak value. The $1/\text{PMF\_}R_3$ function is calculated by multiplying the reference signal, and the normalized range of $1/\text{PMF\_}R_3$ is shown in Fig. 5.38. From Fig. 5.37B, we can know that the rotation radius and the initial phase of the second rotational scattering points are (0.42 m, 4.2 rad) respectively, which are close to the theoretical values.

In order to verify the performance of the HOMF analysis method, the situation of microwave scatter point echo covered by Gaussian white noise is analyzed. Suppose the SNR of the target echo signal is 0 dB. The normalized amplitude value of the $\text{PMF\_}I_3(s_{\text{micro-Doppler}}(t),f,\tau)$ function obtained by the signal $s_{\text{micro-Doppler}}(t)$ is shown in Fig. 5.38A. It can be seen from the figure that the rotation frequency of the rotating

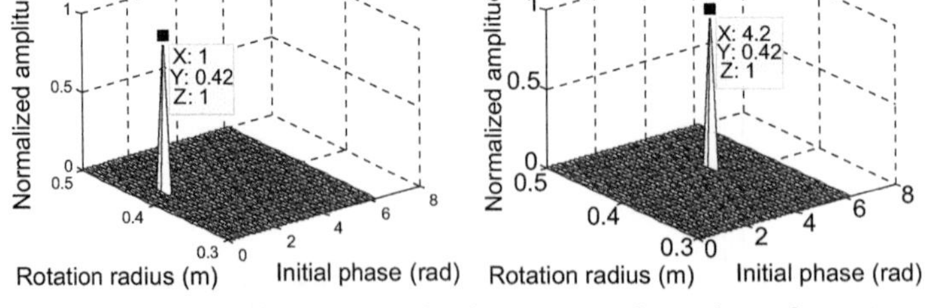

Figure 5.37 High-order moment function method to extract the radius of rotation and the initial phase. (A) The normalized amplitude result of $1/\text{PMF\_}R_3$ after the first calculation; (B) the normalized amplitude result of $1/\text{PMF\_}R_3$ after the second calculation.

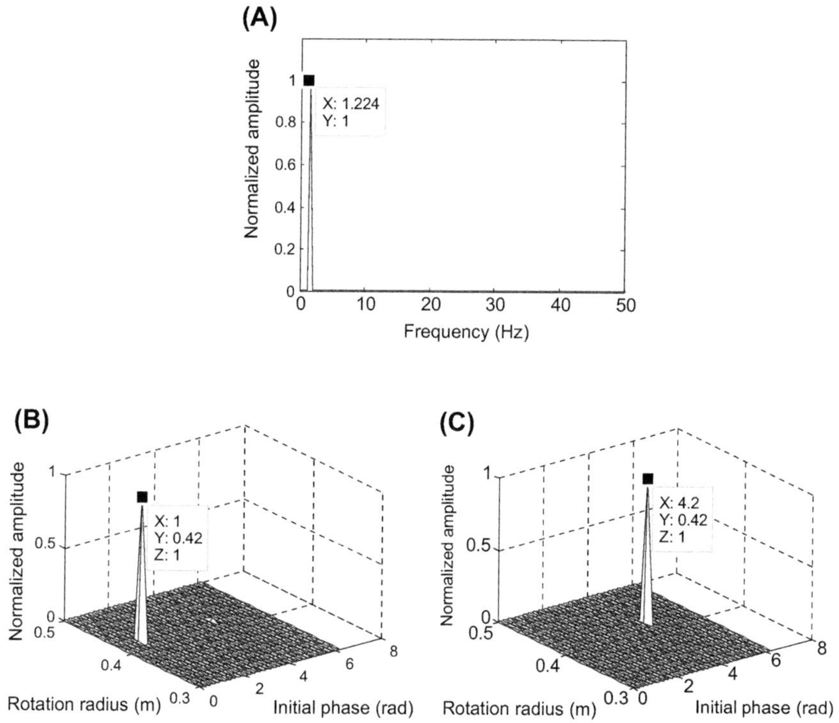

Figure 5.38 The processing result obtained when SNR = 0 dB. (A) The normalized amplitude result of echo signal function MF_I_3; (B) the normalized amplitude result of function 1/PMF_R_3 after the first calculation; (C) the normalized amplitude result of function 1/PMF_R_3 after the second calculation.

scattered point is 1.224 Hz, which is close to the theoretical value. The normalized amplitude of the function $1/\text{PMF\_}R_3\left(s'_{i,j}(\tau,\theta_q),f\right)$ obtained after the first conjugate multiplication with the reference signal is shown in Fig. 5.38B. By extracting the peak position, we can know the rotation radius and the initial phase of the rotation point are respectively (0.42 m, 1 rad). Using the "clean" algorithm to eliminate the influence of the rotational scattering point, the second normalized amplitude of the functions $1/\text{PMF\_}R_3$ is shown in Fig. 5.39C. The rotating radius and the initial phase of the second rotating point are respectively (0.42 m, 4.2 rad). So using the HOMF method, the extraction of micromotion feature can still be accomplished effectively under the condition of SNR = 0 dB.

When the SNR = −5 dB, the normalized amplitude value of function PMF_ $I_3(s_{\text{micro-Doppler}}(t),f,\tau)$ obtained by using signal $s_{\text{micro-Doppler}}(t)$ is shown in Fig. 5.39. It can be seen from the figure, although there is a more obvious peak at the frequency of 1.224 Hz, but because of the influence of noise, the function still has many false peak points, which lead to the error of the rotation frequency extraction.

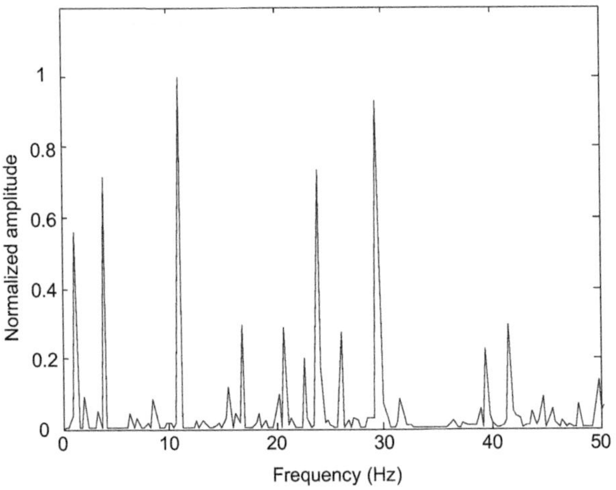

Figure 5.39 The normalized amplitude result of function PMF_l_3 when SNR $= -5$ dB.

5.5.2 Application in Wideband Radar

5.5.2.1 Algorithm

From Eq. (3.26), we can know that after conjugate multiplication with the reference signal, the echo signal of a single rotating scatter point can be written as follows:

$$s_d(t_k, t_m) = \sigma rect\left(\frac{t_k - 2r(t_m)/c}{T_p}\right) \cdot \exp\left(j\left(-\frac{4\pi}{c}\mu\left(t_k - \frac{2R_{ref}(t_m)}{c}\right)\right.\right.$$
$$\left.\left.\times R_\Delta(t_m) - \frac{4\pi}{c}f_c R_\Delta(t_m) + \frac{4\pi\mu}{c^2}R_\Delta(t_m)^2\right)\right) \tag{5.101}$$

According to Eq. (3.29), $R_\Delta(t_m)$ above can be written as

$$R_\Delta(t_m) = A' + r' \cos(\Omega t_m + \theta) \tag{5.102}$$

where $A' = \left\|\overrightarrow{GO''}\right\|$, $r' = r \sin \varepsilon$.

For a given fast time delay τ_k, the first-order and the second-order moment of the fast time are defined as follows, respectively

$$\Psi_{1,1}(s_d(t_k, t_m), (\tau_k, 0)) = s_d(t_k, t_m) \tag{5.103}$$

$$\begin{aligned}\Psi_{2,1}(s_d(t_k, t_m), (\tau_k, 0)) &= \Psi_{1,1}(s_d(t_k, t_m), (\tau_k, 0))\Psi^*_{1,1}(c_n(t_k + \tau_k, t_m), (\tau_k, 0))\\ &= \sigma^2 \mathrm{exrect}_{2,1}\left(\frac{t_k - 2r(t_m)/c}{T_p}, (\tau_k, 0)\right)\exp\left(j\frac{4\pi}{c}\mu\tau_k R_\Delta(t_m)\right)\end{aligned} \tag{5.104}$$

where $\mathrm{exrect}_{2,1}\left(\frac{t_k-2r(t_m)/c}{T_p},(\tau_k,0)\right) = \mathrm{rect}\left(\frac{t_k-2r(t_m)/c}{T_p}\right)\mathrm{rect}\left(\frac{t_k-2r(t_m)/c+\tau_k}{T_p}\right)$.

From Eq. (5.104), we can find that the phase of the second-order moment function $\Psi_{2,1}(s_d(t_k,t_m),(\tau_k,0))$ of the signal $s_d(t_k,t_m)$ about the fast time is irrelevant to the fast time τ_k. The slow time delay τ_m is given, so the second-order moment of the signal $\Psi_{2,1}(s_d(t_k,t_m),(\tau_k,0))$ of the slow time is obtained:

$$\begin{aligned}\Psi_{2,2}(s_d(t_k,t_m),(\tau_k,\tau_m)) &= \Psi_{2,1}(s_d(t_k,t_m),(\tau_k,\tau_m))\Psi^*_{2,1}(s_d(t_k,t_m+\tau_m),(\tau_k,\tau_m))\\ &= \sigma^4\mathrm{exrect}_{2,2}\left(\frac{t_k-2r(t_m)/c}{T_p},(\tau_k,\tau_m)\right)\\ &\times \exp\left(\mathrm{j}\frac{4\pi}{c}\mu\tau_k r'\cdot 2\sin\left(\frac{\Omega}{2}\tau_m\right)\cos\left(\Omega t_m+\theta+\frac{\Omega}{2}\tau_m\right)\right)\end{aligned} \tag{5.105}$$

where

$$\begin{aligned}\mathrm{exrect}_{2,2}\left(\frac{t_k-2r(t_m)/c}{T_p},(\tau_k,\tau_m)\right) &= \mathrm{exrect}_{2,1}\left(\frac{t_k-2r(t_m)/c}{T_p},(\tau_k,0)\right)\\ &\times \mathrm{exrect}_{2,1}\left(\frac{t_k-2r(t_m+\tau_m)/c}{T_p},(\tau_k,0)\right)\end{aligned}$$

By further calculation, we can know that the M-order ($M \geq 3$) moment of the signal $\Psi_{2,1}(s_d(t_k,t_m),(\tau_k,0))$ of the slow time is:

$$\begin{aligned}\Psi_{2,M}(s_d(t_k,t_m),(\tau_k,\tau_m)) &= \Psi_{2,M-1}(s_d(t_k,t_m),(\tau_k,\tau_m))\cdot\Psi^*_{2,M-1}(s_d(t_k,t_m),(\tau_k,\tau_m))\\ &= \sigma^{2^{M-1}}\mathrm{exrect}_{2,M}\left(\frac{t_k-2r(t_m)/c}{T_p},(\tau_k,\tau_m)\right)\\ &\times \exp\left(\mathrm{j}\frac{4\pi}{c}\mu\tau_k r'\left(2\sin\left(\frac{\Omega}{2}\tau_m\right)\right)^{M-1}\right.\\ &\times \left.\sin\left(\Omega t_m+\theta+(M-1)\left(\frac{\Omega}{2}\tau_m+\frac{\pi}{2}\right)\right)\right)\end{aligned} \tag{5.106}$$

where

$$\mathrm{exrect}_{2,M}\left(\frac{t_k - 2r(t_m)/c}{T_p}, (\tau_k, \tau_m)\right) = \mathrm{exrect}_{2,M-1}\left(\frac{t_k - 2r(t_m)/c}{T_p}, (\tau_k, \tau_m)\right)$$
$$\cdot \mathrm{exrect}_{2,M-1}\left(\frac{t_k - 2r(t_m + \tau_m)/c}{T_p}, (\tau_k, \tau_m)\right)$$

Since we only need select several fixed sets of data to make the second moment function of the fast time domain, the influence of $\mathrm{exrect}_{2,M}\left(\frac{t_k - 2r(t_m)/c}{T_p}, (\tau_k, \tau_m)\right)$ can be ignored. Define that

$$a(M) = \frac{4\pi}{c}\mu\tau_k r'\left(2\sin\left(\frac{\Omega}{2}\tau_m\right)\right)^{M-1} \tag{5.107}$$

When $0 < \tau_m < \pi/(3\Omega)$ (because the sampling rate of the slow time is $\mathrm{PRF} \geq \Omega/(2\pi)$ in general, which is easy to satisfy), make the derivation of $a(M)$:

$$\frac{\mathrm{d}a(M)}{\mathrm{d}M} = \frac{4\pi}{c}\mu\tau_k r'\left(2\sin\left(\frac{\Omega}{2}\tau_m\right)\right)^{M-1}\ln\left(2\sin\left(\frac{\Omega}{2}\tau_m\right)\right) \tag{5.108}$$

Since $0 < \tau_m < \pi/(3\Omega)$, $0 < 2\sin\left(\frac{\Omega}{2}\tau_m\right) < 1$. Meanwhile, it is required that $\tau_k > 0$, and then

$$\ln\left(2\sin\left(\frac{\Omega}{2}\tau_m\right)\right) < 0 \tag{5.109a}$$

$$\left(2\sin\left(\frac{\Omega}{2}\tau_m\right)\right)^{M-1} > 0 \tag{5.109b}$$

$$\lim_{M\to+\infty} a(M) = \frac{4\pi}{c}\mu\tau_k r' \lim_{M\to+\infty}\left(2\sin\left(\frac{\Omega}{2}\tau_m\right)\right)^{M-1} = 0 \tag{5.109c}$$

From Eqs. (5.109a) and (5.109b) we can know that

$$\frac{\mathrm{d}a(M)}{\mathrm{d}M} < 0 \tag{5.110}$$

It means $a(M)$ is a decreasing function of M.

Since $\left|\sin\left(\Omega t_m + \theta + (M-1)\left(\frac{\Omega}{2}\tau_m + \frac{\pi}{2}\right)\right)\right| \leq 1$, as for any given slow time $\lim_{M\to+\infty} a(M)\left|\sin\left(\Omega t_m + \theta + (M-1)\left(\frac{\Omega}{2}\tau_m + \frac{\pi}{2}\right)\right)\right| = 0$, it is always true that

$$\lim_{M\to+\infty} a(M)\left|\sin\left(\Omega t_m + \theta + (M-1)\left(\frac{\Omega}{2}\tau_m + \frac{\pi}{2}\right)\right)\right| = 0 \tag{5.111}$$

That is to say, there must exist a value M, which makes $a(M)$ $\sin(2\pi f_n t_m + \theta_n + (M-1)(\pi f_n \tau_m + \pi/2))$ approach zero.

Therefore, the approximation of the imaginary part of $\Psi_{2,M}(s_d(t_k,t_m),(\tau_k,\tau_m))$ can be taken as follows:

$$\begin{aligned}\mathrm{Im}\left[\Psi_{2,M}(s_d(t_k,t_m),(\tau_k,\tau_m))\right] \approx \sigma^{2^{M-1}} \mathrm{j}\frac{4\pi}{c}\mu\tau_k r'\left(2\sin\left(\frac{\Omega}{2}\tau_m\right)\right)^{M-1} \\ \sin\left(\Omega t_m + \theta + (M-1)\left(\frac{\Omega}{2}\tau_m + \frac{\pi}{2}\right)\right)\end{aligned} \tag{5.112}$$

From Eq. (5.112), the imaginary part of $\Psi_{2,M}(s_d(t_k,t_m),(\tau_k,\tau_m))$ is a single frequency signal with the frequency of $\Omega/(2\pi)$.

Make the slow time Fourier transform and modulus of Eq. (5.112),

$$\begin{aligned}\mathrm{MF\_}I_M(s_d(t_k,t_m),(\tau_k,\tau_m),f_m) &= \left|\mathrm{FFT}\left[\mathrm{Im}\left[\Psi_{2,M}(s_d(t_k,t_m),(\tau_k,\tau_m))\right],f_m\right]\right| \\ &= \sigma^{2^{M-1}}\frac{4\pi}{c}\mu\tau_k r'\left(2\sin\left(\frac{\Omega}{2}\tau_m\right)\right)^{M-1}\left(\delta\left(f+\frac{\Omega}{2\pi}\right)\right. \\ &\quad \left. + \delta\left(f-\frac{\Omega}{2\pi}\right)\right)\end{aligned} \tag{5.113}$$

From the characteristics of the impulse function, we can know that the corresponding frequency to the peak point of the positive frequency part of $\mathrm{MF\_}I_M(s_d(t_k,t_m),(\tau_k,\tau_m),f_m)$ is $\Omega/(2\pi)$. Therefore, the detection of the rotation frequency can be accomplished by the peak point detection of the function $\mathrm{MF\_}I_M(s_{\text{micro-Doppler}}(t),f,\tau)$.

However, according to the definition of the HOMF Eqs. (5.104) and (5.106), the cross-terms will exist in the HOMFs, when there are multiple rotating scattering points on the target, no matter they are the HOMFs of the echo signal $s_d(t_k,t_m)$ of the fast time or the slow time.

Because of the interference of the cross-terms, the $MF\_I_M(s_d(t_k,t_m),(\tau_k,\tau_m),f_m)$ value of echo signal $s_d(t_k,t_m)$ is very complex. Similar with the HOMF analysis method in narrowband radar, considering the different time delay values, the rotating diffuse point of the echo signal items will always have a peak value at the corresponding rotating frequency. However, the cross-terms of different echo signals will not always produce a peak at a certain frequency. Therefore, a new function will be obtained, when the corresponding function $MF\_I_M(s_d(t_k,t_m),(\tau_k,\tau_m),f_m)$ of the different fast time delay τ_k and the slow time delay τ_m are accumulated by the cumulative multiplication.

$$PMF\_I_M(s_d(t_k,t_m),(\tau_k,\tau_m),f_m) = \prod_{\tau_m}\prod_{\tau_k} MF\_I_M(s_d(t_k,t_m),(\tau_k,\tau_m),f_m) \quad (5.114)$$

Similarly, when calculating the higher moment functions, the value $MF\_I_M(s_d(t_k,t_m),(\tau_k,\tau_m),f_m)$ of the echo signal of the rotating point in the corresponding peak values can be amplified of the rotating frequency since there is always a peak. Since $PMF\_I_M(s_d(t_k,t_m),(\tau_k,\tau_m),f_m)$ is always amplified because there is a maximum value, which results in a higher peak value of the $PMF\_I_M(s_d(t_k,t_m),(\tau_k,\tau_m),f_m)$ function at the corresponding rotation frequency point. However, for the cross-terms, $MF\_I_M(s_d(t_k,t_m),(\tau_k,\tau_m),f_m)$ cannot always reach its maximum value at a fixed nonrotation frequency point, then the accumulative rising speed of amplitude at the nonrotation frequency point is far slower than the one at the rotation frequency point in $PMF\_I_M(s_d(t_k,t_m),(\tau_k,\tau_m),f_m)$. Therefore, the false peak will be weakened relatively in $PMF\_I_M(s_d(t_k,t_m),(\tau_k,\tau_m),f_m)$.

All in all, the procedures of rotating frequency extraction are as follows:

Step (1) Set the corresponding fast time delay τ_k. Calculate the second-order function $\Psi_{2,1}(s_d(t_k,t_m),(\tau_k,0))$ of the fast time, which is corresponding to echo signal $s_d(t_k,t_m)$ of a micromotion point on certain fast time.

Step (2) Set the corresponding slow time delay τ_m. Obtain the M-order moment function $\Psi_{2,M}(s_d(t_k,t_m),(\tau_k,\tau_m))$, which is corresponding to the slow time $\Psi_{2,1}(s_d(t_k,t_m),(\tau_k,0))$.

Step (3) For the imaginary part of $\Psi_{2,M}(s_d(t_k,t_m),(\tau_k,\tau_m))$, we can make Fourier transforming on slow time t_m and take its mod, then get the function $MF\_I_M(s_d(t_k,t_m),(\tau_k,\tau_m),f_m)$.

Step (4) Reset the value of τ_k and τ_m, repeat Steps (1)—(3), cumulatively multiply the function $MF\_I_M(s_d(t_k,t_m),(\tau_k,\tau_m),f_m)$, and we can get $PMF\_I_M(s_d(t_k,t_m),(\tau_k,\tau_m),f_m)$, until the value of its nonobvious peak point less than the set threshold.

Thus, through the peak detection of function Ω'_n, you can get the rotation frequency of the rotary scatters.

The extraction methods of the rotation radius and the initial phase of the rotation point will be discussed as following. If there are different rotating frequencies of the scattering points on the target, when extracting the rotating radius and the initial phase of each rotation scatterer, it should be analyzed respectively for different rotational frequencies. A new reference signal is constructed:

$$s'_n(t_k, t_m) = \exp\left(\mathrm{j}\frac{4\pi}{c}\mu\tau_k r'_n \sin\left(\Omega'_n t_m + \theta'_n\right)\right) \tag{5.115}$$

where τ_k represents the delay value used in the calculation of the fast time on the second moment function.

We still take a single rotating scatterer for example to analyze. A new signal is obtained by conjugate multiplication Eqs. (5.104) and (5.115):

$$\begin{aligned} s_{nc}\left(t_k, t_m, r'_n, \theta'_n\right) &= \Psi_{2,1}(s_d(t_k, t_m), (\tau_k, 0))s'^{*}_n(t_k, t_m) \\ &= \sigma^2 \exp\left(\mathrm{j}\frac{4\pi}{c}\mu\tau_k\left(A + r'\cos(\Omega t_m + \theta) - r'_n \sin\left(\Omega'_n t_m + \theta'_n\right)\right)\right) \end{aligned} \tag{5.116}$$

Calculate the second-order moment of Eq. (5.116) for the slow time,

$$\begin{aligned} \Psi_2\left(s_{nc}\left(t_k, t_m, r'_n, \theta'_n\right), \tau_m\right) &= s_{nc}\left(t_k, t_m, r'_n, \theta'_n\right) \cdot s^{*}_{nc}\left(t_k, t_m + \tau_m, r'_n, \theta'_n\right) \\ &= \sigma^2 \exp\left(\mathrm{j}\frac{4\pi}{c}\mu\tau_k\left(r'\left(2\sin\left(\frac{\Omega}{2}\tau_m\right)\right)^{M-1}\right.\right. \\ &\quad \times \sin\left(\Omega t_m + \theta + (M-1)\cdot\left(\frac{\Omega}{2}\tau_m + \frac{\pi}{2}\right)\right) \\ &\quad - r'_n\left(2\sin\left(\frac{\Omega'_n}{2}\tau_m\right)\right)^{M-1} \\ &\quad \left.\left. \times \sin\left(\Omega'_n t_m + \theta'_n + (M-1)\cdot\left(\frac{\Omega'_n}{2}\tau_m + \frac{\pi}{2}\right)\right)\right)\right) \end{aligned} \tag{5.117}$$

Analyzing Eq. (5.117) we found that, there must exist value $\mathrm{Re}\left[\Psi_M\left(s_{nc}\left(t_k, t_m, r'_n, \theta'_n\right), \tau_m\right)\right]$, which makes both $\mathrm{Re}\left[\Psi_M\left(s_{nc}\left(t_k, t_m, r'_n, \theta'_n\right), \tau_m\right)\right]$ and $\mathrm{Re}\left[\Psi_M\left(s_{nc}\left(t_k, t_m, r'_n, \theta'_n\right), \tau_m\right)\right]$ approach zero. Thus, the real part of Eq. (5.117) can be approximately written as [The process of the proof is approximate to Eqs. (5.106)−(5.111)]:

$$
\begin{aligned}
&\mathrm{Re}\left[\Psi_M\left(s_{nc}\left(t_k, t_m, r'_n, \theta'_n\right), \tau_m\right)\right]\\
&\quad\approx 1-\left(\sigma^{2^{M-1}}\frac{4\pi}{c}\mu\tau_k\left(r'\left(2\sin\left(\frac{\Omega}{2}\tau_m\right)\right)^{M-1}\right.\right.\\
&\qquad\times\sin\left(\Omega t_m+\theta+(M-1)\cdot\left(\frac{\Omega}{2}\tau_m+\frac{\pi}{2}\right)\right)\\
&\qquad-r'_n\left(2\sin\left(\frac{\Omega'_n}{2}\tau_m\right)\right)^{M-1}\\
&\qquad\left.\left.\times\sin\left(\Omega'_n t_m+\theta'_n+(M-1)\cdot\left(\frac{\Omega'_n}{2}\tau_m+\frac{\pi}{2}\right)\right)\right)\right)^2\\
&\quad= 1-\left(\sigma^{2^{M-1}}2^{M-1}\frac{4\pi}{c}\mu\tau_k\right)^2\left(\frac{r'^2}{2}\left(\sin\left(\frac{\Omega}{2}\tau_m\right)\right)^{2M-2}\right.\\
&\qquad\times\left(1-\cos\left(2\Omega t_m+2\theta+2(M-1)\cdot\left(\frac{\Omega}{2}\tau_m+\frac{\pi}{2}\right)\right)\right)\\
&\qquad+r'r'_n\left(\sin\left(\frac{\Omega}{2}\tau_m\right)\right)^{M-1}\left(\sin\left(\frac{\Omega'_n}{2}\tau_m\right)\right)^{M-1}\\
&\qquad\times\cos\left(\left(\Omega+\Omega'_n\right)t_m+\theta+\theta'_n+(M-1)\cdot\left(\frac{\Omega+\Omega'_n}{2}\tau_m+\pi\right)\right)\\
&\qquad-r'r'_n\left(\sin\left(\frac{\Omega}{2}\tau_m\right)\right)^{M-1}\left(\sin\left(\frac{\Omega'_n}{2}\tau_m\right)\right)^{M-1}\\
&\qquad\times\cos\left(\left(\Omega-\Omega'_n\right)t_m+\theta-\theta'_n+(M-1)\cdot\frac{\Omega-\Omega'_n}{2}\tau_m\right)\\
&\qquad+\frac{\left(r'_n\right)^2}{2}\left(\sin\left(\frac{\Omega'_n}{2}\tau_m\right)\right)^{2M-2}\\
&\qquad\left.\times\left(1-\cos\left(2\Omega'_n t_m+2\theta'_n+2(M-1)\cdot\left(\frac{\Omega'_n}{2}\tau_m+\frac{\pi}{2}\right)\right)\right)\right)
\end{aligned}
\tag{5.118}
$$

Analyzing Eq. (5.118) we found that, the signal $\Omega/\pi$is the sum of a DC signal and a sinusoidal signal at the frequencies of Ω/π, Ω_i'/π, $(\Omega+\Omega_i')/(2\pi)$, and $(\Omega-\Omega_i')/(2\pi)$. Fourier transform process is as follows:

$$
\begin{aligned}
&\mathrm{FFT}\left[\mathrm{Re}\left[\Psi_M\left(s_{nc}\left(t_k,t_m,r_n',\theta_n'\right),\tau_m\right)\right]\right]\\
&=2\pi\delta(f)-\frac{\pi}{2}\left(\sigma^{2^{M-1}}2^{M-1}\frac{4\pi}{c}\mu\tau_k\right)^2\\
&\times\left(r'^2\left(\sin\left(\frac{\Omega}{2}\tau_m\right)\right)^{2M-2}\left(2\delta(f)-\left(\delta\left(f+\frac{\Omega}{\pi}\right)+\delta\left(f-\frac{\Omega}{\pi}\right)\right)\right.\right.\\
&\left.\left.\times\exp\left(\mathrm{j}2\pi f\frac{\theta+(M-1)\cdot\left(\frac{\Omega}{2}\tau_m+\frac{\pi}{2}\right)}{\Omega}\right)\right)-\left(r_n'\right)^2\left(\sin\left(\frac{\Omega_n'}{2}\tau_m\right)\right)^{2M-2}\\
&\times\left(2\delta(f)-\left(\delta\left(f+\frac{\Omega_n'}{\pi}\right)+\delta\left(f-\frac{\Omega_n'}{\pi}\right)\right)\right.\\
&\left.\times\exp\left(\mathrm{j}2\pi f\frac{\theta_n'+(M-1)\cdot\left(\frac{\Omega_n'}{2}\tau_m+\frac{\pi}{2}\right)}{\Omega_n'}\right)\right)\\
&+2r'r_n'\left(\sin\left(\frac{\Omega}{2}\tau_m\right)\right)^{M-1}\left(\sin\left(\frac{\Omega_n'}{2}\tau_m\right)\right)^{M-1}\\
&\times\left(\delta\left(f+\frac{\Omega+\Omega_n'}{\pi}\right)+\delta\left(f-\frac{\Omega+\Omega_n'}{\pi}\right)\right)\\
&\times\exp\left(\mathrm{j}2\pi f\frac{\theta+\theta_n'+(M-1)\cdot\left(\frac{\Omega+\Omega_n'}{2}\tau_m+\pi\right)}{\Omega+\Omega_n'}\right)-2r'r_n'\left(\sin\left(\frac{\Omega}{2}\tau_m\right)\right)^{M-1}\\
&\times\left(\sin\left(\frac{\Omega_n'}{2}\tau_m\right)\right)^{M-1}\cdot\left(\delta\left(f+\frac{\Omega-\Omega_n'}{\pi}\right)+\delta\left(f-\frac{\Omega-\Omega_n'}{\pi}\right)\right)\\
&\left.\times\exp\left(\mathrm{j}2\pi f\frac{\theta-\theta_n'+(M-1)\cdot\frac{\Omega-\Omega_n'}{2}\tau_m}{\Omega-\Omega_n'}\right)\right)
\end{aligned}
\tag{5.119}
$$

Define the function Ω_i'/π as the value at frequency Ω_i'/π of the mod value $\Omega \neq \Omega_n'$ of the Fourier transform result of signal $\Omega \neq \Omega_n'$.The analysis of $\Omega \neq \Omega_n'$ is as follows:

1. if $\text{MF\_}R_M\left(s_{nc}\left(t_k, t_m, r_n', \theta_n'\right), \Omega_n'/\pi, \tau_m\right)$, then

$$\begin{aligned}&\text{MF\_}R_M\left(s_{nc}\left(t_k, t_m, r_n', \theta_n'\right), \Omega_n'/\pi, \tau_m\right)\\&= \frac{\left(r_n'\right)^2}{2}\pi\left(\sigma^{2^{M-1}}2^{M-1}\frac{4\pi}{c}\mu\tau_k\left(\sin\left(\frac{\Omega_n'}{2}\tau_m\right)\right)^{M-1}\right)^2\end{aligned} \tag{5.120}$$

2. if $\text{MF\_}R_M\left(s_{nc}\left(t_k, t_m, r_n', \theta_n'\right), \Omega_n'/\pi, \tau_m\right)$, then

$$\begin{aligned}&\text{MF\_}R_M\left(s_{nc}\left(t_k, t_m, r_n', \theta_n'\right), \Omega_n'/\pi, \tau_m\right)\\&= \frac{\pi}{2}\left(\sigma^{2^{M-1}}2^{M-1}\frac{4\pi}{c}\mu\tau_k\left(\sin\left(\frac{\Omega_n'}{2}\tau_m\right)\right)^{M-1}\right)^2\left|r'^2 + r_n'^2 - 2r'r_n'\cos\frac{\theta-\theta_q}{2}\right|\\&\geq \frac{\pi}{2}\left(\sigma^{2^{M-1}}2^{M-1}\frac{4\pi}{c}\mu\tau_k\left(\sin\left(\frac{\Omega_n'}{2}\tau_m\right)\right)^{M-1}\right)^2\left(r' - r_{i,j}'\right)^2\end{aligned} \tag{5.121}$$

When $r_{i,j}' = r$, symbol "=" in the formula "$r_{i,j}' = r$" sets up.

Analyze Eqs. (5.120) and (5.121) comprehensively, we can know that when $\Omega_i' = \Omega$, $\Omega_i' = \Omega$, and $\theta_q = \theta$, the function $\Omega_i'/(2\pi)$ will obtain the minimum. That is when the estimation of the rotation radius and the initial phase of the scattering point at rotating frequency $\Omega_i'/(2\pi)$ is accurate, the function value of Ω_i'/π at Ω_i'/π will be smaller than that of τ_k when the estimation is wrong. In order to weaken the influence of the cross-terms, we can accumulate function τ_k for different fast time delay τ_k and slow time delay τ_m to get function $\Omega_n'/(2\pi)$. Similarly, in order to avoid the influence of the different fast time delay $\Omega_n'/(2\pi)$ and the slow time delay $\Omega_n'/(2\pi)$ of the weights of $\Omega_n'/(2\pi)$ in the process of calculating function $\Omega_n'/(2\pi)$, the $\Omega_n'/(2\pi)$ can be divided by $\Omega_n'/(2\pi)$ then cumulatively multiply the results, and we can get $\Omega_n'/(2\pi)$.

In conclusion, the estimation of the rotation radius and the initial phase of the scattering point at the rotation frequency $\left(\widehat{r}_n, \widehat{\theta}_n\right) = \underset{\widehat{r}_n, \widehat{\theta}_n}{\arg\min}\, \text{PMF\_}R_M\left(s_{nc}\left(t_k, t_m,\right.\right.$ $\left.\left.r_n', \theta_n'\right), \Omega_n'/\pi\right)$ is:

$$\begin{aligned}\left(\widehat{r}_n, \widehat{\theta}_n\right) &= \underset{\widehat{r}_n, \widehat{\theta}_n}{\arg\min}\, \text{PMF\_}R_M\left(s_{nc}\left(t_k, t_m, r_n', \theta_n'\right), \Omega_n'/\pi\right)\\&= \underset{\widehat{r}_n, \widehat{\theta}_n}{\arg\max}\, \frac{1}{\text{PMF\_}R_M\left(s_{nc}\left(t_k, t_m, r_n', \theta_n'\right), \Omega_n'/\pi\right)}\end{aligned} \tag{5.122}$$

where $\arg\min_{\widehat{r}_n,\widehat{\theta}_n}(\cdot)$ and $\arg\max_{\widehat{r}_n,\widehat{\theta}_n}(\cdot)$ indicate the minimum and the maximum of $(\widehat{r}_n, \widehat{\theta}_n)$ within the brackets, respectively.

If there are no rotating scatterer points detected, then the "clean" algorithm can be used to remove the echo signal of the rotating points from the signal after dechirp processing. Then the extraction of the frequency, phase, and radius will be completed by higher moments function analysis.

These previous discussions are primarily for micromotion feature extraction methods of the rotating form. Actually, the form of echo signals generated by the vibration is approximately the same as that of rotation. Therefore, the methods can also be used to extract the micro-Doppler features induced by vibration.

5.5.2.2 Simulation

The simulation parameters are the same as the simulation parameters in Section 5.3.2.2. The coordinates of the rotation center C in the local coordinate system are (5m,5m,5m). Three scatters rotate around C and all of their angular velocities and radii are $\omega = (0,0,2\pi)^{\mathrm{T}}$ rad/s, the duration of echo signal is $T_s = 2$ s, and the sampling rate is $f_s = 500$ MHz.

Fig. 5.40A shows the normalized amplitude of the echo signal function $\mathrm{PMF\_}I_3(s_d(t_k,t_m),(\tau_k,\tau_m),f_m)$. [The range of the fast time delay τ_k is $(1/f_s,10/f_s)$, sample spacing is $1/f_s$. Meanwhile, the range of the slow time τ_m is (1/PRF,10/PRF), sample spacing is 1/PRF]. In Fig. 5.40A, because of cumulative multiplication, there is only one obvious peak. So the rotation frequency can be determined as 1 Hz. It is clear that using HOMF analysis method can effectively achieve the extraction of the rotation frequency in broadband radar.

Fig. 5.40B shows the normalized amplitude value of function $\mathrm{PMF\_}R_3\left(s_{nc}\left(t_k, t_m, r_n', \theta_n'\right), f_m\right)$ of the signal $s_{nc}\left(t_k, t_m, r_n', \theta_n'\right)$ after conduct conjugate multiplication between echo signal and reference signal.

Setting the frequency corresponding to the rotation radius of the reference signal ranges from 0.3 to 0.5 m, and the sampling interval is 0.01 m, the selection range of the initial phase θ_q is $[0,2\pi)$ rad, and the sampling interval is 0.1 rad.

It can be seen from Fig. 5.40B that there are three peaks. In order to guarantee the veracity of judgment, and detect the peak position, we can know the rotation radius is 2.8 m and the initial phase of the rotation point is 2.8 rad. Then using the "clean" algorithm and calculating again, the normalized amplitude of the functions $1/\mathrm{PMF\_}R_3$ is shown in Fig. 5.40C. There are two peaks in the figure. We can also detect the peak position and get the rotating radius and the initial phase of the second rotating point are, respectively, (2.8 m, 5.1 rad), which is close to theory value.

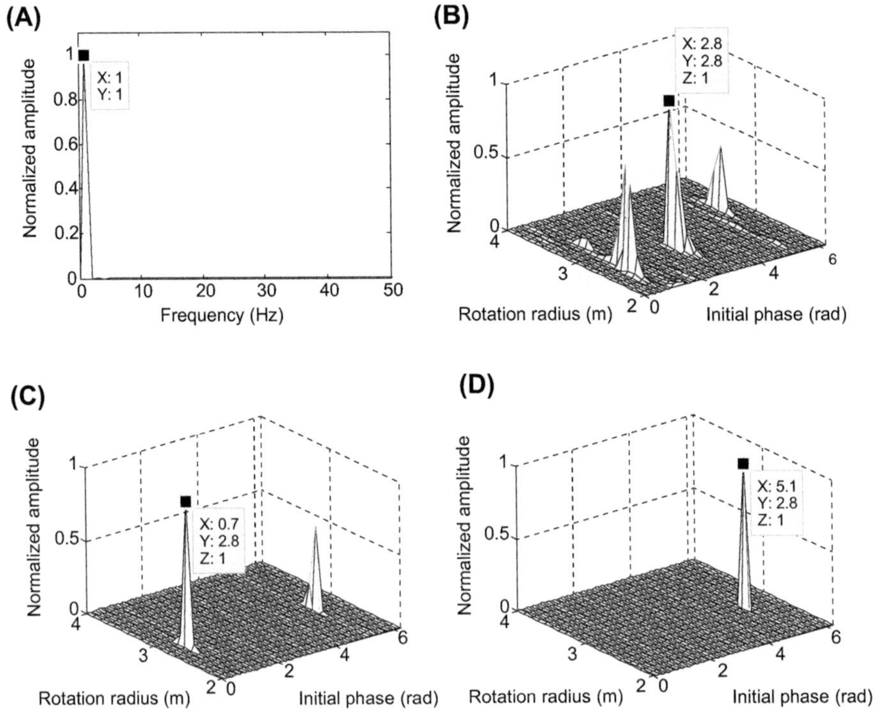

Figure 5.40 The result of HOMF method; (A) Normalized amplitude results of echo signal function PMF_I_3; (B) the first normalized amplitude of the function PMF_R_3; (C) the second normalized amplitude of the function PMF_R_3; (D) the third normalized amplitude of the function PMF_R_3.

It can be seen from Fig. 5.41 that the extraction of micromotion feature can be accomplished effectively under the condition of SNR = 0 dB and broadband radar by the HOMF method.

When reducing SNR to −5 dB, the normalized amplitude results of echo signal function PMF_$I_3(s_d(t_k,t_m),(\tau_k,\tau_m),f_m)$ are shown in Fig. 5.42. Because of the influence of noise, we can see from the figure that the rotational frequency extraction is unable to complete as SNR = −5 dB. In wideband radar system, there exists anti-noise performance reduction of HOMF method compared to image processing method, since the HOMF method only focuses on the processing of certain data of a fixed fast time t_k.

5.6 COMPARISON

In this chapter, five kinds of methods for micro-Doppler feature analysis and extraction are introduced. Due to the advantages and disadvantages of each method, the scopes of their applications are different, and the performances of micro-Doppler feature extraction are also different from each other.

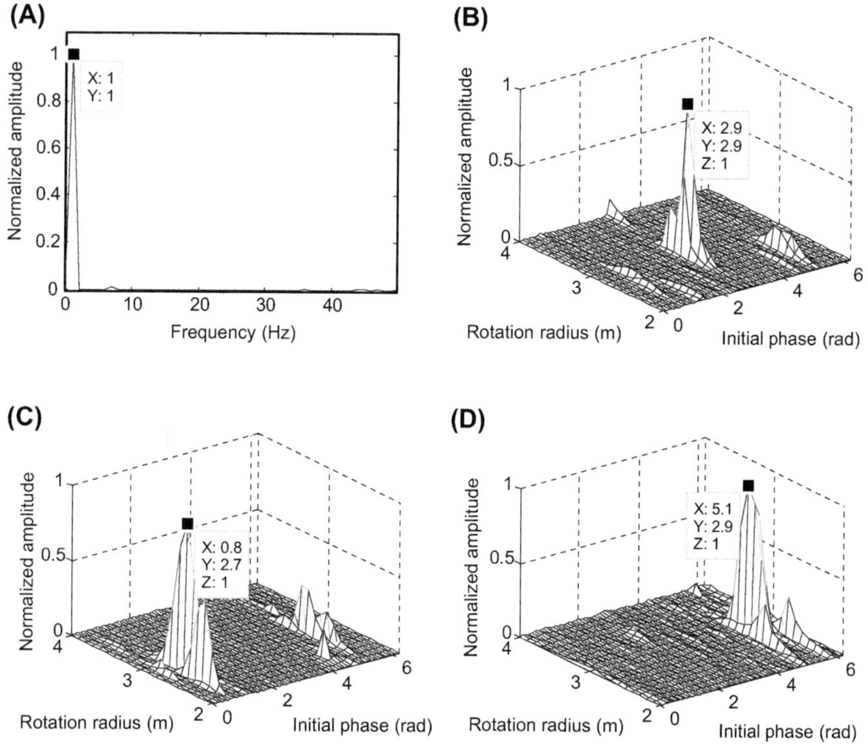

Figure 5.41 The result of the HOMF method when SNR = 0 dB. (A) Normalized amplitude results of echo signal function PMF_I_3; (B) the first normalized amplitude of the function PMF_R_3; (C) the second normalized amplitude of the function PMF_R_3; (D) the third normalized amplitude of the function PMF_R_3.

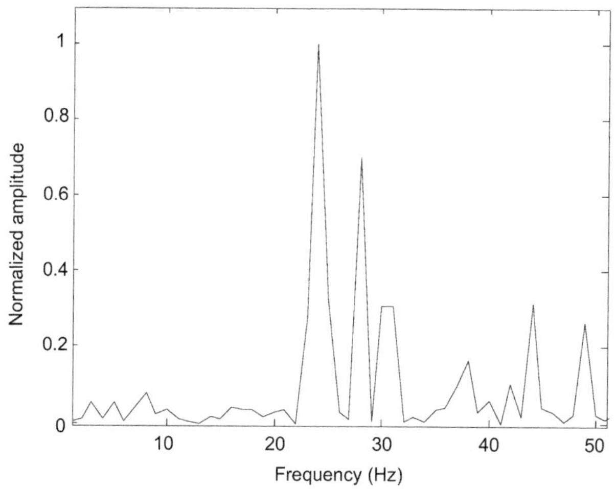

Figure 5.42 The normalized amplitude results of echo signal function PMF_I_3 when SNR = 5 dB.

The image-processing-based micro-Doppler feature extraction methods extract the curve features in range—slow time image or time-frequency image to obtain the micro-Doppler features. The image processing methods have been widely studied, so the image processing—based micro-Doppler feature extraction methods will be easy to engineer implementation, and have very good robustness. It must be noted that, when using the Hough transform to detect patterns, the computer storage and the computational time exponentially grow with the number of dimensions in the parameter space [14]. As for the simulation in Section 5.2.3, the computation time for obtaining the parameter pairs (A',Ω_0,ζ) was approximately 175 s on a personal computer with an Intel Pentium Dual T2330 (1.60 GHz) processor, whereas that for obtaining the parameter pairs (R_Q,A,Ω_0,ζ) was approximately 1311 s. In order to reduce the computation cost for real-time applications, some supplementary processing for speeding up the computation of the algorithm can be used, eg, the fast method presented in Ref. [15] to detect fixed-period sinusoids based on the HT by decomposing the parameters into two groups and estimating them in turn can be extended to detect curves in this paper. The parallel processing technique can also be used to accelerate the computation of the HT.

The OMP-based micro-Doppler feature extraction methods can be utilized in both narrowband radar and wideband radar. In this chapter, the implementation in wideband radar is mainly introduced. Because both the echo amplitude and phase information are used to extract micro-Doppler features, the OMP-based methods possess better robustness compared to the image processing—based methods. When the complex scattering coefficients of micromotional scatterers are time varying, the performance of micro-Doppler feature extraction will be deteriorated. In fact, in this case, many existing micro-Doppler feature extraction methods based on time-frequency analysis are unable to reach a required performance. The computational costs of the OMP-based methods are generally larger than those of the image processing—based methods. Some supplementary processing for speeding up the computation of the algorithm can be used, for example, the stage-wise orthogonal matching pursuit (StOMP) [16], which is a refinement of OMP and more suitable for solving large-scale problems, can be easily extended to decompose the micro-Doppler signals.

EMD decomposition method has advantages in extraction of micro-Doppler signatures of targets with procession. The micro-Doppler effect caused by procession is very complicated. Hough transform and OMP decomposition method both face the problems of overfull search parameters and huge calculation amount. EMD decomposition can decompose complicated micro-Doppler curve into multiple IMF components. From the IMF components, we can get the procession parameters of the targets, which other methods cannot achieve.

Analysis method of HOMF mainly achieves the extraction of micro-Doppler characteristics in rotation and vibration mode. Through the processing of echo signal and calculating its HOMF, we can finish the extraction of rotational frequency via using the characteristic that the image of echo signal can be approximated as simple-frequency signal. And we can finish the estimation of rotational radius via searching the module value accumulation function of real Fourier transform result of M-th order instantaneous torque of searching signal. Compared to image processing method and OMP decomposition method, analysis method of HOMF has lower computational complexity, fast calculation speed, and is easy for engineering realization. In the previous simulation, the computer running time usually is seconds to tens of seconds. But in wideband radar, the antinoise performance of this method is worse than image processing method and OMP decomposition method.

REFERENCES

[1] Xueru B, Mengdao X, Feng Z, et al. Imaging of micromotion targets with rotating parts based on empirical-mode decomposition. IEEE Transactions on Geoscience and Remote Sensing 2008;46(11):3514–23.
[2] Cai C, Liu W, Fu JS, et al. Empirical mode decomposition of micro-Doppler signature. In: Proceedings of International Conference on Radar. Washington, USA: IEEE Press; 2005. p. 895–9.
[3] Auger F, Flandrin P. Improving the readability of time-frequency and time-scale representations by the reassignment method. IEEE Transactions on Signal Processing 1995;43(5):1086–9.
[4] Gonzalez RC, Woods RE. Digital image processing. 2nd ed. Prentice; 2002. p. 519–43.
[5] Zhang Q, Yeo TS, Tan HS, et al. Imaging of a moving target with rotating parts based on the Hough transform. IEEE Transactions on Geoscience and Remote Sensing 2008;46(1):291–9.
[6] Hough PVC. Method and means for recognizing complex patterns. US Patent: 3069654. December 1962.
[7] Luo Y, Zhang Q, Qiu C-W, Liang X-J, Li K-M. Micro-Doppler effect analysis and feature extraction in ISAR imaging with stepped-frequency chirp signals. IEEE Transactions on Geoscience and Remote Sensing 2010;48(4):2087–98.
[8] Mallat SG, Zhang Z. Matching pursuits with time–frequency dictionaries. IEEE Transactions on Signal Processing 1993;41(12):3397–415.
[9] Pati YC, Rezaiifar R, Krishnaprasad PS. Orthogonal matching pursuit: recursive function approximation with applications to wavelet decomposition. In: Proc. 27th Annu. Asilomar Conf. Signals, Systems, and Computers, vol. 1. Pacific Grove, CA; November 1993. p. 40–4.
[10] Tropp JA, Gilbert AC. Signal recovery from random measurements via orthogonal matching pursuit. IEEE Transactions on Information Theory 2007;53(12):4655–66.
[11] Huang NE, Shen Z, Long SR, et al. The empirical mode decomposition and the Hilbert spectrum for nonlinear and nonstationary time series analysis. Proceedings of the Royal Society A Mathematical Physical and Engineering Sciences 1998;454(1971):903–95.
[12] Tsao J, Steinberg BD. Reduction of sidelobe and speckle artifacts in microwave imaging: the CLEAN technique. IEEE Transactions on Antennas and Propagation 1988;36(4):543–56.
[13] Weiss LG. Wavelets and wideband correlation processing. IEEE Signal Processing Magazine 1994;11(1):13–32.
[14] Ballard DH. Generalization the Hough transform to detect arbitrary shapes. Pattern Recognition 1981;13(2):111–22.
[15] Changchun Z, Ge S. A Hough transform-based method for fast detection of fixed period sinusoidal curves in images. In: 6th International Conference on Signal Processing; 2002. p. 909–12.
[16] Donoho DL, Tsaig Y, Drori I, Starck J-L. Sparse solution of underdetermined linear equations by stagewise orthogonal matching pursuit. IEEE Transactions on Information Theory 2012;58(2):1094–121.

APPENDICES

Appendix 5-A

Proof for Eq. (5.70):

$\mathbf{S} = (s_{a,b})$ can be rewritten as

$$\mathbf{S} = \left(s_{a,b}\right) = \begin{bmatrix} s_{1,1} & s_{1,2} & \cdots & s_{1,N_B} \\ s_{2,1} & s_{2,2} & \cdots & s_{2,N_B} \\ \vdots & \vdots & \ddots & \vdots \\ s_{N_A,1} & s_{N_A,2} & \cdots & s_{N_A,N_B} \end{bmatrix}$$

Reshape $\mathbf{S}$ and then a new vector with length of $N_A \times N_B$ can be obtained as follows:

$$\boldsymbol{s}' = \left[s_{1,1}, s_{1,2}, \cdots, s_{1,N_B}, s_{2,1}, s_{2,2}, \cdots, s_{2,N_B}, \cdots, s_{N_A,1}, s_{N_A,2}, \cdots s_{N_A,N_B}\right]^{\mathrm{T}}$$

Similarly, $\mathbf{h}_i$ can be rewritten as

$$\mathbf{h}_i = \left(h_{ia,b}\right) = \begin{bmatrix} h_{i1,1} & h_{i1,2} & \cdots & h_{i1,N_B} \\ h_{i2,1} & h_{i2,2} & \cdots & h_{i2,N_B} \\ \vdots & \vdots & \ddots & \vdots \\ h_{iN_A,1} & h_{iN_A,2} & \cdots & h_{iN_A,N_B} \end{bmatrix}$$

Reshape $\mathbf{h}_i$ and then a new vector with length of $N_A \times N_B$ can be obtained as follows:

$$\boldsymbol{h}_i' = \left[h_{i1,1}, h_{i1,2}, \cdots, h_{i1,N_B}, h_{i2,1}, h_{i2,2}, \cdots, h_{i2,N_B}, \cdots, h_{iN_A,1}, h_{iN_A,2}, \cdots h_{iN_A,N_B}\right]^{\mathrm{T}}$$

Let $\mathbf{H}_g' = \left[\boldsymbol{h}_1', \boldsymbol{h}_2', \cdots, \boldsymbol{h}_g'\right]$. Then Eq. (5.69) is equivalent to

$$\boldsymbol{x} = \arg\min_{\boldsymbol{x}} \left\| \boldsymbol{s}' - \mathbf{H}_g' \boldsymbol{x} \right\|_2^2$$

Apply the least-squares algorithm, and it yields

$$\boldsymbol{x} = \left(\mathbf{H}_g'^{\mathrm{H}} \mathbf{H}_g'\right)^{-1} \mathbf{H}_g'^{\mathrm{H}} \boldsymbol{s}'$$

where

$$\mathbf{H}_g'^{\mathrm{H}}\mathbf{H}_g' = \left[\boldsymbol{h}_1', \boldsymbol{h}_2', \cdots, \boldsymbol{h}_g'\right]^{\mathrm{H}}\left[\boldsymbol{h}_1', \boldsymbol{h}_2', \cdots, \boldsymbol{h}_g'\right]$$

$$= \begin{bmatrix} \boldsymbol{h}_1'^{\mathrm{H}}\boldsymbol{h}_1' & \boldsymbol{h}_1'^{\mathrm{H}}\boldsymbol{h}_2' & \cdots & \boldsymbol{h}_1'^{\mathrm{H}}\boldsymbol{h}_g' \\ \boldsymbol{h}_2'^{\mathrm{H}}\boldsymbol{h}_1' & \boldsymbol{h}_2'^{\mathrm{H}}\boldsymbol{h}_2' & \cdots & \boldsymbol{h}_2'^{\mathrm{H}}\boldsymbol{h}_g' \\ \vdots & \vdots & \ddots & \vdots \\ \boldsymbol{h}_g'^{\mathrm{H}}\boldsymbol{h}_1' & \boldsymbol{h}_g'^{\mathrm{H}}\boldsymbol{h}_2' & \cdots & \boldsymbol{h}_g'^{\mathrm{H}}\boldsymbol{h}_g' \end{bmatrix} = \begin{bmatrix} \langle \mathbf{h}_1, \mathbf{h}_1 \rangle & \langle \mathbf{h}_2, \mathbf{h}_1 \rangle & \cdots & \langle \mathbf{h}_g, \mathbf{h}_1 \rangle \\ \langle \mathbf{h}_1, \mathbf{h}_2 \rangle & \langle \mathbf{h}_2, \mathbf{h}_2 \rangle & \cdots & \langle \mathbf{h}_g, \mathbf{h}_2 \rangle \\ \vdots & \vdots & \ddots & \vdots \\ \langle \mathbf{h}_1, \mathbf{h}_g \rangle & \langle \mathbf{h}_2, \mathbf{h}_g \rangle & \cdots & \langle \mathbf{h}_g, \mathbf{h}_g \rangle \end{bmatrix}$$

$$\begin{aligned} \mathbf{H}_g'^{\mathrm{H}}\boldsymbol{s}' &= \left[\boldsymbol{h}_1', \boldsymbol{h}_2', \cdots, \boldsymbol{h}_g\right]^{\mathrm{H}}\boldsymbol{s}' \\ &= \left[\boldsymbol{h}_1'^{\mathrm{H}}\boldsymbol{s}', \boldsymbol{h}_2'^{\mathrm{H}}\boldsymbol{s}', \cdots, \boldsymbol{h}_g'^{\mathrm{H}}\boldsymbol{s}'\right]^{\mathrm{T}} \\ &= \left[\langle \mathbf{S}, \mathbf{h}_1 \rangle, \langle \mathbf{S}, \mathbf{h}_2 \rangle, \cdots, \langle \mathbf{S}, \mathbf{h}_g \rangle\right]^{\mathrm{T}} \end{aligned}$$

Then it yields

$$\boldsymbol{x} = \begin{bmatrix} \langle \mathbf{h}_1, \mathbf{h}_1 \rangle & \langle \mathbf{h}_2, \mathbf{h}_1 \rangle & \cdots & \langle \mathbf{h}_g, \mathbf{h}_1 \rangle \\ \langle \mathbf{h}_1, \mathbf{h}_2 \rangle & \langle \mathbf{h}_2, \mathbf{h}_2 \rangle & \cdots & \langle \mathbf{h}_g, \mathbf{h}_2 \rangle \\ \vdots & \vdots & \ddots & \vdots \\ \langle \mathbf{h}_1, \mathbf{h}_g \rangle & \langle \mathbf{h}_2, \mathbf{h}_g \rangle & \cdots & \langle \mathbf{h}_g, \mathbf{h}_g \rangle \end{bmatrix}^{-1} \begin{bmatrix} \langle \mathbf{S}, \mathbf{h}_1 \rangle \\ \langle \mathbf{S}, \mathbf{h}_2 \rangle \\ \vdots \\ \langle \mathbf{S}, \mathbf{h}_g \rangle \end{bmatrix}$$

Appendix 5-B

According to $\boldsymbol{x} = \left(\mathbf{H}_g'^{\mathrm{H}}\mathbf{H}_g'\right)^{-1}\mathbf{H}_g'^{\mathrm{H}}\boldsymbol{s}'$, it yields

$$\begin{aligned} \left(\boldsymbol{s}' - \mathbf{H}_g'\boldsymbol{x}\right)^{\mathrm{H}}\mathbf{H}_g' &= \left[\boldsymbol{s}' - \mathbf{H}_g'\left(\mathbf{H}_g'^{\mathrm{H}}\mathbf{H}_g'\right)^{-1}\mathbf{H}_g'^{\mathrm{H}}\boldsymbol{s}'\right]^{\mathrm{H}}\mathbf{H}_g' \\ &= \boldsymbol{s}'^{\mathrm{H}}\mathbf{H}_g' - \boldsymbol{s}'^{\mathrm{H}}\mathbf{H}_g'\left[\left(\mathbf{H}_g'^{\mathrm{H}}\mathbf{H}_g'\right)^{-1}\right]^{\mathrm{H}}\mathbf{H}_g'^{\mathrm{H}}\mathbf{H}_g' \\ &= 0 \end{aligned}$$

Because $\mathbf{H}_g' = \left[\boldsymbol{h}_1', \boldsymbol{h}_2', \cdots, \boldsymbol{h}_g'\right]$, it can be obtained

$$\left\langle \boldsymbol{s}' - \mathbf{H}_g'\boldsymbol{x}, \boldsymbol{h}_i' \right\rangle = \left[\left(\boldsymbol{s}' - \mathbf{H}_g'\boldsymbol{x}\right)^{\mathrm{H}}\boldsymbol{h}_i'\right]^{*} = 0; \quad i = 1, 2, \cdots, g$$

It is equivalent to

$$\left\langle \mathbf{S} - \sum_{i=1}^{g} x_i \mathbf{h}_i, \mathbf{h}_i \right\rangle = 0; \quad i = 1, 2, \cdots, g$$

Therefore, $\mathbf{R}_{\mathbf{S}g}$ is orthogonal to every atom in $\mathbf{H}_g$.

CHAPTER SIX

Three-Dimensional Micromotion Feature Reconstruction

Methods of three-dimensional micromotion feature reconstruction are introduced in this chapter. It is known that on the condition of monostatic radar, the micro-Doppler feature parameters of radar returned signal, by the preceding discussions, are determined by the projection from motion vectors of micromotion components in the direction of radar sight line. So only the structure and motion feature of micromotion components in the direction of radar sight line can be abstracted. For the complexity of posture changes of moving targets, the micro-Doppler features differ markedly in different radar angles of view, which will have an influence on the accuracy of target recognition. For example, in the recognition of missile warhead targets, it is easy to obtain rotation frequency of warhead targets on the condition of monostatic radar but difficult to get the real radius of warhead, as even complex high-resolution imaging algorithms can just obtain the scattering-distribution projection information of warhead in the line of sight (LOS), which definitely increases the difficulty of true-false warhead recognition. So we must manage to reconstruct three-dimensional micromotion information that can reflect the true motion of micromotion components from returned signal in order to overcome postural sensibility of micro-Doppler feature. This chapter introduces methods of three-dimensional micromotion feature reconstruction based on multistatic radar techniques.

6.1 MULTISTATIC RADAR TECHNIQUES

The development history and advantages of bistatic radar are briefly introduced in Chapter 4. And with the fast development and maturation of synchronous control, coherent processing, wideband communication, and other techniques, multistatic radar techniques have received extensive attention especially multiple input, multiple output (MIMO) radar and distributed radar network, which, as two important forms of multistatic radar, have been intensively studied in recent years.

Since Eran Fishler formally introduced the definition of "MIMO radar" in 2004 [1,2], MIMO radar has drawn wide attention from researchers. MIMO is a definition in control system in origin, which represents that there are multiple inputs and multiple outputs in a system. Since the mid-1990s, the project of adopting multiple antennas in base stations and mobile terminals of a wireless communication system was introduced by Bell Laboratories and others, that is to say, there are multiple input signals and multiple

Micro-Doppler Characteristics of Radar Targets
ISBN 978-0-12-809861-5, http://dx.doi.org/10.1016/B978-0-12-809861-5.00006-0

output signals for a mobile channel system (MIMO system). Because a MIMO communication system can receive space diversity gain, it can significantly increase channel capacity of a mobile communication system in a fading channel. Similarly, radar returned signal has some characteristics with mobile communication channel for the notable scintillation properties of radar targets. Both theory and practice show that the radar cross-section (RCS) of target is sensitive to direction, thus a tiny change in posture or direction will lead to severe ups and downs to returned signal, which can lead to 10–25 dB [2]. The ups and downs are very similar to signal fading of a mobile channel, which will greatly influence the detection performance of conventional radar. After extending the definition of MIMO to radar system, MIMO radar system, through adopting multiple transmitting arrays, each array of which radiates mutually orthogonal signal, and multiple receiving arrays, each array of which simultaneously receives multiple signals, makes use of the orthogonality between signals to separate various signals. This kind of working characteristic makes MIMO radar have the advantages of anti-intercepting, antijamming, and so on. Besides, recent research results have shown that MIMO radar has obvious advantages over conventional radar in weak target detection, parameter estimation precision of motion target, target resolution, and so on. In addition, MIMO radar has the advantage of space diversity so it can achieve special parallel sampling to scatter distributed information of target and consequently reduce sampling numbers of pulse in time domain to image rapidly. Thus, MIMO radar imaging techniques have been studied intensively in recent years.

With the fast development of modern radar technology, "distributed radar networks" [3], which were introduced in the 1980s have regained attention in recent years [4–7]. Distributed radar networks collect and transform target information from multiple radars with different locations, different frequency ranges, and different systems in the shape of a "net." An information fusion and processing center will carry out integrated processing for the target information for the purpose of improving system efficiency. The radar in the network can be monostatic radar, bistatic (multistatic) radar, even MIMO radar. Thus, distributed radar networks are a broader category than MIMO radar.

Both MIMO radar and distributed radar networks can get the target information in every perspective, which not only equips the radar with perfect antistealth capability but provides the possibility of reconstructing three-dimensional structure and motion features of a target. Particularly, micromotions of a target have different projection vectors in different radar line sights and micro-Doppler signals received by different antennas have different phase changes so it is possible to reconstruct three-dimensional structure and motion features of targets and improve the ability of target recognition by the changes. In the following chapters and sections, methods of rotating-form three-dimensional micromotion feature reconstruction when transmitting narrowband and wideband signals based on MIMO radar are introduced in order; then methods of precession-form three-dimensional micromotion feature reconstruction when transmitting wideband signal based on distributed radar networks are introduced.

6.2 THREE-DIMENSIONAL MICROMOTION FEATURE RECONSTRUCTION IN NARROWBAND MULTIPLE INPUT MULTIPLE OUTPUT RADAR

6.2.1 Micro-Doppler Effect in Narrowband Multiple Input Multiple Output Radar

MIMO radar includes two types: one is called single-station configuration radar, that is to say that there is not so large an array space between transmitter arrays and receiver arrays that signals received by different arrays are coherent; the other is called multistation configuration radar, that is to say that there is large space in transmitting arrays and different arrays face different directions of target so RCS of target is different to different arrays, then we can promote detection performance to target making use of space diversity. In order to reconstruct three-dimensional micromotion feature of the target, returned signals in different radar line sights are needed, so MIMO radar discussed in this chapter mainly refers to the second type MIMO radar.

A MIMO radar system is shown in Fig. 6.1. (U,V,W) is radar coordinate system. Reference coordinate system (X,Y,Z) is parallel to the radar coordinate system and its origin is O. Target-local coordinate system (x,y,z) shares the same origin O with the reference coordinate system. $O_{T1},O_{T2},\cdots,O_{TM}$ are M radar transmitting arrays and $O_{R1},O_{R2},\cdots,O_{RN}$ are N radar receiving arrays. When undergoing a translation with velocity $\boldsymbol{v}$, the target rotates about the x-axis, y-axis, and z-axis with the angular rotation

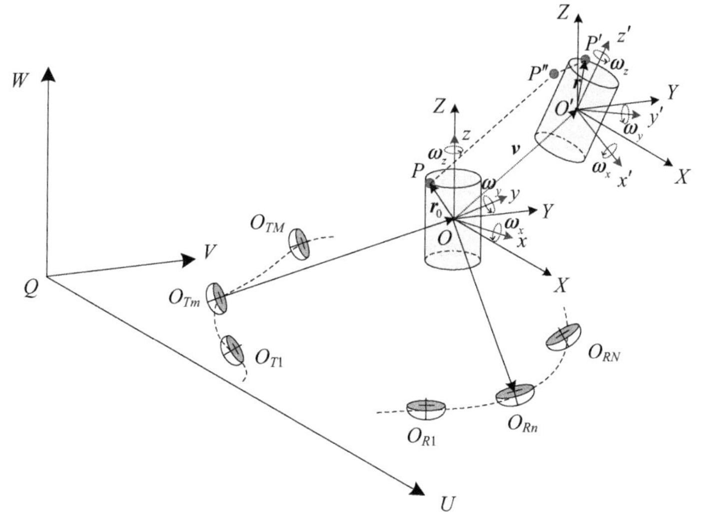

Figure 6.1 Geometry of MIMO radar and a target with three-dimensional rotation.

velocity ω_x, ω_y, and ω_z, respectively, which can be either represented in the target-local coordinate system as $\boldsymbol{\omega} = (\omega_x, \omega_y, \omega_z)^{\mathrm{T}}$ (the superscript "T" means transposition), or represented in the reference coordinate system as $\widehat{\boldsymbol{\omega}} = (\omega_X, \omega_Y, \omega_Z)^{\mathrm{T}}$. A point scatterer P of the target is located at $\boldsymbol{r}_0 = (r_{x0}, r_{y0}, r_{z0})^{\mathrm{T}}$ in the target-local coordinate system and $\widehat{\boldsymbol{r}}_0 = (r_{X0}, r_{Y0}, r_{Z0})^{\mathrm{T}}$ in the reference coordinate system at the initial time.

Suppose that transmitting arrays transmit single-frequency continuous waves with different carrier frequencies. Any transmitting-receiving array $O_{Tm}O_{Rn}$ constructs a group of bistatic radar in fact, thus, according to the derivation in Section 4.2, while f_{cm} denotes the carrier frequency of the transmitted signal by the m-th array, $\boldsymbol{n}_{b0}^{(m,n)} = \left(n_{b0x}^{(m,n)}, n_{b0y}^{(m,n)}, n_{b0z}^{(m,n)}\right)^{\mathrm{T}}$ is the vector opposite to angular bisector vector of angle between target and transmitting-receiving array $O_{Tm}O_{Rn}$. Then it is easy to obtain

$$f_{\text{micro-Doppler}}(t; m, n) = \frac{\Omega f_{cm}}{c}\sqrt{A_{m,n}^2 + B_{m,n}^2}\sin(\Omega t + \phi_{m,n}) \tag{6.1}$$

where

$$\begin{aligned} A_{m,n} = & -n_{b0x}^{(m,n)} r_{X0}\omega_Z'^2 - n_{b0x}^{(m,n)} r_{X0}\omega_Y'^2 + n_{b0x}^{(m,n)} r_{Y0}\omega_X'\omega_Y' + n_{b0x}^{(m,n)} r_{Z0}\omega_X'\omega_Z' \\ & + n_{b0y}^{(m,n)} r_{X0}\omega_X'\omega_Y' - n_{b0y}^{(m,n)} r_{Y0}\omega_Z'^2 - n_{b0y}^{(m,n)} r_{Y0}\omega_X'^2 + n_{b0y}^{(m,n)} r_{Z0}\omega_Y'\omega_Z' \\ & + n_{b0z}^{(m,n)} r_{X0}\omega_X'\omega_Z' + n_{b0z}^{(m,n)} r_{Y0}\omega_Y'\omega_Z' - n_{b0z}^{(m,n)} r_{Z0}\omega_Y'^2 - n_{b0z}^{(m,n)} r_{Z0}\omega_X'^2 \end{aligned}$$

$$\begin{aligned} B_{m,n} = & -n_{b0x}^{(m,n)} r_{Y0}\omega_Z' + n_{b0x}^{(m,n)} r_{Z0}\omega_Y' + n_{b0y}^{(m,n)} r_{X0}\omega_Z' - n_{b0y}^{(m,n)} r_{Z0}\omega_X' - n_{b0z}^{(m,n)} r_{X0}\omega_Y' \\ & + n_{b0z}^{(m,n)} r_{Y0}\omega_X' \end{aligned}$$

$$\phi_{m,n} = \operatorname{atan}(B_{m,n}/A_{m,n})$$

If conducting time-frequency analysis to returned signal, micro-Doppler effect is shown as a sinusoidal curve in time-frequency plane, whose angular frequency is the angular rotation frequency Ω of rotating scatterer points and amplitude $M_a^{(m,n)}$ is determined by the carrier frequency of the transmitted signal, the rotation radius, the angular rotation velocity vector, angular bisector vector of angle between target and transmitting-receiving array, and so on, that is,

$$M_a^{(m,n)} = \Omega f_{cm}\sqrt{A_{m,n}^2 + B_{m,n}^2}\Big/c \tag{6.2}$$

6.2.2 Three-Dimensional Micromotion Feature Reconstruction of Rotating Targets

For the groups of transmitting-receiving array in MIMO radar, the changes of geometry structure parameters between target and transmitting-receiving array lead to difference of returned micro-Doppler signals received by receiving antennas and diversities of amplitudes of sinusoidal curve in time-frequency figure, which provide the possibility for establishing multivariable equations to solve spatial three-dimensional feature of rotating elements using the feature of multiple angles of view of MIMO radar. Frequency, amplitude, and other parameters of sinusoidal curve in time-frequency figure can be abstracted with image processing method introduced in Chapter 5, that is to say, first abstracting the curve frame with mathematical morphology method, then abstracting parameters of the frame with Hough transform; we can also adopt orthogonal matching pursuit decomposition method and derive frequency and amplitude of sinusoidal curve by decomposed atomic parameters.

Based on deriving Ω and $M_a^{(m,n)}$, through analyzing on Eq. (6.2), f_{cm} is known, and after confirming the relative location between radar and target, $\boldsymbol{n}_{b0}^{(m,n)} = \left(n_{b0x}^{(m,n)}, n_{b0y}^{(m,n)}, n_{b0z}^{(m,n)}\right)^{\mathrm{T}}$ is also known. So the unknown parameter is a six-component array $\left(r_{0X}, r_{0Y}, r_{0Z}, \omega'_X, \omega'_Y, \omega'_Z\right)$. Thus six equations are needed to solve the array correctly. By the existing conditions, the following two equations can be derived:

1. $\widehat{\boldsymbol{\omega}}' = \widehat{\boldsymbol{\omega}}/\Omega = \left(\omega'_X, \omega'_Y, \omega'_Z\right)^{\mathrm{T}}$, thus $\omega'^2_X + \omega'^2_Y + \omega'^2_Z = 1$;
2. for rotation, $\widehat{\boldsymbol{r}}_0$ rotating about rotation axis is equivalent to the component of $\widehat{\boldsymbol{r}}_0$, which is perpendicular to angular rotation velocity vector rotating about rotation axis, thus it is impossible to derive a unique solution. But as long as the component of $\widehat{\boldsymbol{r}}_0$ that is perpendicular to angular rotation velocity vector is derived, rotating micromotion information of target is derived too. So we can absolutely substitute $\widehat{\boldsymbol{r}}_0$ for the component of $\widehat{\boldsymbol{r}}_0$, which is perpendicular to angular rotation velocity vector, in which case there is

$$\omega'_X r_{0X} + \omega'_Y r_{0Y} + \omega'_Z r_{0Z} = 0$$

Thus, four equations are needed to obtain $\left(r_{0X}, r_{0Y}, r_{0Z}, \omega'_X, \omega'_Y, \omega'_Z\right)$, that is to say, the product of numbers of transmitting-receiving arrays must meet $MN \geq 4$. It is usually hard to derive analytical solutions of equations like (6.2) but only approximate solutions by numerical calculation methods like Newton method. And to make sure that iteration solving can converge to the accurate solutions and be free from the influence of repeated roots with no physical significance, MIMO radar numbers can be added to increase equations' numbers, which restrain roots of equation more strictly. The simulation shows that when the numbers of transmitting-receiving arrays meet $MN \geq 6$, quite accurate resolutions of equations array can be derived by Newton method.

To sum up, methods of three-dimensional micromotion feature reconstruction of rotating targets in narrowband MIMO radar are specifically expressed as follows:

Step (1) constructing a MIMO radar system whose numbers of transmitting-receiving arrays meet $MN \geq 6$;

Step (2) band-pass filtering for returned signals received by all receiving arrays and separating returned signals with different carrier frequencies, MN groups of returned signal are obtained;

Step (3) executing time-frequency analysis or orthogonal matching pursuit (OMP) decomposition for MN groups of returned signal separately and obtaining the angular frequency Ω amplitude $M_a^{(m,n)}$ of sinusoidal curve in time-frequency plane, $m = 1,2,\cdots,M$, $n = 1,2,\cdots,N$;

Step (4) solving nonlinear equations array

$$\begin{cases} M_a^{(m,n)} = \Omega f_{cm}\sqrt{A_{m,n}^2 + B_{m,n}^2}\Big/c, & m = 1,2,\cdots,M, n = 1,2,\cdots,N \\ \omega_X'^2 + \omega_Y'^2 + \omega_Z'^2 = 1 \\ \omega_X' r_{0X} + \omega_Y' r_{0Y} + \omega_Z' r_{0Z} = 0 \end{cases} \tag{6.3}$$

and deriving the arithmetic resolutions $\left(r_{0X}, r_{0Y}, r_{0Z}, \omega_X', \omega_Y', \omega_Z'\right)$.

Step (5) deriving the rotation radius $\left\|\widehat{\boldsymbol{r}}_0\right\| = \sqrt{r_{0X}^2 + r_{0Y}^2 + r_{0Z}^2}$ of rotating scatterer points and angular rotation velocity vector $\widehat{\boldsymbol{\omega}} = \left(\Omega\omega_X', \Omega\omega_Y', \Omega\omega_Z'\right)^{\mathrm{T}}$.

Through the previous steps, spatial three-dimensional vectors of real rotating radius and angular rotation velocity of rotating components are constructed. And compared with monostatic radar, which only gets the micromotion information of target in the radar line of sight, three-dimensional micromotion feature reconstruction in MIMO radar provides more abundant and accurate feature information for accurate recognition, accurate location of motion posture of target components, and so on.

Simulation and analysis: Suppose that MIMO radar consists of four transmitting arrays and two receiving arrays. The locations of transmitting arrays are (0,0,0), (1000,3000,0), (2000,6000,0), and (3000,9000,0), respectively. And the locations of receiving arrays are (7000,0,0) and (6000,4000,0), respectively. The unit is meter. The carrier frequencies of signals transmitted by transmitting arrays are 6, 7, 8, and 9 GHz, respectively. The location of target center is (3000 m,4000 m,5000 m), and $\boldsymbol{v} = (150\ \mathrm{m/s}, 150\ \mathrm{m/s}, 150\ \mathrm{m/s})^{\mathrm{T}}$. The target consists of three rotating scatterer points that are distributed in target-local cooperate system as shown in Fig. 6.2. The three scatterer points are equally spaced in long circumference and rotate about point (3,4,5) in target-local cooperate system for uniform circular motion with the same angular rotation velocity $\widehat{\boldsymbol{\omega}} = (2\pi\ \mathrm{rad/s}, 4\pi\ \mathrm{rad/s}, 2\pi\ \mathrm{rad/s})^{\mathrm{T}} = (6.2832, 12.5664, 6.2832)^{\mathrm{T}}$, the same rotating radius $\left\|\widehat{\boldsymbol{r}}_0\right\| = 0.6633$ m. The initial rotating radius vectors of the three scatterer points are $(0.2,0.2,-0.6)^{\mathrm{T}}$, $(-0.5950,0.1828,0.2293)^{\mathrm{T}}$, and $(0.3950,-0.3828,0.3707)^{\mathrm{T}}$, respectively. And the unit is meter. After calculation, $\Omega = 15.3906$ rad/s, that is 2.4495 Hz. The duration of sampled returned signal is 0.5 s and sampling frequency is 6 KHz.

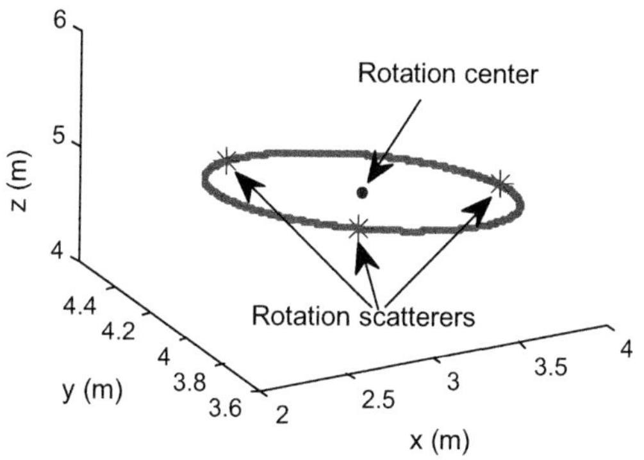

Figure 6.2 Model of target scatterer points ("*" represents scatterer points).

Compensating the Doppler shift induced by translation for returned signal received by radar observation group consisting of transmitting array (0,0,0) and receiving array (7000,0,0) and then executing time-frequency analysis (Gabor transform is adopted), time-frequency distribution shown as Fig. 6.3 is obtained. It is clear that the frequency changes of returned signals are represented as sinusoidal curves.

To eliminate the influence of surrounding noises of curve in Fig. 6.3, the image is smoothed with a Gaussian spatial mask of 25 × 25 pixels. The result is shown in Fig. 6.4A, which shows that the noises in time-frequency plane have been commendably eliminated. And on this basis, abstracted frame curve is shown in Fig. 6.4B. Executing Hough transform for the frame curve in Fig. 6.4B can derive the angular frequency 15.40 rad/s and amplitude 214.6 Hz of the sinusoidal curve. Similarly, for the MIMO radar consisting of four transmitting arrays and two receiving arrays, eight groups of returned signals are isolated at the receiving terminal. Denote that radar observation group consisting of m-th transmitting array and n-th receiving array is (m,n). After eliminating Doppler shift for the eight groups of returned signals, eight different angular

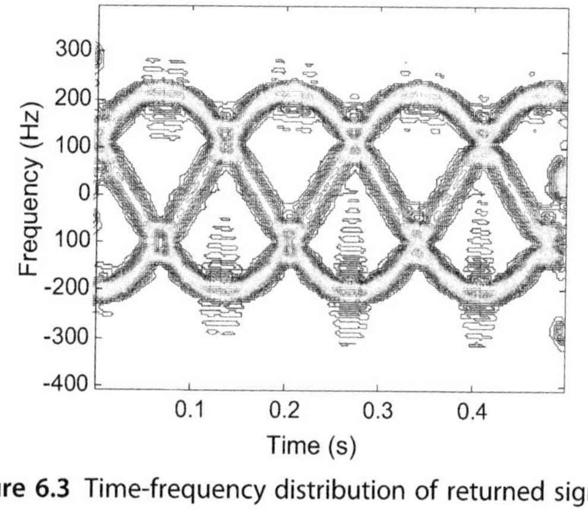

Figure 6.3 Time-frequency distribution of returned signal.

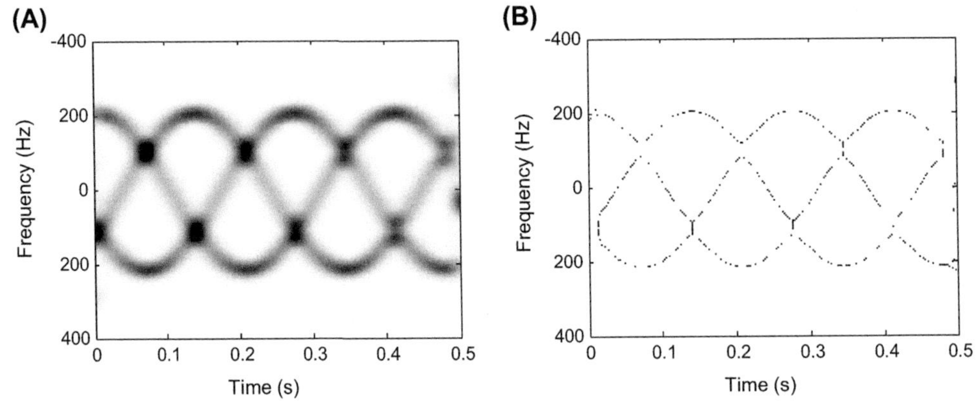

Figure 6.4 Time-frequency image processing of micro-Doppler signal. (A) Smoothing process; (B) Frame abstracting.

frequencies and amplitudes of sinusoidal curves can be derived by Hough transform, which is shown in Table 6.1.

It can be seen from the table that the abstracted angular frequencies of curves lie between 15.39 rad/s and 15.42 rad/s. And angular frequency of curve is equal to angular rotation frequency of rotating scatterer points, as is apparent in previous analysis, and has no relation with geometry configuration of transmitting and receiving radar. Thus, take the mean of angular frequencies of the curves abstracted by the eight groups of transmitting and receiving radars, and regard the calculated result as the estimated value of angular rotation frequency Ω of rotating scatterer points. After calculation, $\Omega = 15.39875$ rad/s.

On the basis of abstracting parameters of curves, adopt command "fsolve" in MATLAB to solve the equations array constructed by (6.3) and we can get

Table 6.1 Amplitudes and Angular Frequencies Abstracted by Hough Transform of Sinusoidal Curves

(m,n)	(1,1)	(1,2)	(2,1)	(2,2)	(3,1)	(3,2)	(4,1)	(4,2)
Theoretical amplitudes (Hz)	213.8667	261.8331	308.3771	379.0710	406.0460	494.1058	448.8819	547.0639
Abstracted amplitudes (Hz)	214.6	264.6	308.0	378.0	411.3	498.0	451.3	548.0
Abstracted angular frequencies (rad/s)	15.40	15.39	15.40	15.40	15.42	15.39	15.40	15.39

$\left(r_{0X}, r_{0Y}, r_{0Z}, \omega'_X, \omega'_Y, \omega'_Z\right) = (-0.1699, 0.3590, -0.5655, 0.3833, 0.8078, 0.3977)$. The initial iterative searching value is (−1,−1,−1,0,0,0). Thus the rotation radius of rotating scatterer points is $\left\|\widehat{\boldsymbol{r}}_0\right\| = \sqrt{r_{0X}^2 + r_{0Y}^2 + r_{0Z}^2} = 0.6911$, angular rotation velocity vector is $\widehat{\boldsymbol{\omega}} = \left(\Omega\omega'_X, \Omega\omega'_Y, \Omega\omega'_Z\right)^{\mathrm{T}} = (5.9025, 12.4408, 6.1253)^{\mathrm{T}}$, which is close to theoretical value.

Restrained by resolution of time-frequency analysis, Hough transform will inevitably bring in errors when abstracting parameters of curve in time-frequency plane, while evaluated errors of Ω and $M_a^{(m,n)}$ will inevitably affect the solving accuracy of equations array (6.3). In order to investigate and analyze the affection, we define the ratio of absolute value of error and true value as normalized error and denote that the error follows the Gaussian distribution. Execute robustness analysis with Monte-Carlo simulation arithmetic to the algorithm and simulation time is 100.

For convenience of analysis, three kinds of cases are divided to simulate:

1. Suppose that angular frequency Ω of curve is estimated accurately and the mean value of error of $M_a^{(m,n)}$ is 0, which is unbiased estimation. Investigate the relation between normalized error of derived angular rotating velocity $\widehat{\boldsymbol{\omega}} = \left(\Omega\omega'_X, \Omega\omega'_Y, \Omega\omega'_Z\right)^{\mathrm{T}}$ together with radius $\left\|\widehat{\boldsymbol{r}}_0\right\|$ and mean square error of $M_a^{(m,n)}$. Fig. 6.5A shows the mean value changing curve of normalized error of $\left(\omega_X, \omega_Y, \omega_Z, \left\|\widehat{\boldsymbol{r}}_0\right\|\right)$ in 100 times of simulations when mean square error of $M_a^{(m,n)}$ lies in [0,20]. It is clear that the normalized error of $\left(\omega_X, \omega_Y, \omega_Z, \left\|\widehat{\boldsymbol{r}}_0\right\|\right)$ basically increases linearly with the increasing of mean square error of $M_a^{(m,n)}$.
2. Suppose that angular frequency Ω of curve is estimated accurately and $M_a^{(m,n)}$, whose mean square error is assumed to be constant, is biased estimation. Investigate the relation between normalized error of derived $\widehat{\boldsymbol{\omega}}$ together with $\left\|\widehat{\boldsymbol{r}}_0\right\|$ and mean error of $M_a^{(m,n)}$. Fig. 6.5B shows the mean value changing curve of normalized error of $\left(\omega_X, \omega_Y, \omega_Z, \left\|\widehat{\boldsymbol{r}}_0\right\|\right)$ in 100 times of simulations when mean error of $M_a^{(m,n)}$ lies in [−10,10] and mean square error is 1. It is clear that the normalized error of $\left(\omega_X, \omega_Y, \omega_Z, \left\|\widehat{\boldsymbol{r}}_0\right\|\right)$ is much smaller the that of case (1), which shows that mean square error of $M_a^{(m,n)}$ has a greater influence on solving accuracy than its mean error to some extent.

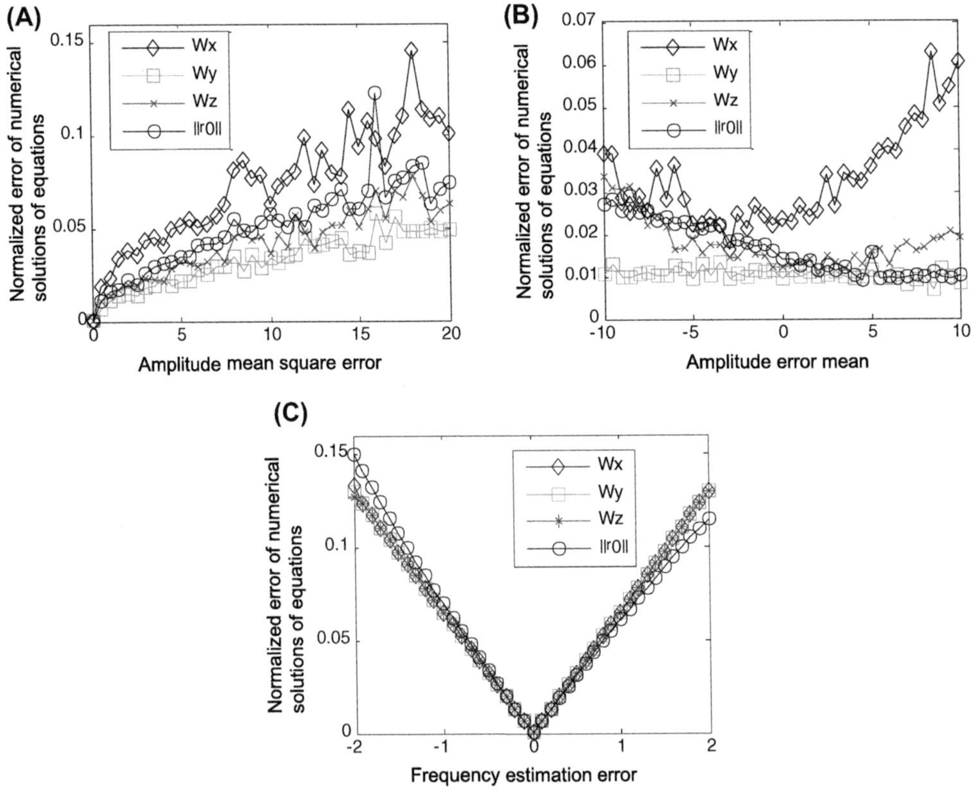

Figure 6.5 Robustness analysis.

3. Suppose that angular frequency Ω of curve is estimated accurately. Investigate the relation between normalized error of derived $\widehat{\boldsymbol{\omega}}$ together with $\left\|\widehat{\boldsymbol{r}}_0\right\|$ and error of Ω. Fig. 6.5C shows the mean value changing curve of normalized error of $\left(\omega_X, \omega_Y, \omega_Z, \left\|\widehat{\boldsymbol{r}}_0\right\|\right)$ in 100 times of simulations when error of Ω lies in $[-1,1]$. It is clear that the normalized error of $\left(\omega_X, \omega_Y, \omega_Z, \left\|\widehat{\boldsymbol{r}}_0\right\|\right)$ basically increases linearly with the increasing of error of Ω.

As is apparent in previous analysis and Fig. 6.5A–C, solving accuracy of equations array is restrained by estimating accuracy of curve parameters to some extent. Angular frequencies are derived from the mean of angular frequencies of curves abstracted by all groups of transmitting-receiving arrays so the evaluating accuracy is high in general. Evaluated errors of curve parameters mainly show as evaluated errors of amplitudes. Thus, accuracy of three-dimensional micromotion feature reconstruction is restrained by accuracy of amplitude abstraction of the sinusoidal curve.

6.3 THREE-DIMENSIONAL MICROMOTION FEATURE RECONSTRUCTION IN WIDEBAND MULTIPLE INPUT MULTIPLE OUTPUT RADAR

6.3.1 Orthogonal Frequency Division Multiplexing Linear Frequency Modulated Signal

Three-dimensional micromotion feature reconstruction in narrowband MIMO radar has been introduced in the previous section. In order to get higher resolution of radar, the wideband MIMO radar has been researched widely in recent years. In wideband MIMO radar, the radiation wave of array antenna element should be designed meticulously to keep ideal orthogonality of them as possible. Orthogonal frequency division multiplexing-linear frequency modulation (OFDM-LFM) which is a wideband signal based on linear frequency modulation (LFM) signal is a good choice and its micro-Doppler effect is similar to the micro-Doppler effect of LFM signal. Therefore, Assuming the OFDM-LFM signals are transmitted by the MIMO radar, the transmitted signal of the m-th transmitter is written as

$$s_m(t_k) = \mathrm{rect}\left(\frac{t_k}{T_p}\right)\exp\left(\mathrm{j}2\pi\left(f_c t_k + m\frac{k_c}{T_p}t_k + \frac{1}{2}\mu t_k^2\right)\right), \quad m = 1, 2, \cdots, M; \quad k_c \in \mathbf{N} \tag{6.4}$$

where t_k is the fast time, T_p is the pulse duration, f_c is the carrier frequency, μ is the chirp rate, M is the total number of transmitters in MIMO radar, and k_c is a natural number. For the signals transmitted by the m_1-th transmitter and m_2-th transmitter, we have

$$\begin{aligned}\int_{-T_p/2}^{T_p/2} s_{m_1}(t_k)s_{m_2}^*(t_k)\mathrm{d}t_k &= \int_{-T_p/2}^{T_p/2}\left(\mathrm{rect}\left(\frac{t_k}{T_p}\right)\cdot\exp\left(\mathrm{j}2\pi\left(f_c t_k + m_1\frac{k_c}{T_p}t_k + \frac{1}{2}\mu t_k^2\right)\right)\right.\\ &\quad \left.\times\,\mathrm{rect}\left(\frac{t_k}{T_p}\right)\cdot\exp\left(-\mathrm{j}2\pi\left(f_c t_k + m_2\frac{k_c}{T_p}t_k + \frac{1}{2}\mu t_k^2\right)\right)\right)\mathrm{d}t_k\\ &= \int_{-T_p/2}^{T_p/2}\exp\left(\mathrm{j}2\pi(m_1 - m_2)\frac{k_c}{T_p}t_k\right)\mathrm{d}t_k\end{aligned} \tag{6.5}$$

If $m_1 = m_2$,

$$\int_{-T_p/2}^{T_p/2} s_{m_1}(t_k)s_{m_2}^*(t_k)\mathrm{d}t_k = \int_{-T_p/2}^{T_p/2}\exp\left(\mathrm{j}2\pi(m_1 - m_2)\frac{k_c}{T_p}t_k\right)\mathrm{d}t_k = \int_{-T_p/2}^{T_p/2}\mathrm{d}t_k = T_p \tag{6.6}$$

If $m_1 \neq m_2$,

$$\begin{aligned}\int_{-T_p/2}^{T_p/2} s_{m_1}(t_k)s^*_{m_2}(t_k)\mathrm{d}t_k &= \int_{-T_p/2}^{T_p/2} \exp\left(\mathrm{j}2\pi(m_1 - m_2)\frac{k_c}{T_p}t_k\right)\mathrm{d}t_k \\ &= \int_{-T_p/2}^{T_p/2}\left(\cos\left(2\pi(m_1 - m_2)\frac{k_c}{T_p}t_k\right)\right. \\ &\quad \left. + \mathrm{j}\sin\left(2\pi(m_1 - m_2)\frac{k_c}{T_p}t_k\right)\right)\mathrm{d}t_k \\ &= \frac{T_p}{2\pi(m_1 - m_2)k}\left(\sin\left(2\pi(m_1 - m_2)\frac{k_c}{T_p}t_k\right)\right. \\ &\quad \left.\left. - \mathrm{j}\cos\left(2\pi(m_1 - m_2)\frac{k_c}{T_p}t_k\right)\right)\right|_{-T_p/2}^{T_p/2} \\ &= \frac{T_p}{2\pi(m_1 - m_2)k_c}(\sin((m_1 - m_2)k_c\pi) - \mathrm{j}\cos((m_1 - m_2)k_c\pi) \\ &\quad - \sin((m_1 - m_2)k_c\pi) + \mathrm{j}\cos((m_1 - m_2)k_c\pi)) = 0\end{aligned} \tag{6.7}$$

ie,

$$\int_{-T_p/2}^{T_p/2} s_{m_1}(t_k)s^*_{m_2}(t_k)\mathrm{d}t_k = \begin{cases} T_p, & m_1 = m_2 \\ 0, & m_1 \neq m_2 \end{cases} \tag{6.8}$$

It demonstrates that the transmitted signals between each two different transmitters are orthogonal to each other. In practice, we should select appropriate k_c to ensure that the operation frequency band of each two transmitters should not be overlapped, ie,

$$\frac{k_c}{T_p} > c_1 \mu T_p, \quad c_1 > 1 \tag{6.9}$$

In the following, we assume the ideal orthogonality between each two transmitted signals for simplicity.

6.3.2 Micro-Doppler Effect

The geometry of the wideband MIMO radar system and the target is shown in Fig. 6.1 when the OFDM-LFM signals are transmitted by the MIMO radar, at the slow time t_m, the received echo of the n-th receiver from the scatterer P transmitted is written as

$$s_n(t_k, t_m) = \sum_{m=1}^{M} \sigma s_m\left(t_k - \frac{r(t_m; m, n)}{c}\right), \quad n = 1, 2, \cdots, N \tag{6.10}$$

where σ is the scattering coefficient of P, c is the speed of light, $r(t_m;m,n)$ is the sum distance form P to (m,n) at the slow time t_m. The method that is shown in Fig. 6.6 is used to separate the signal of each transmitter, ie, we use the transmitted signal $s_n(t_k,t_m)$ of each transmitter for the "dechirp" processing and then get the signal $s_{m,n}(t_k,t_m)$ after low-pass filtering processing.

For simplicity, we assume the translational motion between the target and radars can be compensated accurately. Therefore, the motion trajectory of the target is known, then the echoes from o can be chosen as the reference signals for the "dechirp" processing, ie, $s_m(t_k)$ in Fig. 6.8 takes the place of $s_{m0}(t_k,t_m)$:

$$s_{m0}(t_k, t_m) = s_m\left(t_k - \frac{R_0(t_m; m, n)}{c}\right) \tag{6.11}$$

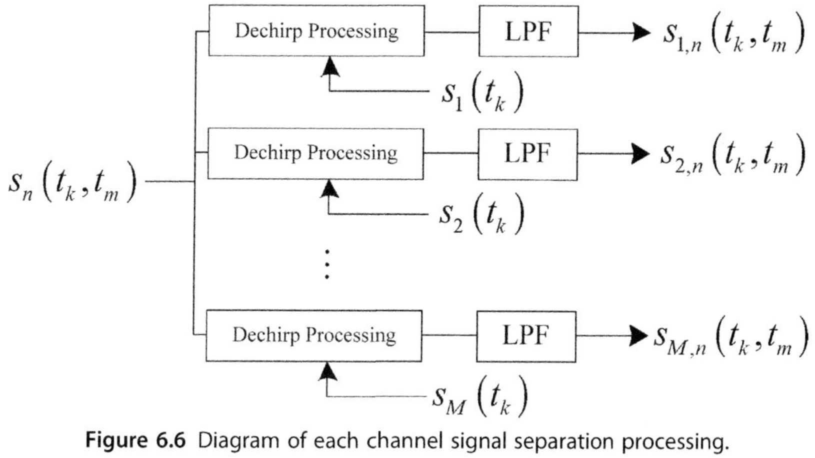

Figure 6.6 Diagram of each channel signal separation processing.

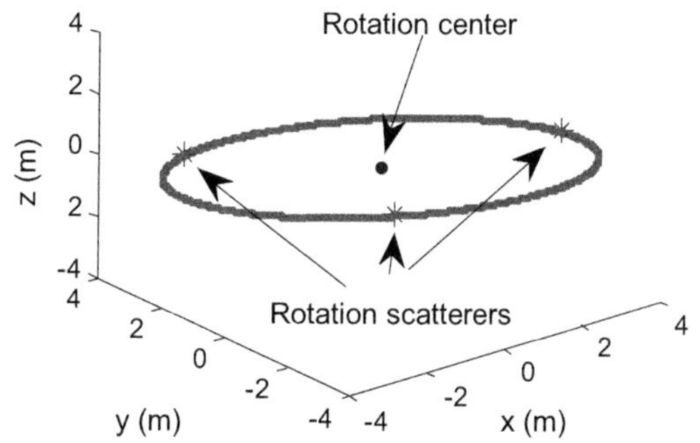

Figure 6.7 Target scatterer model.

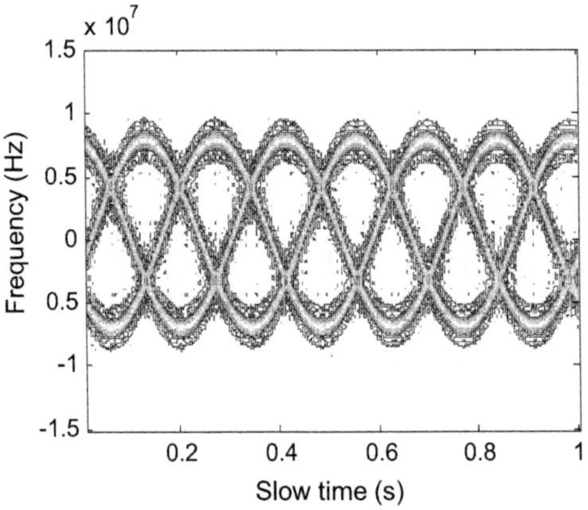

Figure 6.8 The obtained micro-Doppler curve of transmitter–receiver pair (1,1).

where $R_0(t_m;m,n)$ is the sum distance from o to (m,n) at the slow time t_m. After the "dechirp" processing, it yields

$$\begin{aligned}
s_{\mathrm{d}}(t_k,t_m) &= s_{m0}^*(t_k,t_m)s_n(t_k,t_m)\\
&= \sigma\exp\left(-\mathrm{j}2\pi\left(\left(f_0+\frac{mk_c}{T_p}\right)\left(t_k-\frac{R_0(t_m;m,n)}{c}\right)+\frac{\mu}{2}\left(t_k-\frac{R_0(t_m;m,n)}{c}\right)^2\right)\right)\\
&\quad\times\sum_{m'=1}^{M}\mathrm{rect}\left(\frac{t_k-r(t_m;m',n)/c}{T_p}\right)\exp\left(\mathrm{j}2\pi\left(\left(f_0+\frac{m'k_c}{T_p}\right)\left(t_k-\frac{r(t_m;m',n)}{c}\right)\right.\right.\\
&\quad\left.\left.+\frac{\mu}{2}\left(t_k-\frac{r(t_m;m',n)}{c}\right)^2\right)\right)\\
&= \sigma\sum_{m'=1}^{M}\mathrm{rect}\left(\frac{t_k-r(t_m;m',n)/c}{T_p}\right)\exp\left(\mathrm{j}2\pi\left(-\frac{f_0R_\Delta(t_m;m',n)}{c}+\frac{(m'-m)k_c}{T_p}t_k\right.\right.\\
&\quad-\frac{k_c}{T_pc}\left(m'r(t_m;m',n)-mR_0(t_m;m,n)\right)-\frac{\mu R_\Delta(t_m;m',n)}{c}t_k\\
&\quad\left.\left.+\frac{\mu\left(r^2(t_m;m',n)-R_0^2(t_m;m,n)\right)}{2c^2}\right)\right)
\end{aligned} \tag{6.12}$$

where $R_\Delta(t_m;m',n) = r(t_m;m',n) - R_0(t_m;m',n)$. So, after the low-pass filtering processing, the component $s_{m,n}(t_k,t_m)$ of $s_{\mathrm{d}}(t_k,t_m)$ will be when $m' = m$ and its expression is as:

$$s_{\mathrm{d}m,n}(t_k,t_m) = \sigma\mathrm{rect}\left(\frac{t_k - r(t_m;m,n)/c}{T_p}\right)\exp\left(\mathrm{j}2\pi\left(-\left(\frac{f_0}{c}+\frac{mk_c}{T_pc}\right)R_\Delta(t_m;m,n)\right.\right.$$
$$\left.\left.+\frac{\mu\left(r^2(t_m;m',n)-R_0^2(t_m;m,n)\right)}{2c^2}-\frac{\mu R_\Delta(t_m;m,n)}{c}t_k\right)\right) \tag{6.13}$$

Taking the Fourier transform to $s_{\mathrm{d}m,n}(t_k,t_m)$ with respect to t_k, and then removing the residual video phase, it yields

$$S_{m,n}(f_k,t_m) = \sigma T_p\mathrm{sinc}\left(T_p\left(f_k+\frac{\mu R_\Delta(t_m;m,n)}{c}\right)\right)$$
$$\times\exp\left(-\mathrm{j}2\pi\left(\frac{f_c}{c}+\frac{mk_c}{T_pc}\right)R_\Delta(t_m;m,n)\right) \tag{6.14}$$

It can be found from (6.14) that $|S_{m,n}(f_k,t_m)|$ is a sinc function with peaks at $f_k = -\mu R_\Delta(t_m;m,n)/c$. It is consistent with the micromotion feature of the double-base wideband radar in Eq. (4.24), and the difference is that the carrier frequency of different transmitter is different. According to the derivation of Section 4.3.2, the cosine curve equation of range profile of (m,n) can be written as:

$$R_\Delta(t_m;m,n) = r\cos(\Omega t_m)\sin\theta_1^{(m,n)} + r\cos\left(\Omega t_m+\phi^{(m,n)}\right)\sin\theta_2^{(m,n)}$$
$$= \left(r\sin\theta_1^{(m,n)} + r\sin\theta_2^{(m,n)}\cos\phi^{(m,n)}\right)\cos(\Omega t_m) - r\sin\theta_2^{(m,n)}\sin\phi^{(m,n)}\sin(\Omega t_m)$$
$$= r\sqrt{\sin^2\theta_1^{(m,n)}+\sin^2\theta_2^{(m,n)}+2\sin\theta_1^{(m,n)}\sin\theta_2^{(m,n)}\cos\phi^{(m,n)}}\sin\left(\Omega t_m-\varphi^{(m,n)}\right) \tag{6.15}$$

where $\theta_1^{(m,n)}$ is the included angle between $\boldsymbol{\omega}$ and gazing direction of m-th transmitter, $\theta_2^{(m,n)}$ is the included angle between $\boldsymbol{\omega}$ and gazing direction of n-th transmitter, $\phi^{(m,n)}$ is the included angle between the projection of gazing direction of m-th transmitter and the projection of gazing direction of n-th transmitter on the spanning plane, and the expression of $\varphi^{(m,n)}$ is as:

$$\varphi^{(m,n)} = \arctan\left(\frac{\sin\theta_1^{(m,n)}+\sin\theta_2^{(m,n)}\cos\phi^{(m,n)}}{\sin\theta_2^{(m,n)}\sin\phi^{(m,n)}}\right)$$

6.3.3 Three-Dimensional Micromotion Feature Reconstruction of Rotating Targets

Based on the previous analysis, as for (m,n), the amplitude of detected micro-Doppler feature curve is determined by the geometric structure parameters between the target and the transceiver, ie,

$$M_a^{(m,n)} = r\sqrt{\sin^2\theta_1^{(m,n)} + \sin^2\theta_2^{(m,n)} + 2\sin\theta_1^{(m,n)}\sin\theta_2^{(m,n)}\cos\phi^{(m,n)}} \tag{6.16}$$

where $\theta_1^{(m,n)}$, $\theta_2^{(m,n)}$ and $\cos\phi^{(m,n)}$ are calculated as similar as Eqs. (4.29) and (4.30). If the position relationship is certain between target and transmitter and receiver, the line of sight between each transceiver is known in MIMO radar and $M_a^{(m,n)}$ can be obtained. There are four unknown parameters $(\omega_X,\omega_Y,\omega_Z,r)$ in (6.16). Because of $\omega_X^2+\omega_Y^2+\omega_Z^2=\Omega^2$, we can structure the simultaneous equation system to get $(\omega_X,\omega_Y,\omega_Z,r)$ by just using multiview specialty of MIMO radar to extract the amplitude and the angular frequency of three micro-Doppler feature curves. And the micro-Doppler feature curves can be obtained by the methods that are introduced in Chapter 5. Thus we can construct the following equation system:

$$\begin{cases} M_a^{(m,n)} = r\sqrt{\sin^2\theta_1^{(m,n)} + \sin^2\theta_2^{(m,n)} + 2\sin\theta_1^{(m,n)}\sin\theta_2^{(m,n)}\cos\phi^{(m,n)}} \\ \qquad m=1,2,\cdots,M;\, n=1,2,\cdots,N \\ \omega_X^2+\omega_Y^2+\omega_Z^2=\Omega^2 \end{cases} \tag{6.17}$$

$(\omega_X,\omega_Y,\omega_Z,r)$ can be gotten through the previous steps. Compared to monostatic radar, which can only get motion information on the line-of-sight direction, MIMO radar provides more accurate information for realizing the accurate recognition of a special target. If you want to improve the precision, you can add equations into the equation system, by adding transceiver into MIMO radar, to restrain the answer of equation system more strictly.

The following is the example of simulation. Assume the MIMO radar contains two transmitters and four receivers. The two transmitters are located at (−3000,0,0) m and (3000,0,0) m with carrier frequencies 10 and 12 GHz, respectively. The bandwidth of the transmitted signal is 600 MHz and pulse width $T_p=1$ μs. The four receivers are located at (−7000,0,0) m, (−1000,0,0) m, (2000,0,0) m, and (5000,0,0) m, respectively. The range resolution is 0.25 m, the pulse repetition frequency is 1000 Hz, and the echo time is 1 s. The target model is shown in Fig. 6.7. The initial location of the target center is (3000,4000,5000) m and the translational velocity of the target is $\boldsymbol{v}=(100,150,100)^{\mathrm{T}}$ m/s. Three scatterers rotate around the center and all of their angular velocities and radii are $\widehat{\boldsymbol{\omega}}=(2\pi,3\pi,3\pi)^{\mathrm{T}}$ rad/s and $r=4.3716$ m, respectively. The initial rotating radii of three scatterers are $(2,2,-3.3333)^T$, $(-3.9542,1.3387,1.2974)^T$, and $(1.9542,-3.3387,2.0359)^T$, respectively. $\Omega = 14.7354$ rad/s, ie, 2.3452 Hz.

The echo signal of (1,1), which is the result of $f - t_m$ plane after one-dimensional imaging processing, is shown in Fig. 6.8, and it can be learned that three rotation scatterers are respectively corresponding to a sine curve. The angular frequency is 14.7 rad/s and the amplitude is 3.72 m by HT (frequency is 7.46 MHz), and the theoretical amplitude calculated by (6.16) is 3.7336. The same process mode is employed to the other transceiver, and the amplitudes are shown in Table 6.2.

Based on the extracted parameters, we structure the equation system according to (6.17), and we get $(\omega_X,\omega_Y,\omega_Z,r) = (6.3023,9.3762,9.4052,4.4069)$ which is very close to the theoretical value.

Owing to the bandwidth of signal, the resolution of MIMO radar is limited. And because of side lobe of range profile, there is error in extracting micro-Doppler feature curve parameters on the $f - t_m$ plane. The answer precision of (6.17) must be influenced by the estimation error of Ω and $M_a^{(m,n)}$. In order to analyze the impact of this, we define the normalized error as

$$\rho = \frac{\hat{X} - X}{X} \tag{6.18}$$

where X denotes the true value, $\hat{X}$ denotes the corresponding estimated value, and $|\rho|$ denotes the normalized absolute error. Without loss of generality, assume the errors obey the Gaussian distribution. The Monte-Carlo method is utilized to analyze the robustness, where 100 simulations are carried out in each following situation:

1. Assuming Ω is estimated precisely, and the estimation of $M_a^{(m,n)}$ is unbiased estimation (ie, $E\left(\rho_{M_a^{(m,n)}}\right) = 0$), we investigate the relationship between the normalized absolute errors of the solved $(\omega_X,\omega_Y,\omega_Z,r)$ and $D\left(\rho_{M_a^{(m,n)}}\right)$. Fig. 6.9A shows the curves of the normalized absolute errors of $(\omega_X,\omega_Y,\omega_Z,r)$ when $D\left(\rho_{M_a^{(m,n)}}\right)$ changes from 0 to 0.5. It can be found that the normalized absolute errors of $(\omega_X,\omega_Y,\omega_Z,r)$ are linear with $D\left(\rho_{M_a^{(m,n)}}\right)$ approximately.

Table 6.2 The Micro-Doppler Curve Amplitude of Each Transmitter–Receiver Pair Obtained Through Hough Transform

(m,n)	(1,1)	(1,2)	(1,3)	(1,4)	(2,1)	(2,2)	(2,3)	(2,4)
Theoretical amplitude value	3.7336	2.0244	0.8396	1.7907	0.8838	1.6006	3.2194	4.8919
Extracted amplitude	3.72	2.00	0.84	1.80	0.90	1.60	3.23	5.00

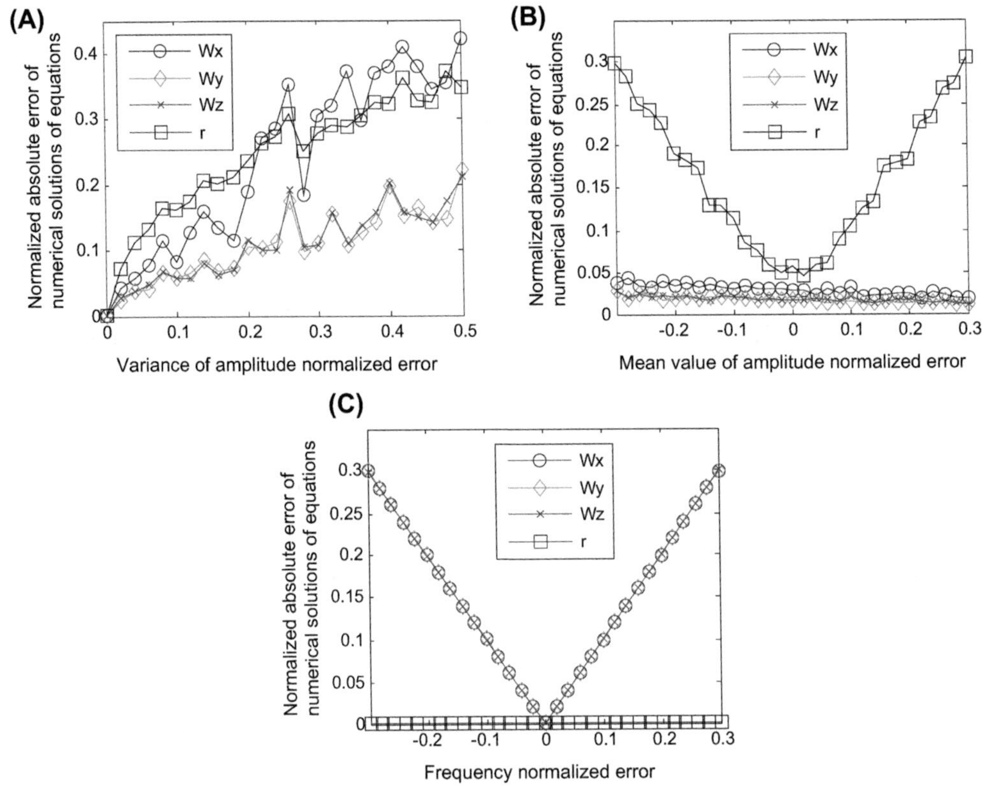

Figure 6.9 Robustness analysis.

2. Assume Ω is estimated precisely, and the estimation of $M_a^{(m,n)}$ is biased estimation (ie, $E\left(\rho_{M_a^{(m,n)}}\right) \neq 0$). For the given $D\left(\rho_{M_a^{(m,n)}}\right)$, we investigate the relationship between the normalized absolute errors of $(\omega_X,\omega_Y,\omega_Z,r)$ and $E\left(\rho_{M_a^{(m,n)}}\right)$. Fig. 6.9B shows the curves of the normalized absolute errors of $(\omega_X,\omega_Y,\omega_Z,r)$ when $D\left(\rho_{M_a^{(m,n)}}\right) = 0.01$, $E\left(\rho_{M_a^{(m,n)}}\right)$ changes from -0.3 to 0.3. It can be found that the normalized absolute errors of r are linear with $\left|E\left(\rho_{M_a^{(m,n)}}\right)\right|$ approximately, while the normalized absolute errors of $(\omega_X,\omega_Y,\omega_Z,r)$ are nearly independent to $E\left(\rho_{M_a^{(m,n)}}\right)$, and the error is smaller than (1) deeply. It indicates that the influence of $E\left(\rho_{M_a^{(m,n)}}\right)$ on the solutions of $(\omega_X,\omega_Y,\omega_Z)$ is weaker than that of $D\left(\rho_{M_a^{(m,n)}}\right)$.
3. Assume $M_a^{(m,n)}$ is estimated precisely, we investigate the relationship between the normalized absolute errors of $(\omega_X,\omega_Y,\omega_Z,r)$ and the normalized error of Ω (ie, ρ_Ω).

Fig. 6.9C shows the curves of the normalized absolute errors of $(\omega_X, \omega_Y, \omega_Z, r)$ when ρ_Ω changes from -0.3 to 0.3. It can be found that the normalized absolute errors of r are close to zero, while the normalized absolute errors of $(\omega_X, \omega_Y, \omega_Z, r)$ are linear with ρ_Ω. This phenomenon can be explained as follows: in (6.17), the value of $M_a^{(m,n)}$ is determined by r, $\theta_1^{(m,n)}$, $\theta_2^{(m,n)}$, and $\phi^{(m,n)}$. Because the values of $\theta_1^{(m,n)}$, $\theta_2^{(m,n)}$, and $\phi^{(m,n)}$ are related to the direction of $\widehat{\boldsymbol{\omega}} = (\omega_X, \omega_Y, \omega_Z)^{\mathrm{T}}$ and independent to the Ω, the estimation precision of Ω does not affect the solution of r. Whereas according to $\omega_X^2 + \omega_Y^2 + \omega_Z^2 = \Omega^2$, when ρ_Ω increases linearly, the normalized absolute errors of $(\omega_X, \omega_Y, \omega_Z)$ will also increase linearly.

According to the previous analysis, the answer precision of the equation system is affected by the estimation precision of curve parameters. Therefore, if you want to improve the precision of $(\omega_X, \omega_Y, \omega_Z, r)$, you should improve the estimation precision of Ω and $M_a^{(m,n)}$ possibly.

6.4 THREE-DIMENSIONAL MICROMOTION FEATURE RECONSTRUCTION OF TARGETS WITH PRECESSION

The micro-Doppler effect of precession motion in monostatic radar has been discussed in Chapters 2 and 3. Due to the complexity of the theoretical expressions, it is difficult to construct a three-dimensional micromotion feature reconstruction algorithm for targets with precession. To simplify the discussion, the following reconstruction algorithm is deduced based on distributed wideband radar networks composed with monostatic miniradars.

Many artificial space targets have regular shapes, such as spherical, cylindrical, conical, etc. In this section, we focus on the 3-D precession feature extraction of cone-shaped space targets. The cone-shaped target is one important kind of space object; for example, ballistic warheads are usually cone shaped. Because of their regular shape, the 3-D precession feature extraction of cone-shaped targets is relatively simpler than that of the objects with irregular shapes. The feature extraction algorithm for irregular-shaped objects needs more in-depth work in the future.

In this section, although the micro-Doppler effect of precession motion has been discussed before, the geometry of precession target and distributed radar networks is established and the echo model is analyzed again for clarity. Then, a 3-D micromotion feature reconstruction algorithm is introduced.

6.4.1 Echo Model of Targets With Precession in Distributed Radar Networks

The scatterer model of a cone-shaped space target after translational motion compensation is shown in Fig. 6.10. In the scatterer model, we assume that the target contains three main scatterers, where scatterer A is the conical point, and B and C are the scatterers located at the tails of the target, which are usually on the edge of the cone bottom,

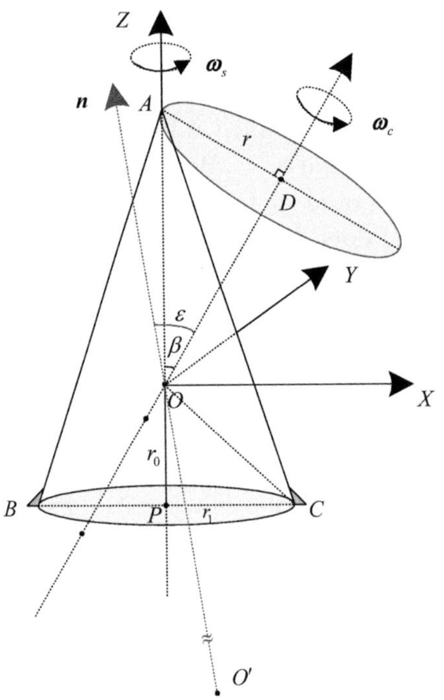

Figure 6.10 The cone-shaped space target model.

as shown in Fig. 6.10. Generally, the tails have relatively high scattering coefficients due to the surface curvature discontinuity. In the figure, just two tail scatterers are drawn for clarity. The target is spinning around the symmetry axis of the target at an angular velocity $\boldsymbol{\omega}_s$, and is simultaneously coning around a certain fixed axis with an included angle β with respect to the spinning axis at an angular velocity $\boldsymbol{\omega}_c$, which forms the precession motion. Taking the intersection point O of the spinning axis and coning axis as the origin, a Cartesian coordinate system $OXYZ$ can be defined as shown in the figure, where the Z-axis is the symmetry axis of the target at the initial time, the X-axis is generated by clockwise rotating the Z-axis for $\pi/2$ rad on the plane spanned by the spinning axis and coning axis, and the Y-axis is determined by the right-hand rule. The distance between point O and the center of the cone bottom P is r_0, and the radius of the cone bottom is r_1.

Assume the miniradar in distributed radar networks are simple transceiver units, ie, each unit is in fact a monostatic radar. The transmitted waves of the miniradar are LFM signals, and the miniradar operate at different frequency bands to avoid mutual interferences. Assume one of the miniradars is located at the point O' in far field, and the unit vector of the LOS is $\boldsymbol{n}$ with an included angle ε to the coning axis. According to the analyses in Chapter 3, the micro-Doppler curves are determined by $R_\Delta(t_m)$.

For scatterer A, since it is located at the spinning axis, only the coning motion can be observed from the radar echoes. According to the geometry shown in Fig. 6.10, it can be found that

$$R_\Delta(t_m) \approx \overrightarrow{O'A} \cdot \boldsymbol{n} - \overrightarrow{O'O} \cdot \boldsymbol{n} = \boldsymbol{r}_A^{\mathrm{T}}(t_m)\boldsymbol{n} = \left(\mathbf{R}_{\text{coning}}\boldsymbol{r}_{A0}\right)^{\mathrm{T}}\boldsymbol{n} \tag{6.19}$$

where $\boldsymbol{r}_{A0}$ is the vector $\overrightarrow{OA}$ at the initial time, $\boldsymbol{r}_A^{\mathrm{T}}(t_m)$ is the vector $\overrightarrow{OA}$ at slow time t_m, and $\mathbf{R}_{\text{coning}}$ is the rotation matrix corresponding to the coning motion.

Different from the conical point A, the micromotion of scatterer B and C is synthesized by spinning and coning. Taking scatterer C as an example and ignoring the scatterers' shadowing effect, the distance between point C and point O along the LOS can be rewritten as follows:

$$R_\Delta(t_m) \approx \overrightarrow{O'C} \cdot \boldsymbol{n} - \overrightarrow{O'O} \cdot \boldsymbol{n} = \left(\mathbf{R}_{\text{coning}}\mathbf{R}_{\text{spinning}}\boldsymbol{r}_{C0}\right)^{\mathrm{T}}\boldsymbol{n} \tag{6.20}$$

where $\boldsymbol{r}_{C0}$ is the vector $\overrightarrow{OC}$ at the initial time and $\mathbf{R}_{\text{spinning}}$ is the rotating matrix corresponding to the spinning motion. We can see that the micro-Doppler curves of scatterers B and C do not follow the simple sinusoidal pattern anymore. It is more complicated than that of scatterer A. However, the amplitude of the curve is also modulated by the unit vector of LOS, which is similar to that of scatterer A.

6.4.2 Three-Dimensional Micromotion Feature Reconstruction of Precession

In order to reconstruct the 3-D micro-Doppler features of the target, the micro-Doppler curve parameters should be extracted from the range—slow time images obtained by miniradar. Fig. 6.11 shows a range—slow time image of the target in Fig. 6.10. It is easy to obtain the curve features as follows.

The coning period T_c: can be calculated by the autocorrelation analysis of the curve A, or by EMD algorithm;

The amplitude M_c of the coning curve: is equivalent to half of the difference between the maximum and the minimum of the curve A.

The precession period T_{pr}: can be calculated by the autocorrelation analysis of the curve B or C.

The range span M_p of the precession curve: is equivalent to the difference between the maximum and the minimum of the curve B or C.

The range span L_{max} of all the micro-Doppler curves: is equivalent to the difference between the maximum and the minimum of curves A, B, and C.

The number N_p of local maximums of the precession curve during a precession period: after T_{pr} is calculated, it is easy to count the number of local maximums of the precession curve during a precession period.

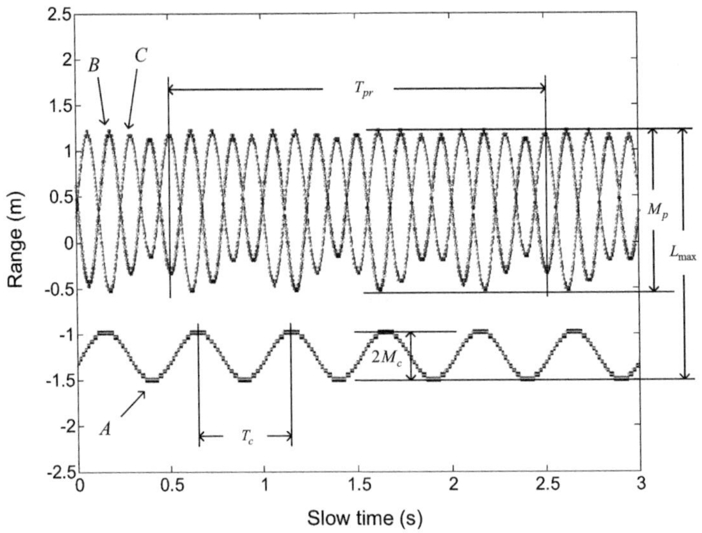

Figure 6.11 Range—slow time image of a precession target.

Based on the above micro-Doppler curve parameters, the 3-D precession feature reconstruction method is given as follows.

6.4.2.1 Solving for the Three-Dimensional Coning Vector

As for the conical point A, only its coning motion can be observed from the radar echoes, so the coning vector could be solved by analyzing the micro-Doppler signatures of A. According to (6.19), after the micro-Doppler curve parameters extraction, six unknown parameters, ie, $\boldsymbol{\omega}_c = (\omega_{cx}, \omega_{cy}, \omega_{cz})^{\mathrm{T}}$ and $\boldsymbol{r}_{A0} = (r_{A0x}, r_{A0y}, r_{A0z})^{\mathrm{T}}$, must be solved when the unit vector of LOS $\boldsymbol{n}$ is known. Therefore, at least six equations are required to solve for the unknown parameters, that is to say, at least six miniradar are required to observe the target at different viewing angles. However, solving a matrix equation as in (6.19) is very complicated. Next, we attempt to simplify the solving process and reduce the number of unknowns.

For better clarity, the geometry of the coning motion of conical scatterer A shown in Fig. 6.10 is redrawn in Fig. 6.12. In the figure, the trajectory of A appears as a circle on the plane with a radius of r. The center of the circle locates at point D, which is the projection of A on the coning axis. $'$ is the projection of $\boldsymbol{n}$ on the plane. According to the geometry shown in the figure, (6.19) can be rewritten as

$$
\begin{aligned}
R_\Delta(t_m) &= \left(\mathbf{R}_{\text{coning}} \boldsymbol{r}_{A0}\right)^{\mathrm{T}} \boldsymbol{n} = \overrightarrow{OD} \cdot \boldsymbol{n} + \left(\mathbf{R}_{\text{coning}} \overrightarrow{DA}\right)^{\mathrm{T}} \boldsymbol{n} \\
&= \overrightarrow{OD} \cdot \boldsymbol{n} + r \cos(\Omega_c t_m + \theta) \cdot \sin \boldsymbol{\varepsilon}
\end{aligned} \tag{6.21}
$$

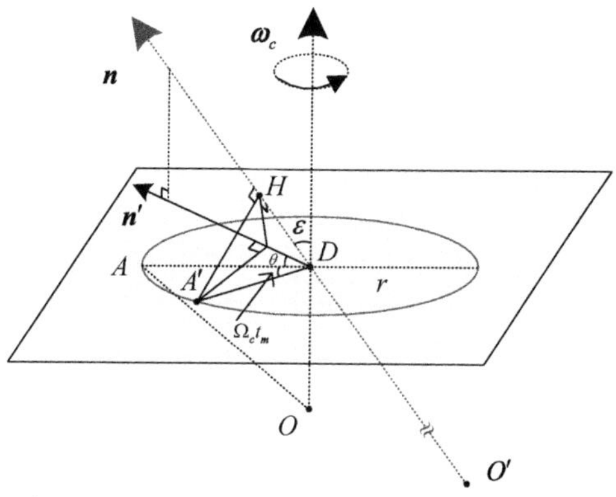

Figure 6.12 The geometry of the coning motion of the conical point A.

where θ is the initial phase. When $\boldsymbol{n} = (n_x, n_y, n_z)^{\mathrm{T}}$, we have

$$\varepsilon = \arccos\left(\frac{\omega_c \boldsymbol{n}}{\Omega_c}\right) = \arccos\left(\frac{n_x\omega_{cx} + n_y\omega_{cy} + n_z\omega_{cz}}{\Omega_c}\right) \tag{6.22}$$

It can be found from (6.21) that the amplitude of the micro-Doppler curve on the range—slow time plane is $r\sin\varepsilon$, and ε is related to the LOS of the miniradar. Assume that the distributed radar networks consist of N miniradars, after the coning period T_c and the amplitude M_{ci} ($i = 1,2,\cdots,N$ and i indicates the index of miniradars) are extracted from the micro-Doppler curve, a multivariable equation system can be constructed:

$$\begin{cases} M_{ci} = r\sin\varepsilon_i, & i = 1,2,\cdots,N \\ \omega_{cx}^2 + \omega_{cy}^2 + \omega_{cz}^2 = \Omega_c^2 \end{cases} \tag{6.23}$$

where the unknown parameters to be solved are $(\omega_{cx}, \omega_{cy}, \omega_{cz}, r)$. It is difficult to derive the analytical solution to this equation system. Therefore, numerical methods such as the Newton method and quasi-Newton method can be used to solve the approximate roots. Since there are four parameters to be solved, at least three miniradars are necessary to solve the roots of (6.23). The accuracy of the solution could be further improved by increasing the number of equations for stricter constraints by utilizing more miniradars in the network.

It must be mentioned that, because of $\sin\varepsilon = \sin(\pi - \varepsilon)$, double roots exist when solving the equation system (6.23), ie, $(\omega_{cx}, \omega_{cy}, \omega_{cz}, r)$ and $(-\omega_{cx}, -\omega_{cy}, -\omega_{cz}, r)$. The vector $(\omega_{cx}, \omega_{cy}, \omega_{cz})^{\mathrm{T}}$ is the converse of the vector $(-\omega_{cx}, -\omega_{cy}, -\omega_{cz})^{\mathrm{T}}$, which means the rotation direction corresponding to $(\omega_{cx}, \omega_{cy}, \omega_{cz})^{\mathrm{T}}$ (eg, clockwise rotation) is opposite to that corresponding to $(-\omega_{cx}, -\omega_{cy}, -\omega_{cz})^{\mathrm{T}}$ (eg, counterclockwise rotation). In practice, the rotation direction is not a valuable feature for space target recognition, therefore either root is acceptable as the solution to (6.23).

6.4.2.2 Solving for the Spinning Period

According to the analyses in Section 3.2.5, the right-hand side of (6.20) contains four frequency components: $\Omega_c + \Omega_s$, Ω_c, Ω_s, and $\Omega_c - \Omega_s$. The component with the highest frequency $\Omega_c + \Omega_s$ makes the amplitude of the curve of $R_\Delta(t_m)$ oscillate at the frequency $\Omega_c + \Omega_s$, and the other components with lower frequencies form the envelope of the curve of $R_\Delta(t_m)$ together. In fact, the component with frequency $\Omega_c + \Omega_s$ can be rewritten as

$$\begin{aligned}
&\frac{A_4'}{2}\sin((\Omega_c+\Omega_s)t_m+\phi_4)-\frac{A_3'}{2}\cos((\Omega_c+\Omega_s)t_m+\phi_3)\\
&=\frac{A_4'}{2}\sin((\Omega_c+\Omega_s)t_m+\phi_4)-\frac{A_3'}{2}\cos(\phi_3-\phi_4)\cos((\Omega_c+\Omega_s)t_m+\phi_4)\\
&\quad+\frac{A_3'}{2}\sin(\phi_3-\phi_4)\sin((\Omega_c+\Omega_s)t_m+\phi_4)\\
&=\left(\frac{A_4'}{2}+\frac{A_3'}{2}\sin(\phi_3-\phi_4)\right)\sin((\Omega_c+\Omega_s)t_m+\phi_4)-\frac{A_3'}{2}\cos(\phi_3-\phi_4)\\
&\quad\times\cos((\Omega_c+\Omega_s)t_m+\phi_4)=\frac{A_5'}{2}\sin((\Omega_c+\Omega_s)t_m+\phi_4+\phi_5)
\end{aligned}\tag{6.24}$$

where

$$A_5' = \sqrt{(A_4' + A_3'\sin(\phi_3-\phi_4))^2 + (A_3'\cos(\phi_3-\phi_4))^2},\ \phi_5 = \operatorname{atan}\left(-\frac{A_3'\cos(\phi_3-\phi_4)}{A_4' + A_3'\sin(\phi_3-\phi_4)}\right)$$

Hence, the number of local maximums of $R_\Delta(t_m)$ during a precession period T_{pr} is $N_p = T_{pr}(\Omega_c + \Omega_s)/(2\pi)$, then the spinning period can be solved as

$$T_s = \frac{T_{pr}T_c}{N_p T_c - T_{pr}} \tag{6.25}$$

6.4.2.3 Solving for the Precession Angle and the Radius of the Cone Bottom

The micro-Doppler signatures of scatterers on the edge of the cone bottom are modulated by the precession angle β and the radius r_1 of the cone bottom. Therefore, the values of β and r_1 can be solved from (6.20) theoretically. However, it is difficult to obtain the two parameters from (6.20) directly. Similar to the analysis in Section 6.4.2.1, the geometrical analysis is carried out to solve β and r_1 as follows.

To ensure flight stability, most artificial cone-shaped space targets have a relatively small precession angle β (usually from several degrees to 10 or more degrees). As shown in Fig. 6.13, the scatterers B and C are taking a spinning motion around the major axis of the target and a simultaneous coning motion around the coning axis. Generally, rotating a circular disk around an axis, whose projection to the disk hits the center of the circle, yields a body whose surface partially coincides with a sphere, shown as the body taking B, E, Q, C, B', and G as the edge points in Fig. 6.13A, or taking B, G, B', E, Q, and C as the edge points in Fig. 6.13B. Point G and Q are the projection of point B and C on the coning axis, respectively. $\overline{CQ}$ and $\overline{BG}$ are the radius of the higher surface and the lower surface of the body, respectively. Therefore, the projection length on the LOS of this body is equal to the range span M_p of the precession curve (as shown in Fig. 6.11).

The geometry of the precession motion of the scatterers on the edge of the cone bottom contains the following two cases:

Case 1: The location of point K is within the cone bottom, where K is the intersection point of the coning axis and the plane spanned by the cone bottom. The

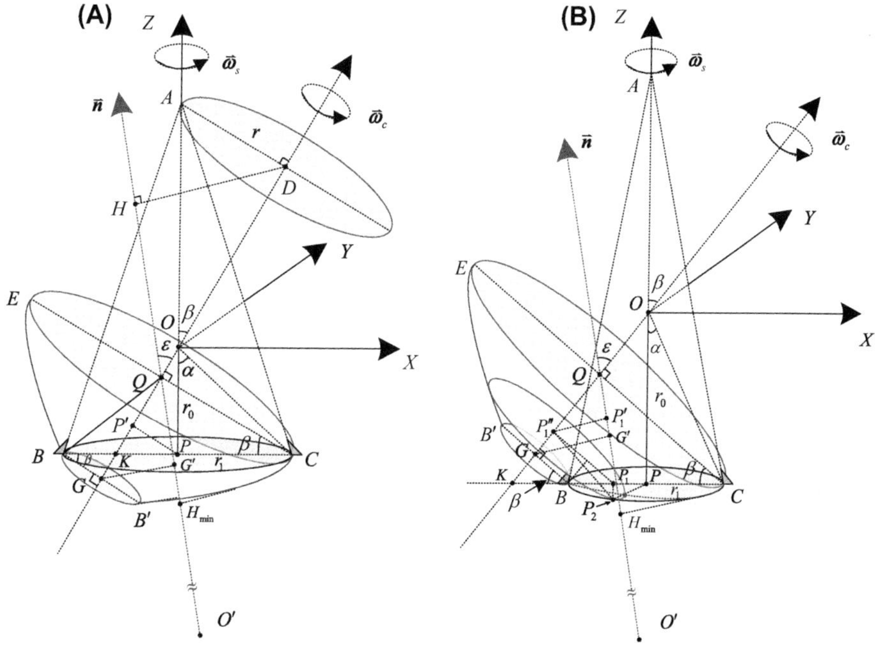

Figure 6.13 The geometry of the precession motion of the scatterers on the edge of the cone bottom. K is the intersection point of the coning axis and the plane spanned by the cone bottom. (A) Case 1: the location of K is within the cone bottom; (B) Case 2: the location of K is on the outside of the cone bottom.

geometry of this case is shown in Fig. 6.13A, where the point K locates on the segment $\overline{BP}$. Because $\angle KOP = \beta$ and $\overline{OP} = r_0$, we have $\overline{KP} = r_0 \tan\beta$, therefore the condition of location of K within the cone bottom is equivalent to $r_0 \tan\beta \le \overline{BP} = r_1$.

Case 2: Point K is located outside the cone bottom, ie, $r_0\tan\beta > r_1$. This case is shown in Fig. 6.13B, where K is located on the elongation line of $\overline{PB}$.

Both in *Case 1* and *Case 2*, for the i-th miniradar, the range span M_p of the precession curve can be expressed as follows (please refer to Appendix 6-A)

$$M_{pi} = \begin{cases} (r_0 \sin\beta + r_1 \cos\beta)\sin\varepsilon_i + 2r_1 \sin\beta|\cos\varepsilon_i| + |r_0 \sin\beta - r_1\cos\beta|\sin\varepsilon_i, \\ \qquad \text{when } l'_{x1} > 2r_1 \text{ and } l'_{x2} < 0 \quad (\textit{Case I}) \\ 2\sqrt{r_0^2 + r_1^2}, \qquad \text{when } 0 \le l'_{x1} \le 2r_1 \text{ and } 0 \le l'_{x2} \le 2r_1 \quad (\textit{Case II}) \\ (r_0 \sin\beta + r_1\cos\beta)\sin\varepsilon_i - (r_0\cos\beta - r_1\sin\beta)|\cos\varepsilon_i| + \sqrt{r_0^2 + r_1^2}, \\ \qquad \text{when } l'_{x1} > 2r_1 \text{ and } 0 \le l'_{x2} \le 2r_1 \quad (\textit{Case III}) \\ 2(r_0\sin\beta + r_1\cos\beta)\sin\varepsilon_i, \quad \text{when } l'_{x1} > 2r_1 \text{ and } l'_{x2} > 2r_1 \quad (\textit{Case IV}) \end{cases} \tag{6.26}$$

where

$$l'_{x1} = \frac{1}{\sin\beta}\left(r_0\cos\beta + r_1\sin\beta + |\cos\varepsilon_i|\sqrt{r_0^2 + r_1^2}\right) \tag{6.27}$$

$$l'_{x2} = \frac{1}{\sin\beta}\left(r_0\cos\beta + r_1\sin\beta - |\cos\varepsilon_i|\sqrt{r_0^2 + r_1^2}\right) \tag{6.28}$$

and $I = 1,2,\cdots,N$. The value of M_{pi} is obtained from Fig. 6.11, and ε_i can be calculated from (6.22) after the coning vector is solved for. Therefore, the unknown parameters in (6.26) are (r_0, r_1, β). When the number of miniradar N is greater than 2, it is possible to obtain the values of (r_0, r_1, β) by solving the equation system of (6.26). However, when $r_0\sin\beta \le r_1\cos\beta$ (ie, $r_0\tan\beta \le r_1$ in *Case 1*), *Case 1* in (6.26) is equal to

$$M_{pi} = 2r_1\cos\beta\sin\varepsilon_i + 2r_1\sin\beta|\cos\varepsilon_i|, \tag{6.29}$$

which is independent of r_0. Therefore, in this case, the correct value of r_0 cannot be obtained by solving the equation system. In *Case 1*, the set of inequalities $l'_{x1} > 2r_1$ and $l'_{x2} < 0$ is equivalent to the inequality

$$|\cos\varepsilon_i| > (r_0\cos\beta + r_1\sin\beta)\Big/\sqrt{r_0^2 + r_1^2} \tag{6.30}$$

In order to determine whether the obtained value of r_0 is correct, we use the obtained (r_0,r_1,β) to judge whether both (6.30) and the inequality $r_0\tan\beta \le r_1$ are satisfied for each $M_{pi}(i = 1,2,\cdots,N)$. If the number of M_{pi} that does not simultaneously satisfy these two inequalities is greater than 2, the obtained value of r_0 can be considered as correct. Otherwise, the obtained value of r_0 is inaccurate and only (r_1,β) could be obtained by solving (6.26).

Furthermore, in *Case 2* of (6.26), M_{pi} is independent of $\boldsymbol{\varepsilon}_i$. It means that if the parameters of several miniradars simultaneously satisfy with *Case 2*, the corresponding equations of these miniradars in (6.26) would be the same. Therefore, if *Case 2* occurs, two or more miniradars with parameters belonging to the other three cases in (6.26) are necessary to solve (r_0,r_1,β).

In fact, there is little probability that the parameters of several miniradars simultaneously satisfy *Case 2*. The set of inequalities $0 \le l'_{x1} \le 2r_1$ and $0 \le l'_{x2} \le 2r_1$ in *Case 2* is equivalent to the inequality as following:

$$|\cos \boldsymbol{\varepsilon}_i| \le (r_1 \sin\beta - r_0 \cos\beta) \Big/ \sqrt{r_0^2 + r_1^2} \tag{6.31}$$

Since $|\cos \boldsymbol{\varepsilon}_i| \ge 0$, $r_1\sin\beta$ should be equal to or greater than $r_0\cos\beta$, ie, $r_0 \le r_1\tan\beta$. Defining $\angle COP = \alpha$, it turns out that $\tan\alpha = r_1/r_0$, and then we have $\tan\alpha \ge \tan(\pi/2 - \beta)$. Because $\alpha \in (0,\pi/2)$, it yields $\pi/2 - \beta \le \alpha \le \pi/2$, and it is equivalent to $\pi/2 \le \alpha + \beta \le \pi/2 + \beta$. Then, it can be obtained that $-\sin\beta \le \cos(\alpha+\beta) \le 0$. Therefore, (6.31) becomes

$$|\cos \boldsymbol{\varepsilon}_i| \le \sin\alpha \sin\beta - \cos\alpha\cos\beta = -\cos(\alpha+\beta) \le \sin\beta \tag{6.32}$$

Then it can be obtained that

$$\pi/2 - \beta \le \boldsymbol{\varepsilon}_i \le \pi/2 + \beta \tag{6.33}$$

Because β is usually very small, the value set of $\boldsymbol{\varepsilon}_i$ is very limited. Hence, there is little probability that the $\boldsymbol{\varepsilon}_i$ of several miniradars simultaneously satisfy with (6.33).

The least-squares method is used to search the approximate solution of (r_0,r_1,β) since analytic solution to (6.26) is infeasible. After the precession angle β is obtained, the distance from the conical point to the intersection point of spinning axis and coning axis (ie, the length of segment $\overline{OA}$ in Fig. 6.13) can be calculated as $r/\sin\beta$ (r has been solved for in Section 6.4.2.1).

6.4.2.4 Estimation of Object Length

The length of the object (ie, the length of $\overline{PA}$ in Fig. 6.10) is calculated as $r_0 + r/\sin\beta$. When the obtained value of r_0 is correct in the previous subsection, it is very easy to calculate the length of the object. However, if the obtained value of r_0 is incorrect, it is difficult to calculate the accurate value of the object's length even though the length of $\overline{OA}$ can be obtained precisely. In this case, to estimate the length of the object, some approximate calculations have to be adopted. As shown in Fig. 6.13A, point D is the projection of A on the coning axis, H is the projection of D on the LOS, and P' is the projection of P on the coning axis, where P is the center of the cone bottom. The length of the object is

$$\begin{aligned}\overline{AP} &= \overline{P'D}\Big/\cos\beta = \left(\overline{GD}-\overline{GP'}\right)\Big/\cos\beta \\ &= \left(\overline{G'H}\Big/|\cos\varepsilon_i| - \overline{BP}\sin\beta\right)\Big/\cos\beta \end{aligned} \tag{6.34}$$

In (6.34), β, ε_i and $\overline{BP} = r_1$ have all been solved for while $\overline{G'H}$ is unknown. Considering the relationship between the curve features shown in Fig. 6.11, we have

$$\overline{H_{\min}H} = L_{\max} - M_c = L_{\max} - r\sin\varepsilon_i \tag{6.35}$$

where $H_{\min}$ is the point on the $\boldsymbol{n}$-axis with the minimum coordinate of the projection of the body formed by the precession of scatterers B and C. Considering the approximation $\overline{G'H}\approx\overline{H_{\min}H}$, we can obtain the estimated length of the object

$$\overline{PA}\approx((L_{\max} - r\sin\varepsilon_i)/|\cos\varepsilon_i| - r_1\sin\beta)/\cos\beta \tag{6.36}$$

Obviously, the estimation error induced by the replacement of $\overline{G'H}$ with $\overline{H_{\min}H}$ is

$$e_l = \left|\overline{H_{\min}G'}\Big/(\cos\beta|\cos\varepsilon_i|)\right| = \left|\left(\overline{QG'} - \dot{H}_{\min}\right)\Big/(\cos\beta|\cos\varepsilon_i|)\right| \tag{6.37}$$

where

$$\begin{aligned}\overline{QG'} &= -\overline{QG}\cdot|\cos\varepsilon_i| = -\left(\overline{QK}+\overline{KG}\right)|\cos\varepsilon_i| \\ &= -\left(\overline{KC}\cdot\sin\beta + \overline{BK}\cdot\sin\beta\right)|\cos\varepsilon_i| = -2r_1\sin\beta|\cos\varepsilon_i| \end{aligned} \tag{6.38}$$

According to (6.49)–(6.51) in Appendix 6-A, if $l'_{x2} < 0$, (6.37) becomes

$$e_{l1} = |r_1 - r_0\tan\beta|\cdot|\tan\varepsilon_i| \tag{6.39}$$

If $0 \le l'_{x2} \le 2r_1$, it becomes

$$e_{l2} = \frac{1-\cos(\alpha-\beta)|\cos\varepsilon_i|}{\cos\beta|\cos\varepsilon_i|}\sqrt{r_0^2+r_1^2} \tag{6.40}$$

If $l'_{x2} > 2r_1$, it becomes

$$e_{l3} = |-2r_1\tan\beta + (r_1 + r_0\tan\beta)|\tan\varepsilon_i|| \tag{6.41}$$

It can be found that, the estimation error e_{l1} is in direct proportion to $|\tan \varepsilon_i|$, e_{l2} is in inverse proportion to $|\cos \varepsilon_i|$, and as in (6.41), the term $-2r_1 \tan \beta$ is usually relatively small when β is no larger than 10°, hence e_{l3} is also directly proportional to $|\tan \varepsilon_i|$ approximately. When $\varepsilon \in [0,\pi)$, the less $|\tan \varepsilon_i|$ is, the greater $|\cos \varepsilon_i|$ is. Therefore, to minimize the estimation error, radar echo of the miniradar with the maximum value of $|\cos \varepsilon_i|$ is preferred to estimate the object's length.

6.4.2.5 The Chart of the Three-Dimensional Precession Feature Extraction

For better understanding, we present the flowchart of the proposed three-dimensional precession feature extraction algorithm in Fig. 6.14. Due to the advantage of multi-view of miniradars in the distributed radar networks, the main features of the target's micromotions and structures can be extracted from the radar echoes by following the processing procedure listed in the flowchart. The following features can be obtained from the algorithm: (1) the coning period T_c, (2) the precession period T_{pr}, (3) the spinning period T_s, (4) the coning vector $\boldsymbol{\omega}_c = (\omega_{cx},\omega_{cy},\omega_{cz})^{\mathrm{T}}$, (5) the precession angle β,

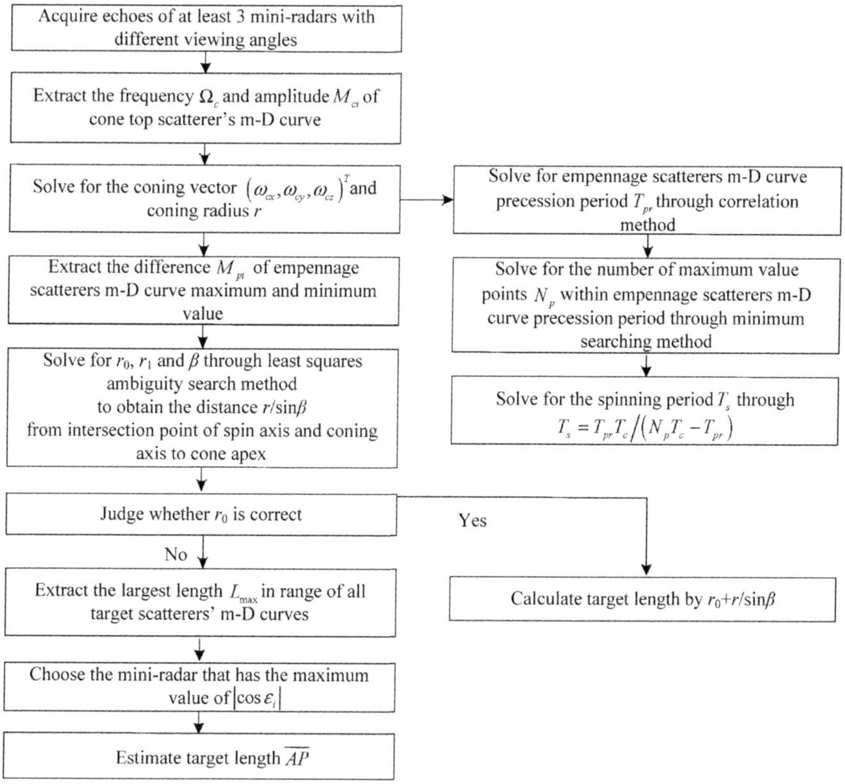

Figure 6.14 The flowchart of the proposed 3-D precession feature extraction algorithm.

(6) the radius r_1 of cone bottom, (7) the distance $r/\sin\beta$ from the conical point to the intersection point of spinning axis and coning axis, and (8) the estimated length of the target. All these features play very important roles in improving the accuracy of automatic target recognition.

6.4.2.6 Simulation

The parameters of the simulation model are as follow: $r_0 = 0.6$ m, $r_1 = 1$ m, $\overline{OA} = 2$ m, the length of the target $\overline{PA} = 2.6$ m, $\Omega_s = 5\pi$ rad/s, $\omega_c = (7.6756, -2.3930, 9.6577)^{\mathrm{T}}$ rad/s, $\Omega_c = 4\pi$ rad/s $= 12.5644$ rad/s, the precession angle $\beta = \pi/18$ rad $= 0.1745$ rad. Six miniradars operating at 20.5, 24, 10, 13.5, 17, and 27.5 GHz are located at (100 km, 300 km, −500 km) (indexed as miniradar *1*), (−300 km, −40 km, −500 km) (indexed as miniradar *2*), (−300 km, −100 km, −500 km) (indexed as miniradar *3*), (0 km, −100 km, −500 km) (indexed as miniradar *4*), (−100 km, 200 km, −500 km) (indexed as miniradar *5*) and (−220 km, −100 km, −500 km) (indexed as miniradar *6*), respectively. The bandwidth of the transmitted LFM signals is 3 GHz and $T_p = 50$ μs, the obtained range resolution is 0.05 m. The pulse repetition frequency is 500 Hz, and the coherent integration time is 6 s. Fig. 6.15 shows the simulation results of miniradar *1* and miniradar *2* with different viewing angles. In the figure, only the micro-Doppler signatures in the first 3 s are drawn for clarity. From the figure, we can see that: the micro-Doppler signature of scatterer A appears as a sinusoidal curve on the range–slow time plane (ie, the $f_k - t_m$ plane after range scaling) and the period of the curve is equivalent to the coning period. As for scatterers B and C, their micro-Doppler signatures appear as relatively complicated curves. It can also be found that the amplitudes of the corresponding curves are different from each other when observed from different viewing angles.

By utilizing the feature extraction method, the amplitude of the micro-Doppler curves of the conical scatterer are calculated as $M_{ci} = 0.2683$ m, 0.0935 m, 0.1287 m,

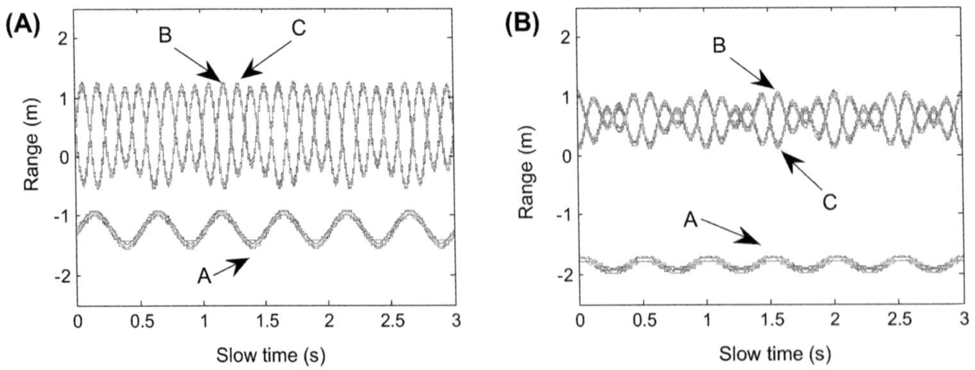

Figure 6.15 The micro-Doppler curves observed by miniradars. (A) Miniradar 1; (B) miniradar 2.

0.2458 m, 0.1635 m, and 0.1517 m, respectively. Then, we use the command *fsolve* in MATLAB to solve the Eq. (6.23). Setting the initial guess to be (1, 1, 1, 1), the calculated solution to $(\omega_{cx},\omega_{cy},\omega_{cz},r)$ is (7.8944, −2.4866, 9.6034, 0.3496), which are close to the true values. As for the micro-Doppler curves in Fig. 6.11, where $T_c = 0.5$ s, $T_{pr} = 2$ s, and $N_p = 9$, the calculated spinning period is $T_s = 0.4$ s, which is equivalent to the setting value of the simulation.

The range span M_{pi} of the precession curve B or C is 1.7290 m, 0.8887 m, 1.0750 m, 1.6368 m, 1.2155 m, and 1.1688 m, respectively. Using the solved coning vector, the value of $\boldsymbol{\varepsilon}_i$ corresponding to each miniradar can be obtained by (6.22) and the value of $|\cos \boldsymbol{\varepsilon}_i|$ is calculated to be 0.6429, 0.9580, 0.9271, 0.7163, 0.8826, and 0.8997, respectively. Because the set value of β is $\pi/18$ rad, all of $|\cos \boldsymbol{\varepsilon}_i|$ are greater than $\sin \beta$, therefore *Case 2* in (6.26) has not occurred in this simulation. Then using the least-squares method, the approximate solution to (6.26) is obtained as $r_0 = 0.4$ m, $r_1 = 1$ m and $\beta = 0.17$ rad. It can be found the values of r_1 and β are close to the set values. The distance from the conical point to the intersection point of spinning axis and coning axis can be also calculated as $\overline{OA} = r/\sin \beta = 2.0664$ m, which is very close to the theoretical value of 2 m as mentioned earlier. However, it can be calculated that $r_0 \tan \beta = 0.0687 < r_1$ and the value of the right term in (6.30) is 0.5270, that means all of $|\cos \boldsymbol{\varepsilon}_i|$ satisfy with (6.30). Therefore, the obtained value of r_0 is incorrect. The echo of the miniradar *2* is chosen to estimate the object's length because it has the greatest value of $|\cos \boldsymbol{\varepsilon}_i|$. Feature extraction of the micro-Doppler curves yields $L_{\max} = 2.9930$ m and $M_c = 0.0935$ m. According to (6.36), we have $\overline{PA} \approx 2.8921$ m with a relative error of 11.2% compared with the predesigned value of 2.6 m.

When setting the simulation parameters as $r_0 = 1$ m, $r_1 = 0.5$ m, and the other parameters are same to those in the previous simulation, the solution to (6.26) is obtained as $r_0 = 0.98$ m, $r_1 = 0.5$ m, and $\beta = 0.17$ rad. It can be calculated that $r_0 \tan \beta = 0.1728 < r_1$ and the value of the right term in (6.30) is 0.9561. The number of $|\cos \boldsymbol{\varepsilon}_i|$ that does not satisfy with (6.30) is 5, so the obtained value of r_0 can be considered as correct. In fact, it is close to the set value.

It is known that the accuracy of the extracted micro-Doppler curve features is limited by the radar resolution. Therefore, errors are inevitable in the solution of three-dimensional precession features. The micro-Doppler curve features include the coning period T_c, the amplitude M_c of the coning curve, the precession period T_{pr}, the range span M_p of the precession curve, the range span $L_{\max}$ of all the micro-Doppler curves, and the number N_p of local maximums of the precession curve during a precession period. As for the coning period T_c and the precession period T_{pr}, it could be imagined that their errors are mainly dependent on the sampling interval of slow time (ie, the pulse repetition interval), since they are calculated by the autocorrelation analysis of curves A and B, respectively. Usually, the pulse repetition interval is very small (up to several milliseconds), thus the errors of T_c and T_{pr} are accordingly very small. Furthermore, because the values of T_c and T_{pr} extracted from the micro-Doppler curves are independent of the viewing angle of miniradars, the accuracy of the estimated T_c and T_{pr} can be improved by averaging the coning periods and precession periods extracted from all miniradars respectively. In this case, the errors of the obtained T_c and T_{pr} are usually negligible. And then, the estimation of the spinning period T_s can be very precise because it only depends on T_c, T_{pr}, and N_p according to (6.25), where the value of N_p could be accurately calculated. As for the other features such as M_c, M_p, and $L_{\max}$, their accuracies are mainly limited by the range resolution of the miniradars, while the range resolution is determined by the bandwidth of the transmitted waveforms and usually are up to several centimeters. Furthermore, the accuracies of these parameters cannot be improved by averaging the corresponding feature values obtained from different miniradars since their values vary if the viewing angle changes. Therefore, it is necessary to analyze the robustness of the proposed algorithm when the errors of the obtained M_c, M_p, and $L_{\max}$ exist.

For the convenience of the analysis, we define the normalized error as

$$\rho = \frac{\widehat{X} - X}{X}$$

where X denotes the true value of M_c, M_p, $L_{\max}$, ω_{cx}, ω_{cy}, ω_{cz}, r, β, r_1, and $r/\sin\beta$, and $\widehat{X}$ denotes the corresponding estimated value. Then we define $|\rho|$ as the normalized absolute error.

For simplicity, in the following Monte Carlo simulations, we assume that the normalized error ρ of M_c and M_p obey a uniform distribution in $[-a,a]$, and the errors of the estimated T_c, T_{pr}, and N_p are neglected. The radar parameters and target parameters are the same as the previous simulations. Fig. 6.16 shows the variation curves of $|\rho|$ of ω_{cx}, ω_{cy}, ω_{cz}, r, r_1, β, and $r/\sin\beta$ with respect to a when a varies from 0.01 to 0.1. In the figure, $|\rho|$ is the average value of 100 times of simulation experiments. It can be found

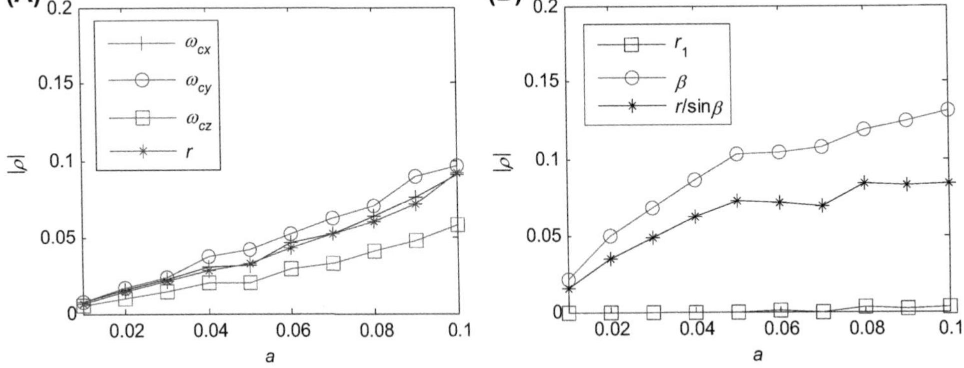

Figure 6.16 The variation curves of $|\rho|$ of ω_{cx}, ω_{cy}, ω_{cz}, r, r_1, β, and $r/\sin\beta$ with respect to a when a varies from 0.01 to 0.1.

that, although the accuracies of the solution ω_{cx}, ω_{cy}, ω_{cz}, r, r_1, β, and $r/\sin\beta$ are restricted by the estimation errors of M_c and M_p, the accuracies are reasonable and acceptable when $a < 0.1$. To achieve higher accuracy, larger bandwidth of transmitted waveforms is required to decrease the estimation errors of M_c and M_p.

APPENDIX 6-A

The scatterers B and C with precession motion yield a body taking B, E, Q, C, B', and G as the edge points in Fig. 6.13A, or taking B, G, B', E, Q, and C as the edge points in Fig. 6.13B. Because M_p is the projection length of the body projected to the LOS, we calculate the projection length to analyze the relationship between M_p and the parameters of the target model. Set Q to be the origin of the $\boldsymbol{n}$-axis, the direction of the positive axis is the same as the $\boldsymbol{n}$-axis, the length of segment on the $\boldsymbol{n}$-axis is positive when the corresponding vector of the segment has the same direction with $\boldsymbol{n}$, and vice versa.

Taking *Case 2* ($r_0\tan\beta > r_1$), for example, the geometry of the target is shown in Fig. 6.13B. Consider a point P_1 on the segment $\overline{BC}$ with $\overline{BP_1} = l_x$ $(0 \le l_x \le 2r_1)$ and a point P_2 on the edge of the cone bottom with $P_1P_2 \perp BP_1$, the projection of P_2 on the coning axis is shown as the point P_1'' and the projection of P_1'' on the $\vec{\boldsymbol{n}}$-axis is shown as the point P_1'. It is obvious that $\left|\overline{GP_1''}\right| / \left|\overline{GQ}\right| = \left|\overline{BP_1}\right| / \left|\overline{BC}\right| = l_x/(2r_1)$. Since $\left|\overline{GQ}\right| = 2r_1\sin\beta$, we have $\left|\overline{GP_1''}\right| = l_x\sin\beta$ and $\left|\overline{QP_1''}\right| = 2r_1\sin\beta - l_x\sin\beta$. Then, it can be calculated that $\left|\overline{QP_1'}\right| = (2r_1\sin\beta - l_x\sin\beta)|\cos\boldsymbol{\varepsilon}_i|$ and $\overline{QP_1'} = -\left|\overline{QP_1'}\right| = (l_x\sin\beta - 2r_1\sin\beta)|\cos\boldsymbol{\varepsilon}_i|$. According to the geometry of the triangle $\Delta P_1P_2P_1''$ and ΔP_1P_2P, it can be obtained

$$
\begin{aligned}
\overline{P_1''P_2}^2 &= \overline{P_1P_2}^2 + \overline{P_1''P_1}^2 = \left(\overline{PP_2}^2 - \overline{PP_1}^2\right) + \left(\overline{BG} + l_x\cos\beta\right)^2 \\
&= \left[r_1^2 - (r_1 - l_x)^2\right] + \left[(r_0\tan\beta - r_1)\cos\beta + l_x\cos\beta\right]^2 \\
&= -\sin^2\beta\cdot l_x^2 + 2\left(r_0\sin\beta\cos\beta + r_1\sin^2\beta\right)l_x + \left(r_0\sin\beta - r_1\cos\beta\right)^2 \\
&= r_0^2 + r_1^2 - \left[l_x\sin\beta - \left(r_0\cos\beta + r_1\sin\beta\right)\right]^2
\end{aligned}
\tag{6.42}
$$

Because $\left|\overline{P_1''P_2}\right|$ is the radius of the circle with center P_1'', the maximum and minimum coordinate of the projection of the circle on the $\boldsymbol{n}$-axis are $\dot{H}_1(l_x)$ and $\dot{H}_2(l_x)$, respectively, where

$$\dot{H}_1(l_x) = \overline{QP_1'} + \left|\overline{P_1''P_2}\right|\sin\boldsymbol{\varepsilon}_i \tag{6.43}$$

$$\dot{H}_2(l_x) = \overline{QP_1'} - \left|\overline{P_1''P_2}\right|\sin\boldsymbol{\varepsilon}_i \tag{6.44}$$

It can be proved that the maximum value of $\dot{H}_1(l_x)$, ie, $\dot{H}_{\max}$, occurs when $d\dot{H}_1(l_x)/dl_x = 0$, that is

$$l'_{x1} = \frac{1}{\sin\beta}\left(r_0 \cos\beta + r_1 \sin\beta + |\cos\varepsilon_i|\sqrt{r_0^2 + r_1^2}\right) \tag{6.45}$$

Obviously, $l'_{x1} \geq 0$. When $l_x \leq l'_{x1}$, $\dot{H}_1(l_x)$ is a monotonic increasing function; otherwise, $\dot{H}_1(l_x)$ is a monotonic decreasing function. Due to the limitation of $0 \leq l_x \leq 2r_1$, if $0 \leq l'_{x1} \leq 2r_1$, substitute (6.45) into (6.43) and then it can be obtained that

$$\dot{H}_{\max} = \dot{H}_1(l'_{x1}) = (r_0 \cos\beta - r_1 \sin\beta)|\cos\varepsilon_i| + \sqrt{r_0^2 + r_1^2} \tag{6.46}$$

Otherwise, if $l'_{x1} > 2r_1$, the maximum value of $\dot{H}_1(l_x)$ occurs when $l_x = 2r_1$, and then it yields

$$\dot{H}_{\max} = \dot{H}_1(2r_1) = (r_0 \sin\beta + r_1 \cos\beta)\sin\varepsilon_i \tag{6.47}$$

Similarly, it can be proved that the minimum value of $\dot{H}_2(l_x)$, ie, $\dot{H}_{\min}$, occurs when $d\dot{H}_2(l_x)/dl_x = 0$, that is

$$l'_{x2} = \frac{1}{\sin\beta}\left(r_0 \cos\beta + r_1 \sin\beta - |\cos\varepsilon_i|\sqrt{r_0^2 + r_1^2}\right) \tag{6.48}$$

When $l_x \leq l'_{x2}$, $\dot{H}_2(l_x)$ is a monotonic decreasing function; otherwise, $\dot{H}_2(l_x)$ is a monotonic increasing function. If $0 \leq l'_{x2} \leq 2r_1$, substitute (6.48) into (6.44) and then it can be obtained that

$$\dot{H}_{\min} = \dot{H}_2(l'_{x2}) = (r_0 \cos\beta - r_1 \sin\beta)|\cos\varepsilon_i| - \sqrt{r_0^2 + r_1^2} \tag{6.49}$$

If $l'_{x2} < 0$, the minimum value of $\dot{H}_2(l_x)$ occurs when $l_x = 0$, then it yields

$$\dot{H}_{\min} = \dot{H}_2(0) = -2r_1 \sin\beta|\cos\varepsilon_i| - |r_0 \sin\beta - r_1 \cos\beta|\sin\varepsilon_i \tag{6.50}$$

Otherwise, if $l'_{x2} > 2r_1$, the minimum value of $\dot{H}_2(l_x)$ occurs when $l_x = 2r_1$, and then it yields

$$\dot{H}_{\min} = \dot{H}_2(2r_1) = -(r_0 \sin\beta + r_1 \cos\beta)\sin\varepsilon_i \tag{6.51}$$

Then, M_p can be calculated as $\dot{H}_{\max} - \dot{H}_{\min}$. However, it is impossible to obtain the values of l'_{x1} and l'_{x2} because the values of r_0, r_1, and β are unknown. As a result, the expression of M_p will contain several cases. It is obvious that $l'_{x1} \geq l'_{x2}$, and the inequality $0 \leq l'_{x1} \leq 2r_1$ is equivalent to

$$|\cos\varepsilon_i| \leq (r_1 \sin\beta - r_0 \cos\beta)\Big/\sqrt{r_0^2 + r_1^2}, \tag{6.52}$$

while the inequality $l'_{x2} < 0$ is equivalent to

$$|\cos \varepsilon_i| > (r_1 \sin \beta + r_0 \cos \beta) \Big/ \sqrt{r_0^2 + r_1^2} > (r_1 \sin \beta - r_0 \cos \beta) \Big/ \sqrt{r_0^2 + r_1^2} \quad (6.53)$$

That means the inequalities of $0 \le l'_{x1} \le 2r_1$ and $l'_{x2} < 0$ cannot be simultaneously satisfied. Therefore, M_p can be expressed as

$$M_{pi} = \begin{cases} (r_0 \sin \beta + r_1 \cos \beta)\sin \varepsilon_i + 2r_1 \sin \beta|\cos \varepsilon_i| + |r_0 \sin \beta - r_1 \cos \beta|\sin \varepsilon_i, \\ \qquad \text{when } l'_{x1} > 2r_1 \text{ and } l'_{x2} < 0 \quad (\textit{Case I}) \\ 2\sqrt{r_0^2 + r_1^2}, \qquad \text{when } 0 \le l'_{x1} \le 2r_1 \text{ and } 0 \le l'_{x2} \le 2r_1 \quad (\textit{Case II}) \\ (r_0 \sin \beta + r_1 \cos \beta)\sin \varepsilon_i - (r_0 \cos \beta - r_1 \sin \beta)|\cos \varepsilon_i| + \sqrt{r_0^2 + r_1^2}, \\ \qquad \text{when } l'_{x1} > 2r_1 \text{ and } 0 \le l'_{x2} \le 2r_1 \quad (\textit{Case III}) \\ 2(r_0 \sin \beta + r_1 \cos \beta)\sin \varepsilon_i, \quad \text{when } l'_{x1} > 2r_1 \text{ and } l'_{x2} > 2r_1 \quad (\textit{Case IV}) \end{cases}, \quad (6.54)$$

which is the equation in (6.26).

Taking *Case 1* ($r_0\tan \beta \le r_1$) for analysis, a similar conclusion can be obtained.

REFERENCES

[1] Fishler E, Haimovich A, Blum R, et al. MIMO radar: an idea whose time has come. In: Proceeding of the IEEE Radar Conference, Philadelphia, PA; 2004. p. 71–8.

[2] Fishler E, Haimovich A, Blum R, et al. Performance of MIMO radar systems: advantages of angular diversity. In: Conference Record of the 38th Asilomar Conference on signals, systems and Computers, vol. 1; 2004. p. 305–9.

[3] Heimiller RC, Belyea JE, Tomlinson PG. Distributed array radar. IEEE Transactions on Aerospace and Electronic Systems 1983;19(6):831–9.

[4] Daher R, Adve R. A notion of diversity order in distributed radar networks. IEEE Transactions on Aerospace and Electronic Systems 2010;46(2):818–31.

[5] Subotic NS, Thelen B, Cooper K, et al. Distributed radar waveform design based on compressive sensing considerations. In: IEEE Radar Conference (radar '2008), Rome; May 2008. p. 1–6.

[6] Daher R, Adve R. Diversity order of joint detection in distributed radar networks. In: IEEE Radar Conference (radar '2008), Rome; May 2008. p. 1–6.

[7] Yang Z, Liao G, Zeng C. Reduced-dimensional processing for ground moving target detection in distributed space-based radar. IEEE Geoscience and Remote Sensing Letters 2007;4(2):256–9.

CHAPTER SEVEN

Review and Prospects

Research in recent years at home and abroad has shown that micro-Doppler effects have opened up a new area for radar target characteristics analysis and provided new characteristics with good stability and high separability for target recognition. This book, combining with the recent research achievements of the authors, comprehensively discusses the generation mechanism of micro-Doppler effects of radar targets and the corresponding extraction methods of characteristics. Micro-Doppler effects analysis and extraction methods of micro-Doppler characteristics are the two main deep-expatiated aspects, which include thorough analysis on micro-Doppler effects of various micromotion targets in different radar systems and expounds on various extraction methods of micro-Doppler characteristics along with reconstruction methods of micromotion in three dimensions. Restricted by the research scope, the authors cannot refer to every aspect of micro-Doppler effects in this book. In the later research, the following aspects can be further investigated:

1. Micro-Doppler effects analysis of complex motion targets. The recent work is mainly aimed at relatively simple micromotions such as rotation, vibration, precession, and so on. So micro-Doppler effects of complex motion targets like compound motion targets, human body targets, and other creature targets remain for further study.
2. Three-dimensional reconstruction of characteristics of complex motion targets. The targets change the posture very intricately, which will result in differences in micro-Doppler characteristic when in different radar visual angles, thus the accuracy of target recognition will be affected. In order to overcome the sensitiveness to radar posture of micro-Doppler characteristics, three-dimensional micromotion information reflecting the real motion of micromotion components in targets must be extracted from returned signal of targets. Reconstruction methods of micromotion in three dimensions are expounded on in the book, but for more complicated motions, such as tracking vehicles, creature targets, and so on, the reconstruction methods remain to be studied intensively. Besides, in view of the eclipsing effects of target scatterers, there are differences in the distributions of observed scattering center of target, the three-dimensional micromotion reconstruction issue, which needs further study.
3. Micro-Doppler characteristic extraction of group targets. To implement penetration, helicopters and fighters usually fly with an intensive formation, thus a group of targets take shape in the wave beam of radar; missile targets usually release objects like light and heavy bait, fuel tank, metal chaff cloud, and so on, which form an intensive target

Micro-Doppler Characteristics of Radar Targets
ISBN 978-0-12-809861-5, http://dx.doi.org/10.1016/B978-0-12-809861-5.00007-2

group while flying in midcourse. In this condition, the returned signals from different targets are difficult to separate, and the micro-Doppler effects induced by different micromotions of all the targets are rather complicated. Thus, it remains to be solved how to extract and analyze the micro-Doppler characteristics of target groups.

4. High-resolution imaging of micromotion target in three dimensions. Three-dimensional imaging is an effective approach to obtain three-dimensional characteristics of targets. In the figures of time—frequency and range—slow time, micromotion characteristics are all shown as curves, the parameters of which are closely connected with the spatial position of scatterers in the targets. That relationship provides a new view for reconstructing the scatter-distributed micromotion targets in three dimensions. Some reports in the literature have studied the three-dimensional imaging of simple spinning targets, but the three-dimensional imaging of micromotion targets in more complicated situations remains for further study, for example, imaging of compound micromotion targets, imaging with eclipsing effects of target scatterers, and so on.
5. Target recognition techniques based on micromotion characteristics. Though the important role of micromotion characteristics in radar target recognition has been widely recognized internationally, further research is still needed to apply it into target recognition. Many aspects like how to fuse the micromotion characteristics with other characteristics, how to build a reasonable classifier, and how to improve the online study ability of the classifier still remain for further study.

INDEX

'*Note:* Page numbers followed by "f" indicate figures and "t" indicate tables.'

O

P

Printed and bound by CPI Group (UK) Ltd, Croydon, CR0 4YY
08/05/2025
01864800-0005